Differential Equations in Action: From Modeling to Theory

Differential Equations in Action: From Modeling to Theory

Edited by
Amelie Roberts

Willford Press
www.willfordpress.com

Published by Willford Press,
118-35 Queens Blvd., Suite 400,
Forest Hills, NY 11375, USA

ISBN: 978-1-64728-511-1

Cataloging-in-Publication Data

Differential equations in action : from modeling to theory / edited by Amelie Roberts.
p. cm.
Includes bibliographical references and index.
ISBN 978-1-64728-511-1
1. Differential equations. 2. Calculus. 3. Mathematical models. 4. Group theory. I. Roberts, Amelie.
QA371 .D54 2023
515.35--dc23

For information on all Willford Press publications
visit our website at www.willfordpress.com

Willford Press

Contents

Preface

This book was inspired by the evolution of our times; to answer the curiosity of inquisitive minds. Many developments have occurred across the globe in the recent past which has transformed the progress in the field.

A differential equation is an equation that contains one or more functions with its derivatives. The derivatives of the function define the rate of change of the function at a point. There are different types of differential equations including ordinary differential equations, linear differential equations, partial differential equations, homologous differential equations, and nonlinear differential equations. Differential equations can also be classified based on the order and coefficients of the derivatives, which may be either constants or functions of the independent variable. These equations are applicable in a continuous system. Differential equations are used to model real-world situations involving change. Such equations usually involve derivatives with respect to time. In most cases, a real-life problem involves the rate of change of a variable being proportional to some function of the variable. Such situations can be conveniently modeled using differential equations. This book provides a detailed analysis of the theory and modeling of differential equations. It is appropriate for students seeking detailed information in this area of mathematics as well as for experts.

This book was developed from a mere concept to drafts to chapters and finally compiled together as a complete text to benefit the readers across all nations. To ensure the quality of the content we instilled two significant steps in our procedure. The first was to appoint an editorial team that would verify the data and statistics provided in the book and also select the most appropriate and valuable contributions from the plentiful contributions we received from authors worldwide. The next step was to appoint an expert of the topic as the Editor-in-Chief, who would head the project and finally make the necessary amendments and modifications to make the text reader-friendly. I was then commissioned to examine all the material to present the topics in the most comprehensible and productive format.

I would like to take this opportunity to thank all the contributing authors who were supportive enough to contribute their time and knowledge to this project. I also wish to convey my regards to my family who have been extremely supportive during the entire project.

Editor

1

Ordinary Differential Equation with Left and Right Fractional Derivatives and Modeling of Oscillatory Systems

Liana Eneeva *, Arsen Pskhu and Sergo Rekhviashvili

Institute of Applied Mathematics and Automation, Kabardino-Balkarian Scientific Center of Russian Academy of Sciences, 89-A Shortanov Street, 360000 Nalchik, Russia; pskhu@list.ru (A.P.); rsergo@mail.ru (S.R.)

* Correspondence: eneeva72@list.ru;

Abstract: We consider the principle of least action in the context of fractional calculus. Namely, we derive the fractional Euler–Lagrange equation and the general equation of motion with the composition of the left and right fractional derivatives defined on infinite intervals. In addition, we construct an explicit representation of solutions to a model fractional oscillator equation containing the left and right Gerasimov–Caputo fractional derivatives with origins at plus and minus infinity. We derive a representation for the composition of the left and right derivatives with origins at plus and minus infinity in terms of the Riesz potential, and introduce special functions with which we give solutions to the model fractional oscillator equation with a complex coefficient. This approach can be useful for describing dissipative dynamical systems with the property of heredity.

Keywords: principle of least action; fractional oscillator; equation with left and right fractional derivative; Gerasimov–Caputo fractional derivative; Riesz potential

1. Introduction

The basic framework for Lagrangian and Hamiltonian mechanics is the principle of least action. In problems of mechanics, the principle of stationary action is most commonly used, and is the most important among all extreme principles. The equations of motion obtained using these principles are called the Euler–Lagrange equations. The Euler–Lagrange equation with the Lagrange function can describe any mechanical motion.

In recent years, fractional integro-differentiation has been widely used to model the properties of dynamic systems [1–3]. Fractional calculus enables one to naturally take into account the heredity, time non-locality and memory effects in dynamic processes [2].

The principle of least action in the context of the fractional calculus is studied in [4–11]. In [4,5], it was shown that applying the Lagrangian with fractional derivatives allows one to describe the dynamics of nonconservative systems. In [6], two fractional variational problems were considered. A generalized Euler–Lagrange equation with fractional differentiation was derived. The classical Euler–Lagrange equation is a special case of the latter. In addition, practical examples of the application of the results obtained were considered. It was shown that this approach would make it possible to simulate dynamic systems with greater accuracy. In [8], the Lagrange formalism with fractional derivatives with infinite limits was considered for the first time. In [9], the Hamiltonian and Lagrangian formulations with fractional differentiation were used to describe constrained systems. It was shown that both treatments for systems with linear velocities are equivalent. In [7,12,13] the principle resulted in very exotic differential equations of fractional order, with composition of the Riemann–Liouville or Caputo operators with different origins. These equations and their possible application to modeling dynamic processes can be found in [12–17].

We point out the major challenges related to the equations with different origins. First, methods for obtaining solutions to these equations, both analytically and numerically, are not yet well developed. Next, physical interpretation of the fractional Euler–Lagrange equations is currently not well understood. The first-order Euler–Lagrange equation does not contain the limits of integration in the action functional, unlike the fractional Euler–Lagrange equation. Therefore, the following question arises: is it possible to use a fractional Euler–Lagrange equation if the time variable falls outside the interval within which the action functional is considered? Any value can be chosen as a time interval, including a small physical time (for example, the Planck time, which is approximately 5.4×10^{-44} s). In this case, studying any physical process in this interval is believed to be incorrect.

This ambiguity in a sense can be eliminated by using the Gerasimov–Caputo fractional derivatives (fractional operators defined on an infinite interval [18,19]) in the Euler–Lagrange equation, and we offer this in the present article.

In the present paper, we derive the fractional Euler–Lagrange equation and the general equation of motion with the composition of the left and right fractional derivatives defined on infinite intervals, and construct an explicit representation of solutions to a model fractional oscillator equation containing the left and right Gerasimov–Caputo fractional derivatives with origins at plus and minus infinity.

The rest of the paper is structured as follows: Section 2 contains the definitions of fractional integrals and derivatives; in Section 3, we derive the Euler–Lagrange equation and the equation of motion with fractional derivatives defined on infinite intervals; and, in Section 4, we construct solutions of a model one-dimensional equation for the fractional oscillator; Section 5 presents the main findings of this work.

2. Fractional Differentiation

The fractional derivatives of order α $(0 < \alpha < 1)$ with respect to t, having origins at $t = a$ $(-\infty \le a \le \infty)$, in the Riemann–Liouville and Gerasimov–Caputo senses, are given by ([18–21])

$$D^{\alpha}_{at}g(t) = \text{sign}(t-a)\frac{d}{dt}D^{\alpha-1}_{at}g(t) \qquad \text{and} \qquad \partial^{\alpha}_{at}g(t) = \text{sign}(t-a)D^{\alpha-1}_{at}\frac{d}{dt}g(t),$$

respectively. Here, $D^{\alpha-1}_{at}$ stands for the Riemann–Liouville fractional integral:

$$D^{\alpha-1}_{at}g(t) = \text{sign}(t-a)\int_a^t g(\eta)\frac{|t-\eta|^{-\alpha}}{\Gamma(1-\alpha)}\,d\eta.$$

3. Euler–Lagrange Equation

Introduce a velocity vector for a material point motion in a certain medium:

$$\mathbf{v}(t) = \frac{1}{\tau\Gamma(1-\alpha)}\int_{-\infty}^{t}(t-\eta)^{-\alpha}\frac{d}{d\eta}\mathbf{r}(\eta)\,d\eta = \frac{1}{\tau}\partial^{\alpha}_{-\infty t}\mathbf{r}(t) \qquad (0 < \alpha \le 1), \tag{1}$$

where $\mathbf{r}(t)$ is the radius vector, t is the dimensionless (related to τ) time and τ is some characteristic time for the process. On account of Equation (1), the generalized action functional can be written as

$$S = \int_{-\infty}^{\infty} L[t, \mathbf{r}(t), \mathbf{v}(t)]dt. \tag{2}$$

In Equation (2) infinite limits for integration are established. To determine the velocity vector of Equation (1), it is necessary to take into account the entire time axis. Assume $\mathbf{r}_0$ (t) is a solution for the extremum problem. Then by Equation (2) and the conditions

$$\mathbf{r}(t) = \mathbf{r}_0(t) + \varepsilon\mathbf{h}(t),$$

$$\mathbf{h}(-\infty) = \mathbf{h}(\infty) = 0$$

(ε is a small parameter) we get

$$S = \int_{-\infty}^{\infty} L\left[t, \mathbf{r}_0(t) + \varepsilon\mathbf{h}(t), \frac{1}{\tau}\partial^{\alpha}_{-\infty t}\mathbf{r}_0(t) + \frac{\varepsilon}{\tau}\partial^{\alpha}_{-\infty t}\mathbf{h}(t)\right] dt.$$

The extremum of the functional satisfies the condition

$$\frac{dS}{d\varepsilon} = \int_{-\infty}^{\infty}\left(\frac{dL}{d\varepsilon}\right) dt = \int_{-\infty}^{\infty}\mathbf{h}(t)\left(\frac{\partial}{\partial\mathbf{r}} + \frac{1}{\tau}D^{\alpha}_{\infty t}\frac{\partial}{\partial\mathbf{v}}\right) L[t, \mathbf{r}(t), \mathbf{v}(t)]dt = 0. \tag{3}$$

Using Equation (3) we obtain the Euler–Lagrange equation

$$\left(\frac{\partial}{\partial\mathbf{r}} + \frac{1}{\tau}D^{\alpha}_{\infty t}\frac{\partial}{\partial\mathbf{v}}\right) L[t, \mathbf{r}(t), \mathbf{v}(t)] = 0. \tag{4}$$

It should be noted that the same problem for classical fields was first considered in [8]. In our case, Equation (4) defines the action function for a mechanical system. Moreover, the other techniques can be applied to get Equation (4), for example, using the results of [22], obtained by exploiting the Riesz fractional derivative.

Next, assume that the space of the considered motion is homogeneous and isotropic. Then the scalar function L should not be explicitly dependent on time. The function L should undergo power series expansion that involves only even powers $\mathbf{v}(t)$. Up to the second term of the expansion, we have

$$L = \frac{m\mathbf{v}^2(t)}{2} - U[\mathbf{r}(t), t]. \tag{5}$$

Expansion coefficients m and $U[\mathbf{r}(t), t]$ are independent of the derivative fractional order α. Therefore, there is no indication that these coefficients are different from the mass and potential energy of an object. Substituting Equation (5) into Equation (4), we obtain the equation of motion

$$-\frac{m}{\tau^2}D^{\alpha}_{\infty t}\partial^{\alpha}_{-\infty t}\mathbf{r}(t) = \mathbf{F}[\mathbf{r}(t), t], \tag{6}$$

where $\mathbf{F}[\mathbf{r}(t), t] = -(\partial/\partial\mathbf{r})\, U[\mathbf{r}(t), t]$. The right-hand side of Equation (6) does not explicitly depend on α, and, consequently, $\mathbf{F}[\mathbf{r}(t), t]$ determines the conservative force acting on a material point. The parameter α, according to [23,24] takes into account energy dissipation. Therefore, the equation of motion in the form of Equation (6) is apparently obtained for the first time.

As $\alpha = 1$, by Equation (6) we have the classical equation of motion

$$\frac{m}{\tau^2}\frac{d^2\mathbf{r}(t)}{dt^2} = \mathbf{F}[\mathbf{r}(t), t].$$

4. Solution of Fractional Oscillator Equation

Here, inspired by Equation (6), we are going to solve the equation

$$D^{\alpha}_{\infty t}\partial^{\alpha}_{-\infty t}x(t) + \lambda x(t) = f(t) \qquad (\lambda \in \mathbb{C}), \tag{7}$$

which is a model one-dimensional equation for the fractional oscillator.

Definition 1. *By a solution of Equation (7), we mean a function $x(t)$ such that $x'(t) \in L^1(\mathbb{R})$, $D^{\alpha-1}_{\infty t}D^{\alpha-1}_{-\infty t}x(t) \in C^1(\mathbb{R})$ and $x(t)$ satisfies Equation (7).*

4.1. Riesz Potentials

Now, we will rewrite Equation (7) in terms of the Riesz potential.

Definition 2. *Let $\mu \in (0,1)$. The operator K^{μ} is called the Riesz potential, and, for an integrable function $x(t)$, it is defined by*

$$K^{\mu}x(t) = (K^{\mu}x)(t) = c_{\mu}\int_{-\infty}^{\infty}|t-s|^{\mu-1}x(s)\,ds, \qquad c_{\mu} = \frac{\Gamma(1-\mu)}{\Gamma\left(\frac{\mu}{2}\right)\Gamma\left(1-\frac{\mu}{2}\right)}.$$

Lemma 1. *Let $\mu \in (0,1/2)$. If $x(t) \in L^1(\mathbb{R})$, then*

$$D_{\infty t}^{-\mu}D_{-\infty t}^{-\mu}x(t) = K^{2\mu}x(t). \tag{8}$$

Proof. Let $M = \max\{s,t\}$ and $m = \min\{s,t\}$. It is easy to check that

$$\begin{aligned}
[\Gamma(\mu)]^2\, D_{\infty t}^{-\mu}D_{-\infty t}^{-\mu}x(t) &= \int_t^{\infty}(\xi-t)^{\mu-1}\int_{-\infty}^{\xi}(\xi-s)^{\mu-1}x(s)\,ds\,d\xi \\
&= \int_{-\infty}^{\infty}x(s)\int_{-\infty}^{m}(M-\xi)^{\mu-1}(m-\xi)^{\mu-1}x(s)\,d\xi\,ds \\
&= \int_{-\infty}^{\infty}x(s)\int_0^{\infty}(M-m+\eta)^{\mu-1}\eta^{\mu-1}d\eta\,ds \\
&= \int_{-\infty}^{\infty}x(s)(M-m)^{2\mu-1}ds\cdot\int_0^{\infty}(1+z)^{\mu-1}z^{\mu-1}dz.
\end{aligned}$$

Combining this with the relations

$$M-m = |t-s| \qquad \text{and} \qquad \int_0^{\infty}(1+z)^{\mu-1}z^{\mu-1}dz = \frac{\Gamma(\mu)\Gamma(1-2\mu)}{\Gamma(1-\mu)},$$

we get Equation (8). □

By Equation (8) we have

$$D_{\infty t}^{\alpha}\partial_{-\infty t}^{\alpha}x(t) = -\frac{d}{dt}D_{\infty t}^{\alpha-1}D_{-\infty t}^{\alpha-1}\frac{d}{dt}x(t) = -\frac{d}{dt}K^{2-2\alpha}\frac{d}{dt}x(t),$$

and, consequently, Equation (7) can be rewritten in the form

$$-\frac{d}{dt}K^{2-2\alpha}\frac{d}{dt}x(t) + \lambda x(t) = f(t).$$

4.2. Special Functions

Consider the functions

$$C_{\alpha,\lambda}(t) = \frac{1}{\pi}\int_0^{\infty}\frac{\cos\xi t}{\xi^{2\alpha}+\lambda}\,d\xi$$

and

$$S_{\alpha,\lambda}^{\mu}(t) = \frac{1}{\pi}\int_0^{\infty}\frac{\xi^{\mu}\sin\xi t}{\xi^{2\alpha}+\lambda}\,d\xi$$

for $t \in \mathbb{R}$, $\lambda \in \mathbb{C}\setminus\{0\}$, $\alpha \in (1/2,1)$ and $\mu \in (-2,2\alpha)$.

Lemma 2. *Let $\lambda \in \mathbb{C}\setminus\{0\}$ and $\alpha \in (1/2,1)$. Then*

$$\frac{d}{dt}C_{\alpha,\lambda}(t) = -S_{\alpha,\lambda}^{1}(t), \qquad \frac{d}{dt}S_{\alpha,\lambda}^{-1}(t) = C_{\alpha,\lambda}(t), \tag{9}$$

and

$$K^{2-2\alpha} S^{1}_{\alpha,\lambda}(t) = \frac{\operatorname{sign} t}{2} - \lambda\, S^{-1}_{\alpha,\lambda}(t). \tag{10}$$

Moreover, if $|\arg\lambda| < (1-\alpha)\pi - \varepsilon$ *for some* $\varepsilon > 0$*, then*

$$\lim_{|t|\to\infty} t^2 C_{\alpha,\lambda}(t) < \infty \qquad \text{and} \qquad \lim_{|t|\to\infty} t^3 S^{1}_{\alpha,\lambda}(t) < \infty. \tag{11}$$

Proof. The formulas in Equation (9) are derived by direct differentiation. In the same way, an easy computation gives

$$K^{2-2\alpha} S^{1}_{\alpha,\lambda}(t) = S^{2\alpha-1}_{\alpha,\lambda}(t)$$

$$= \frac{1}{\pi}\int_0^\infty \frac{\xi^{2\alpha-1}\sin\xi t}{\xi^{2\alpha}+\lambda}\, d\xi = \frac{1}{\pi}\int_0^\infty \frac{\sin\xi t}{\xi}\, d\xi - \lambda S^{-1}_{\alpha,\lambda}(t).$$

This proves Equation (10).

Using the properties of the Mittag–Leffler function (see e.g., [25]), we can write

$$C_{\alpha,\lambda}(t) = \frac{1}{\pi}\int_0^\infty \cos\xi t \int_0^\infty e^{-s\xi} s^{2\alpha-1} E_{2\alpha,2\alpha}\left(-\lambda s^{2\alpha}\right) ds\, d\xi$$

$$= \frac{1}{\pi}\int_0^\infty s^{2\alpha-1} E_{2\alpha,2\alpha}\left(-\lambda s^{2\alpha}\right)\int_0^\infty e^{-s\xi}\cos\xi t\, d\xi\, ds = \frac{1}{\pi}\int_0^\infty E_{2\alpha,2\alpha}\left(-\lambda s^{2\alpha}\right)\frac{s^{2\alpha}}{s^2+t^2}\, ds.$$

This gives Equation (11). □

4.3. Solution Representation

Theorem 1. *Let* $\alpha \in (1/2, 1)$, $|\lambda| > 0$, *and* $|\arg\lambda| < (1-\alpha)\pi - \varepsilon$ *for some* $\varepsilon > 0$. *If* $f(t) \in C(\mathbb{R}) \cap L^1(\mathbb{R})$, *then the function*

$$x(t) = \int_{-\infty}^{\infty} f(s) C_{\alpha,\lambda}(t-s)\, ds$$

is a regular solution of Equation (7).

Proof. Lemmas 1 and 2 allow us to write

$$D^{\alpha}_{\infty t}\partial^{\alpha}_{-\infty t} x(t) = \frac{d}{dt} K^{2-2\alpha}\int_{-\infty}^{\infty} f(s)\, S^{1}_{\alpha,\lambda}(t-s)\, ds$$

$$= \frac{d}{dt}\int_{-\infty}^{\infty} f(s)\left[\frac{\operatorname{sign}(t-s)}{2} - \lambda\, S^{-1}_{\alpha,\lambda}(t-s)\right] ds = f(t) - \lambda\int_{-\infty}^{\infty} f(s) C_{\alpha,\lambda}(t-s)\, ds.$$

□

5. Conclusions

The principle of least action is often used to describe dissipative systems, in particular in the case of linear oscillators with viscous friction [26]. In this paper, we used this principle in the context of fractional calculus. In particular, we considered the concept of effective velocity for a moving material point, which was defined in terms of the Gerasimov–Caputo fractional derivative. This allowed us to take into account the heredity and nonlocality in dynamic processes of a material point with respect to the time. Using the fractional Euler–Lagrange equation, we derived the general equation of motion with Gerasimov–Caputo fractional derivatives. The equation of motion of such a form (see Equation (6)), has apparently been obtained for the first time in this paper.

In addition, we expressed the composition of the left and right fractional derivatives with origins at plus and minus infinities in terms of the Riesz potential, and constructed an explicit representation of solutions to a model equation of motion for a fractional oscillator.

It is of interest to give a detailed physical interpretation for Equations (4) and (6), and apply these equations to real physical processes. Undoubtedly, it is relevant to develop algorithms for the numerical solution of equations in the form of Equation (7). These issues should be the subjects of separate works.

Author Contributions: Conceptualization, L.E., A.P., and S.R.; methodology, L.E., A.P., and S.R.; validation, L.E., A.P., and S.R.; formal analysis, L.E., A.P., and S.R.; investigation, L.E., A.P., and S.R.; writing–original draft preparation, L.E. (Sections 1–5), A.P. (Sections 1, 2, 4, 5), and S.R. (Sections 1–3, 5); writing–review and editing, L.E., A.P., and S.R. All authors have read and agreed to the published version of the manuscript.

References

1. Tarasov, V.E. *Fractional Dynamics: Applications of Fractional Calculus to Dynamics of Particles, Fields and Media*; Springer: Berlin/Heidelberg, Germany, 2010.
2. Uchaikin, V.V. *Fractional Derivatives for Physicists and Engineers*; Springer: Berlin/Heidelberg, Germany, 2012.
3. Herrmann, R. *Fractional Calculus—An Introduction for Physicists*; World Scientific: Singapore, 2018.
4. Riewe, F. Non-conservative Lagrangian and Hamiltonian mechanics. *Phys. Rev. E* **1996**, *53*, 1890–1899. [CrossRef] [PubMed]
5. Riewe, F. Mechanics with fractional derivatives. *Phys. Rev. E* **1997**, *55*, 3581–3592. [CrossRef]
6. Agrawal, O.P. Formulation of Euler-Lagrange equations for fractionalvariational problems. *J. Math. Anal. Appl.* **2002**, *72*, 368–379. [CrossRef]
7. Rekhviashvili, S.S. The Lagrange formalism with fractional derivatives in problems of mechanics. *Tech. Phys. Lett.* **2004**, *30*, 33–37. [CrossRef]
8. Baleanu, D.; Muslih, S.I. Lagrangian formulation of classical fields within Riemann-Liouville fractional derivatives. *Phys. Scr.* **2005**, *72*, 119–121. [CrossRef]
9. Muslih, S.I.; Baleanu, D. Hamiltonian formulation of systems withlinear velocities within Riemann-Liouville fractional derivatives. *J. Math. Anal. Appl.* **2005**, *304*, 599–606. [CrossRef]
10. Rabei, E.M.; Nawafleh, K.I.; Hijjawi, R.S.; Muslih, S.I.; Baleanu, D. The Hamilton formalism with fractional derivatives. *J. Math. Anal. Appl.* **2007**, *327*, 891–897. [CrossRef]
11. Atanacković, T.M.; Konjik, S.; Pilipovic, S. Variational Problems with Fractional Derivatives: Euler-Lagrange Equations. *J. Phys. Math. Theor.* **2008**, *41*, 095201. [CrossRef]
12. Atanacković, T.M.; Stanković, B. On a differential equation with left and right fractional derivatives. *Fract. Calc. Appl. Anal.* **2007**, *10*, 139–150. [CrossRef]
13. Atanacković, T.M.; Stanković, B. On a class of differential equations with left and right fractional derivatives. *ZAMM Z. Angew. Math. Mech.* **2007**, *87*, 537–564. [CrossRef]
14. Eneeva, L.M. Boundary value problem for differential equation with fractional order derivatives with different origins. *Vestn. KRAUNC. Fiz. Mat. Nauk.* **2015**, *2*, 39–44.
15. Eneeva, L.M. On Neumann problem for equation with fractional derivatives with different starting points. *Vestn. KRAUNC. Fiz. Mat. Nauk.* **2018**, *4*, 61–65.
16. Eneeva, L.M. Lyapunov inequality for an equation with fractional derivatives with different origins. *Vestn. KRAUNC. Fiz. Mat. Nauk.* **2019**, *3*, 32–39.
17. Eneeva, L.M. A priori estimate for an equation with fractional derivatives with different origins. *Vestn. KRAUNC. Fiz. Mat. Nauk.* **2019**, *4*, 41–47.
18. Gerasimov, A.N. A generalization of linear laws of deformation and its application to internal friction problem. *Prikl. Mat. Mekh.* **1948**, *12*, 251–260.
19. Samko, S.G.; Kilbas, A.A.; Marichev, O.I. *Fractional Integrals and Derivatives: Theory and Applications*; Taylor & Francis Books: Didcot, UK, 1993.
20. Nakhushev, A.M. *Fractional Calculus and Its Applications*; Fizmatlit: Moscow, Russia, 2003.
21. Kilbas, A.A.; Srivastava, H.M.; Trujillo, J.J. *Theory and Applications of Fractional Differential Equations*; North-Holland Math. Stud.; Elsevier: Amsterdam, The Netherlands, 2006; Volume 204.
22. Agrawal, O.P. Fractional variational calculus in terms of Riesz fractional derivatives. *J. Phys. A: Math. Theor.* **2007**, *40*, 6287–6303. [CrossRef]
23. Rekhviashvili, S.S.; Pskhu, A.V.; Agarwal, P.; Jain, S. Application of the fractional oscillator model to describe damped vibrations. *Turk. J. Phys.* **2019**, *43*, 236–242. [CrossRef]

24. Rekhviashvili, S.S.; Pskhu, A.V. New Method for Describing Damped Vibrations of a Beam with a Built-in End. *Tech. Phys.* **2019**, *64*, 1237–1241. [CrossRef]
25. Gorenflo, R.; Kilbas, A.A.; Mainardi, F.; Rogosin, S.V. *Mittag-Leffler Functions, Related Topics and Applications*; Springer: Berlin/Heidelberg, Germany, 2014.
26. Shalashov, A.G. Can we refer to Hamilton equations for an oscillator with friction? *Phys. Usp.* **2018**, *61*, 1082–1088. [CrossRef]

2

Improved Conditions for Oscillation of Functional Nonlinear Differential Equations

Omar Bazighifan [1,2,†] and **Mihai Postolache** [3,4,5,*,†]

1 Department of Mathematics, Faculty of Science, Hadhramout University, Hadhramout 50512, Yemen; o.bazighifan@gmail.com
2 Department of Mathematics, Faculty of Education, Seiyun University, Hadhramout 50512, Yemen
3 Center for General Education, China Medical University, Taichung 40402, Taiwan
4 Department of Mathematics and Informatics, University Politehnica of Bucharest, 060042 Bucharest, Romania
5 Romanian Academy, Gh. Mihoc-C. Iacob Institute of Mathematical Statistics and Applied Mathematics, 050711 Bucharest, Romania
* Correspondence: mihai@mathem.pub.ro
† These authors contributed equally to this work.

Abstract: The aim of this work is to study oscillatory properties of a class of fourth-order delay differential equations. New oscillation criteria are obtained by using generalized Riccati transformations. This new theorem complements and improves a number of results reported in the literature. Some examples are provided to illustrate the main results.

Keywords: oscillatory solutions; nonoscillatory solutions; fourth-order; delay differential equations; riccati transformation

---

1. Introduction

In this article, we investigate the asymptotic behavior of solutions of the fourth-order differential equation

$$\left(b(x)\left(w'''(x)\right)^{\kappa}\right)' + \sum_{i=1}^{j} q_i(x) f\left(w\left(\vartheta_i(x)\right)\right) = 0, \; x \geq x_0. \tag{1}$$

Throughout this paper, we assume the following conditions hold:

(Z_1) κ are quotient of odd positive integers;

(Z_2) $b \in C^1([x_0,\infty),\mathbb{R})$, $b(x) > 0$, $b'(x) \geq 0$ and under the condition

$$\int_{x_0}^{\infty} \frac{1}{b^{1/\kappa}(x)} \mathrm{d}x = \infty. \tag{2}$$

(Z_3) $q_i \in C[x_0,\infty)$, $q(x) > 0$, $i = 1,2,\ldots,j$,

(Z_4) $\vartheta_i \in C[x_0,\infty)$, $\vartheta_i(x) \leq x$, $\lim_{x\to\infty} \vartheta_i(x) = \infty$; $i = 1,2,..,j$,

(Z_5) $f \in C(\mathbb{R},\mathbb{R})$ such that

$$f(x)/x^{\kappa} \geq \ell > 0, \text{ for } x \neq 0. \tag{3}$$

Definition 1. *The function $y \in C^3[\nu_y,\infty)$, $\nu_y \geq \nu_0$, is called a solution of equation (1), if $b(x)(w'''(x))^{\kappa} \in C^1[x_w,\infty)$, and $w(x)$ satisfies (1) on $[x_w,\infty)$.*

Definition 2. *A solution of (1) is called oscillatory if it has arbitrarily large zeros on* $[x_w, \infty)$, *and otherwise is called to be nonoscillatory.*

Definition 3. *Equation (1) is said to be oscillatory if all its solutions are oscillatory.*

Differential equations arise in modeling situations to describe population growth, biology, economics, chemical reactions, neural networks, and in aeromechanical systems, etc.; see [1].

More and more scholars pay attention to the oscillatory solution of functional differential equations, see [2–5], especially for the second/third-order, see [6–8], or higher-order equations see [9–17]. With the development of the oscillation for the second-order equations, researchers began to study the oscillation for the fourth-order equations, see [18–25].

In the following, we show some previous results in the literature which related to this paper: Moaaz et al. [21] studied the fourth-order nonlinear differential equations with a continuously distributed delay

$$\left(b(x)\left((w(x))'''\right)^{\alpha}\right)' + \int_a^c q(x,\xi) f(w(g(x,\xi)))\, d\xi = 0, \tag{4}$$

by means of the theory of comparison with second-order delay equations, the authors established some oscillation criteria of (4) under the condition

$$\int_{x_0}^{\infty} \frac{1}{b^{1/\kappa}(x)} \mathrm{d}x < \infty. \tag{5}$$

Cesarano and Bazighifan [22] considered Equation (4), and established some new oscillation criteria by means of Riccati transformation technique.

Agarwal et al. [9] and Baculikova et al. [10] studied the equation

$$\left(\left(w^{(n-1)}(x)\right)^{\kappa}\right)' + q(x) f(w(\vartheta(x))) = 0 \tag{6}$$

and established some new sufficient conditions for oscillation.

Theorem 1 (See [9]). *If there exists a positive function* $g \in C^1([x_0,\infty),(0,\infty))$, *and* $\theta > 1$ *is a constant such that*

$$\limsup_{x\to\infty} \int_{x_0}^{x} \left(g(s) q(s) - \lambda\theta \frac{(g'(s))^{\kappa+1}}{\left(g(s)\vartheta^{n-2}(s)\vartheta'(s)\right)^{\kappa}} \right) \mathrm{d}s = \infty, \tag{7}$$

where $\lambda := (1/(\kappa+1))^{\kappa+1} (2(n-1)!)^{\kappa}$, *then every solution of (6) is oscillatory.*

Theorem 2 (See [10]). *Let* $f\left(x^{1/\kappa}\right)/x \geq 1$ *for* $0 < x \leq 1$ *such that*

$$\lim_{x\to\infty}\inf \int_{\vartheta_i(x)}^{x} q(s) f\left(\frac{\varsigma}{(n-1)!} \frac{\vartheta^{n-1}(s)}{b^{1/\kappa}(\vartheta(s))} \right) \mathrm{d}s > \frac{1}{\mathrm{e}} \tag{8}$$

for some $\varsigma \in (0,1)$, *then every solution of (6) is oscillatory.*

To prove this, we apply the previous results to the equation

$$w^{(4)}(x) + \frac{c_0}{x^4} w\left(\frac{9}{10}x\right) = 0,\ x \geq 1, \tag{9}$$

then we get that (9) is oscillatory if

The condition	(7)	(8)
The criterion	$c_0 > 60$	$c_0 > 28.7$

From above, we see that [10] improved the results in [9].

The motivation in studying this paper is complementary and improves the results in [9,10].

The paper is organized as follows. In Section 2, we state some lemmas, which will be useful in the proof of our results. In Section 3, by using generalized Riccati transformations, we obtain a new oscillation criteria for (1). Finally, some examples are considered to illustrate the main results.

For convenience, we denote

$$\delta(x) := \int_x^\infty \frac{1}{b^{1/\kappa}(s)} \mathrm{d}s,\ F_+(x) := \max\{0, F(x)\},$$

$$\psi(x) := g(x)\left(\ell \sum_{i=1}^{j} q_i(x)\left(\frac{\vartheta_i^3(x)}{x^3}\right)^\kappa + \frac{\varepsilon \beta_1^{(1+\kappa)/\kappa} x^2 - 2\beta_1\kappa}{2b^{\frac{1}{\kappa}}(x)\,\delta^{\kappa+1}(x)}\right),$$

$$\phi(x) := \frac{g'_+(x)}{g(x)} + \frac{(\kappa+1)\beta_1^{1/\kappa}\varepsilon x^2}{2b^{\frac{1}{\kappa}}(x)\,\delta(x)},\ \phi^*(x) := \frac{\xi'_+(x)}{\xi(x)} + \frac{2\beta_2}{\delta(x)},$$

and

$$\psi^*(x) := \xi(x)\left(\int_x^\infty \left(\frac{\ell}{b(v)}\int_v^\infty \sum_{i=1}^{j} q_i(s)\frac{\vartheta_i^\kappa(s)}{s^\kappa}\mathrm{d}s\right)^{1/\kappa} \mathrm{d}v + \frac{\beta_2^2 - \beta_2 b^{\frac{-1}{\kappa}}(x)}{\delta^2(x)}\right),$$

where β_1, β_2 are constants and $g, \xi \in C^1([x_0, \infty), (0, \infty))$.

Remark 1. *We define the generalized Riccati substitutions*

$$\pi(x) := g(x)\left(\frac{b(x)\,(w''')^\kappa(x)}{w^\kappa(x)} + \frac{\beta_1}{\delta^\kappa(x)}\right), \tag{10}$$

and

$$\varpi(x) := \xi(x)\left(\frac{w'(x)}{w(x)} + \frac{\beta_2}{\delta(x)}\right). \tag{11}$$

2. Some Auxiliary Lemmas

Next, we begin with the following lemmas.

Lemma 1 ([8]). *Let β be a ratio of two odd numbers, $V > 0$ and U are constants. Then,*

$$P^{(\beta+1)/\beta} - (P-Q)^{(\beta+1)/\beta} \le \frac{1}{\beta} Q^{1/\beta}\left[(1+\beta)P - Q\right],\ PQ \ge 0,\ \beta \ge 1$$

and

$$Uw - Vw^{(\beta+1)/\beta} \le \frac{\beta^\beta}{(\beta+1)^{\beta+1}}\frac{U^{\beta+1}}{V^\beta}.$$

Lemma 2 ([15]). *Suppose that $h \in C^n([x_0, \infty), (0, \infty))$, $h^{(n)}$ is of a fixed sign on $[x_0, \infty)$, $h^{(n)}$ not identically zero, and there exists a $x_1 \ge x_0$ such that*

$$h^{(n-1)}(x)\,h^{(n)}(x) \le 0,$$

for all $x \geq x_1$. *If we have* $\lim_{x\to\infty} h(x) \neq 0$, *then there exists* $x_\beta \geq x_1$ *such that*

$$h(x) \geq \frac{\beta}{(n-1)!} x^{n-1} \left| h^{(n-1)}(x) \right|,$$

for every $\beta \in (0,1)$ *and* $x \geq x_\beta$.

Lemma 3 ([19]). *If the function* u *satisfies* $u^{(j)} > 0$ *for all* $j = 0, 1, ..., n$, *and* $u^{(n+1)} < 0$, *then*

$$\frac{n!}{x^n} u(x) - \frac{(n-1)!}{x^{n-1}} \frac{d}{dx} u(x) \geq 0.$$

3. Oscillation Criteria

In this section, we shall establish some oscillation criteria for Equation (1).

Upon studying the asymptotic behavior of the positive solutions of (1), there are only two cases:

$$\begin{array}{llll} \text{Case (1)}: & w^{(r)}(x) > 0 & \text{for} & r = 0,1,2,3. \\ \text{Case (2)}: & w^{(r)}(x) > 0 & \text{for} & r = 0,1,3 \text{ and } w''(x) < 0. \end{array}$$

Moreover, from Equation (1) and condition (3), we have that $\left(b(x)(w'''(x))^\kappa\right)'$. In the following, we will first study each case separately.

Lemma 4. *Assume that* w *be an eventually positive solution of (1) and* $w^{(r)}(x) > 0$ *for all* $r = 1, 2, 3$. *If we have the function* $\pi \in C^1[x, \infty)$ *defined as (10), where* $g \in C^1([x_0, \infty), (0, \infty))$, *then*

$$\pi'(x) \leq -\psi(x) + \phi(x)\pi(x) - \frac{\kappa \varepsilon x^2}{2(b(x)g(x))^{1/\kappa}} \pi^{\frac{\kappa+1}{\kappa}}(x), \tag{12}$$

for all $x > x_1$, *where* x_1 *is large enough.*

Proof. Let w be an eventually positive solution of (1) and $w^{(r)}(x) > 0$ for all $r = 1, 2, 3$. Thus, from Lemma 2, we get

$$w'(x) \geq \frac{\varepsilon}{2} x^2 w'''(x), \tag{13}$$

for every $\varepsilon \in (0,1)$ and for all large x. From (10), we see that $\pi(x) > 0$ for $x \geq x_1$, and

$$\begin{aligned} \pi'(x) &= g'(x)\left(\frac{b(x)(w''')^\kappa(x)}{w^\kappa(x)} + \frac{\beta_1}{\delta^\kappa(x)}\right) + g(x)\frac{\left(b(w''')^\kappa\right)'(x)}{w^\kappa(x)} \\ &\quad -\kappa g(x)\frac{w^{\kappa-1}(x)w'(x)b(x)(w''')^\kappa(x)}{w^{2\kappa}(x)} + \frac{\kappa\beta_1 g(x)}{b^{\frac{1}{\kappa}}(x)\delta^{\kappa+1}(x)}. \end{aligned}$$

Using (13) and (10), we obtain

$$\begin{aligned} \pi'(x) &\leq \frac{g'_+(x)}{g(x)}\pi(x) + g(x)\frac{\left(b(x)(w'''(x))^\kappa\right)'}{w^\kappa(x)} \\ &\quad -\kappa g(x)\frac{\varepsilon}{2}x^2\frac{b(x)(w'''(x))^{\kappa+1}}{w^{\kappa+1}(x)} + \frac{\kappa\beta_1 g(x)}{b^{\frac{1}{\kappa}}(x)\delta^{\kappa+1}(x)} \\ &\leq \frac{g'(x)}{g(x)}\pi(x) + g(x)\frac{\left(b(x)(w'''(x))^\kappa\right)'}{w^\kappa(x)} \\ &\quad -\kappa g(x)\frac{\varepsilon}{2}x^2 b(x)\left(\frac{\pi(x)}{g(x)b(x)} - \frac{\beta_1}{b(x)\delta^\kappa(x)}\right)^{\frac{\kappa+1}{\kappa}} + \frac{\kappa\beta_1 g(x)}{b^{\frac{1}{\kappa}}(x)\delta^{\kappa+1}(x)}. \end{aligned} \tag{14}$$

Using Lemma 1 with $P = \pi(x) / (g(x) b(x))$, $Q = \beta_1 / (b(x) \delta^{\kappa}(x))$ and $\beta = \kappa$, we get

$$\left(\frac{\pi(x)}{g(x) b(x)} - \frac{\beta_1}{b(x) \delta^{\kappa}(x)}\right)^{\frac{\kappa+1}{\kappa}} \geq \left(\frac{\pi(x)}{g(x) b(x)}\right)^{\frac{\kappa+1}{\kappa}} - \frac{\beta_1^{1/\kappa}}{\kappa b^{\frac{1}{\kappa}}(x) \delta(x)} \left((\kappa+1) \frac{\pi(x)}{g(x) b(x)} - \frac{\beta_1}{b(x) \delta^{\kappa}(x)}\right). \tag{15}$$

From Lemma 3, we have that $w(x) \geq \frac{x}{3} w'(x)$ and hence

$$\frac{w(\vartheta_i(x))}{w(x)} \geq \frac{\vartheta_i^3(x)}{x^3}. \tag{16}$$

From (1), (14), and (15), we obtain

$$\begin{aligned}\pi'(x) \leq{}& \frac{g'_+(x)}{g(x)} \pi(x) - \ell g(x) \sum_{i=1}^{j} q_i(x) \left(\frac{\vartheta_i^3(x)}{x^3}\right)^{\kappa} - \kappa g(x) \frac{\varepsilon}{2} x^2 b(x) \left(\frac{\pi(x)}{g(x) b(x)}\right)^{\frac{\kappa+1}{\kappa}} \\ & -\kappa g(x) \frac{\varepsilon}{2} x^2 b(x) \left(\frac{-\beta_1^{1/\kappa}}{\kappa b^{\frac{1}{\kappa}}(x) \delta(x)} \left((\kappa+1) \frac{\pi(x)}{g(x) b(x)} - \frac{\beta_1}{b(x) \delta^{\kappa}(x)}\right)\right) + \frac{\kappa \beta_1 g(x)}{b^{\frac{1}{\kappa}}(x) \delta^{\kappa+1}(x)}.\end{aligned}$$

This implies that

$$\begin{aligned}\pi'(x) \leq{}& \left(\frac{g'_+(x)}{g(x)} + \frac{(\kappa+1) \beta_1^{1/\kappa} \varepsilon x^2}{2 b^{\frac{1}{\kappa}}(x) \delta(x)}\right) \pi(x) - \frac{\kappa \varepsilon x^2}{2 b^{1/\kappa}(x) g^{1/\kappa}(x)} \pi^{\frac{\kappa+1}{\kappa}}(x) \\ & -g(x) \left(\ell \sum_{i=1}^{j} q_i(x) \left(\frac{\vartheta_i^3(x)}{x^3}\right)^{\kappa} + \frac{\varepsilon \beta_1^{(1+\kappa)/\kappa} x^2 - 2\beta_1 \kappa}{2 b^{\frac{1}{\kappa}}(x) \delta^{\kappa+1}(x)}\right).\end{aligned}$$

Thus,

$$\pi'(x) \leq -\psi(x) + \phi(x) \pi(x) - \frac{\kappa \varepsilon x^2}{2 (b(x) g(x))^{1/\kappa}} \pi^{\frac{\kappa+1}{\kappa}}(x).$$

The proof is complete. □

Lemma 5. *Assume that w is an eventually positive solution of (1), $w^{(r)}(x) > 0$ for $r = 1, 3$ and $w''(x) < 0$. If we have the function $\varpi \in C^1[x, \infty)$ defined as (11), where $\xi \in C^1([x_0, \infty), (0, \infty))$, then*

$$\varpi'(x) \leq -\psi^*(x) + \phi^*(x) \varpi(x) - \frac{1}{\xi(x)} \varpi^2(x), \tag{17}$$

for all $x > x_1$, where x_1 is large enough.

Proof. Let w be an eventually positive solution of (1), $w^{(r)} > 0$ for $r = 1, 3$ and $w''(x) < 0$. From Lemma 3, we get that $w(x) \geq x w'(x)$. By integrating this inequality from $\vartheta_i(x)$ to x, we get

$$w(\vartheta_i(x)) \geq \frac{\vartheta_i(x)}{x} w(x).$$

Hence, from (3), we have

$$f(w(\vartheta_i(x))) \geq \ell \frac{\vartheta_i^{\kappa}(x)}{x^{\kappa}} w^{\kappa}(x). \tag{18}$$

Integrating (1) from x to u and using $w'(x) > 0$, we obtain

$$\begin{aligned} b(u)\left(w'''(u)\right)^{\kappa} - b(x)\left(w'''(x)\right)^{\kappa} &= -\int_x^u \sum_{i=1}^{j} q_i(s) f(w(\vartheta_i(s)))\, ds \\ &\leq -\ell w^{\kappa}(x) \int_x^u \sum_{i=1}^{j} q_i(s) \frac{\vartheta_i^{\kappa}(s)}{s^{\kappa}} ds. \end{aligned}$$

Letting $u \to \infty$, we see that

$$b(x)\left(w'''(x)\right)^{\kappa} \geq \ell w^{\kappa}(x) \int_x^{\infty} \sum_{i=1}^{j} q_i(s) \frac{\vartheta_i^{\kappa}(s)}{s^{\kappa}} ds$$

and so

$$w'''(x) \geq w(x) \left(\frac{\ell}{b(x)} \int_x^{\infty} \sum_{i=1}^{j} q_i(s) \frac{\vartheta_i^{\kappa}(s)}{s^{\kappa}} ds \right)^{1/\kappa}.$$

Integrating again from x to ∞, we get

$$w''(x) \leq -w(x) \int_x^{\infty} \left(\frac{\ell}{b(v)} \int_v^{\infty} \sum_{i=1}^{j} q_i(s) \frac{\vartheta_i^{\kappa}(s)}{s^{\kappa}} ds \right)^{1/\kappa} dv. \tag{19}$$

From the definition of $\varpi(x)$, we see that $\varpi(x) > 0$ for $x \geq x_1$. By differentiating, we find

$$\varpi'(x) = \frac{\xi'(x)}{\xi(x)} \varpi(x) + \xi(x) \frac{w''(x)}{w(x)} - \xi(x) \left(\frac{\varpi(x)}{\xi(x)} - \frac{\beta_2}{\delta(x)} \right)^2 + \frac{\xi(x)\beta_2}{b^{1/\kappa}(x)\delta^2(x)}. \tag{20}$$

Using Lemma 1 with $P = \varpi(x)/\xi(x)$, $Q = \beta_2/\delta(x)$ and $\beta = 1$, we get

$$\left(\frac{\varpi(x)}{\xi(x)} - \frac{\beta_2}{\delta(x)} \right)^2 \geq \left(\frac{\varpi(x)}{\xi(x)} \right)^2 - \frac{\beta_2}{\delta(x)} \left(\frac{2\varpi(x)}{\xi(x)} - \frac{\beta_2}{\delta(x)} \right). \tag{21}$$

From (1), (20), and (21), we obtain

$$\begin{aligned} \varpi'(x) \leq\ & \frac{\xi'(x)}{\xi(x)} \varpi(x) - \xi(x) \int_x^{\infty} \left(\frac{\ell}{b(v)} \int_v^{\infty} \sum_{i=1}^{j} q_i(s) \frac{\vartheta_i^{\kappa}(s)}{s^{\kappa}} ds \right)^{1/\kappa} dv \\ & -\xi(x) \left(\left(\frac{\varpi(x)}{\xi(x)} \right)^2 - \frac{\beta_2}{\delta(x)} \left(\frac{2\varpi(x)}{\xi(x)} - \frac{\beta_2}{\delta(x)} \right) \right) + \frac{\beta_2 \xi(x)}{b^{\frac{1}{\kappa}}(x)\delta^2(x)}. \end{aligned}$$

This implies that

$$\begin{aligned} \varpi'(x) \leq\ & \left(\frac{\xi'_+(x)}{\xi(x)} + \frac{2\beta_2}{\delta(x)} \right) \varpi(x) - \frac{1}{\xi(x)} \varpi^2(x) \\ & -\xi(x) \left(\int_x^{\infty} \left(\frac{\ell}{b(v)} \int_v^{\infty} \sum_{i=1}^{j} q_i(s) \frac{\vartheta_i^{\kappa}(s)}{s^{\kappa}} ds \right)^{1/\kappa} dv + \frac{\beta_2^2 - \beta_2 b^{\frac{-1}{\kappa}}(x)}{\delta^2(x)} \right). \end{aligned}$$

Thus,

$$\varpi'(x) \leq -\psi^*(x) + \phi^*(x)\varpi(x) - \frac{1}{\xi(x)} \varpi^2(x).$$

The proof is complete. □

Lemma 6. *Assume that w is an eventually positive solution of (1). If there exists a positive function $g \in C([x_0, \infty))$ such that*

$$\int_{x_0}^{\infty} \left(\psi(s) - \left(\frac{2}{\varepsilon s^2} \right)^{\kappa} \frac{b(s) g(s) (\phi(s))^{\kappa+1}}{(\kappa+1)^{\kappa+1}} \right) ds = \infty, \tag{22}$$

for some $\varepsilon \in (0, 1)$, then w does not fulfill Case (1).

Proof. Assume that w is an eventually positive solution of (1). From Lemma 4, we get that (12) holds. Using Lemma 1 with

$$U = \phi(x), \; V = \kappa \varepsilon x^2 / \left(2 (b(x) g(x))^{1/\kappa} \right) \text{ and } x = \pi,$$

we get

$$\pi'(x) \leq -\psi(x) + \left(\frac{2}{\varepsilon x^2} \right)^{\kappa} \frac{b(x) g(x) (\phi(x))^{\kappa+1}}{(\kappa+1)^{\kappa+1}}. \tag{23}$$

Integrating from x_1 to x, we get

$$\int_{x_1}^{x} \left(\psi(s) - \left(\frac{2}{\varepsilon s^2} \right)^{\kappa} \frac{b(s) g(s) (\phi(s))^{\kappa+1}}{(\kappa+1)^{\kappa+1}} \right) ds \leq \pi(x_1),$$

for every $\varepsilon \in (0, 1)$, which contradicts (22). The proof is complete. □

Lemma 7. *Assume that w is an eventually positive solution of (1), $w^{(r)}(x) > 0$ for $r = 1, 3$ and $w''(x) < 0$. If there exists a positive function $\xi \in C([x_0, \infty))$ such that*

$$\int_{x_0}^{\infty} \left(\psi^*(s) - \frac{1}{4} \xi(s) (\phi^*(s))^2 \right) ds = \infty, \tag{24}$$

then w does not fulfill Case (2).

Proof. Assume that w is an eventually positive solution of (1). From Lemma 5, we get that (17) holds. Using Lemma 1 with

$$U = \phi^*(x), \; V = 1/\xi(x), \kappa = 1 \text{ and } x = \varpi,$$

we get

$$\pi'(x) \leq -\psi^*(x) + \frac{1}{4} \xi(x) (\phi^*(x))^2. \tag{25}$$

Integrating from x_1 to x, we get

$$\int_{x_1}^{x} \left(\psi^*(s) - \frac{1}{4} \xi(s) (\phi^*(s))^2 \right) ds \leq \pi(x_1),$$

which contradicts (24). The proof is complete. □

Theorem 3. *Assume that there exist positive functions $g, \xi \in C([x_0, \infty))$ such that (22) and (24) hold, for some $\varepsilon \in (0, 1)$. Then, every solution of (1) is oscillatory.*

When putting $g(x) = x^3$ and $\xi(x) = x$ into Theorem 3, we get the following oscillation criteria:

Corollary 1. *Let (2) hold. Assume that*

$$\limsup_{x\to\infty}\int_{x_1}^{x}\left(\varphi(s)-\left(\frac{2}{\varepsilon s^2}\right)^{\kappa}\frac{b(s)\,g(s)\,(\tilde{\varphi}(s))^{\kappa+1}}{(\kappa+1)^{\kappa+1}}\right)ds=\infty, \tag{26}$$

for some $\varepsilon\in(0,1)$. *If*

$$\limsup_{x\to\infty}\int_{x_1}^{x}\left(\varphi_1(s)-\frac{1}{4}\xi(s)\,(\tilde{\varphi}_1(s))^2\right)ds=\infty, \tag{27}$$

where

$$\varphi(x) \; : \; = x^3\left(\ell\sum_{i=1}^{j}q_i(x)\left(\frac{\vartheta_i^3(x)}{x^3}\right)^{\kappa}+\frac{\varepsilon\beta_1^{(1+\kappa)/\kappa}x^2-2\beta_1\kappa}{2b^{\frac{1}{\kappa}}(x)\,\delta^{\kappa+1}(x)}\right)$$

$$\tilde{\varphi}(x) \; : \; = \frac{3}{x}+\frac{(\kappa+1)\,\beta_1^{1/\kappa}\varepsilon x^2}{2b^{\frac{1}{\kappa}}(x)\,\delta(x)},\quad \tilde{\varphi}_1(x):=\frac{1}{x}+\frac{2\beta_2}{\delta(x)}$$

and

$$\varphi_1(x):=x\left(\int_x^{\infty}\left(\frac{\ell}{b(v)}\int_v^{\infty}\sum_{i=1}^{j}q_i(s)\frac{\vartheta_i^{\kappa}(s)}{s^{\kappa}}ds\right)^{1/\kappa}dv+\frac{\beta_2^2-\beta_2 b^{\frac{-1}{\kappa}}(x)}{\delta^2(x)}\right),$$

then every solution of (1) is oscillatory.

Example 1. *Consider a differential equation*

$$w^{(4)}(x)+\frac{c_0}{x^4}w\left(\frac{1}{2}x\right)=0,\ x\geq 1, \tag{28}$$

where $c_0>0$ *is a constant. Note that* $\kappa=b(x)=1$, $q(x)=c_0/x^4$ *and* $\vartheta(x)=x/2$. *Hence, we have*

$$\delta(x_0)=\infty,\ \varphi(s)=\frac{c_0}{8s}.$$

If we set $\ell=\beta_1=1$, *then condition (26) becomes*

$$\begin{aligned}\limsup_{x\to\infty}\int_{x_1}^{x}\left(\varphi(s)-\left(\frac{2}{\varepsilon s^2}\right)^{\kappa}\frac{b(s)\,g(s)\,(\tilde{\varphi}(s))^{\kappa+1}}{(\kappa+1)^{\kappa+1}}\right)ds &= \limsup_{x\to\infty}\int_{x_1}^{x}\left(\frac{c_0}{8s}-\frac{9}{2s}\right)ds\\ &= \infty\ \text{if}\ c_0>36.\end{aligned}$$

Therefore, from Corollary 1, the solutions of Equation (28) are all oscillatory if $c_0>36$.

Remark 2. *We compare our result with the known related criteria for oscillations of this equation as follows:*

1. *By applying Condition (7) in [9] on Equation (28) where* $\theta=2$, *we get*

$$c_0>432.$$

2. *By applying Condition (8) in [10] on Equation (28) where* $\varsigma=1/2$, *we get*

$$c_0>51.$$

Therefore, our result improves results [9,10].

Remark 3. *By applying Condition (26) in Equation (9), we find*

$$c_0 > 6.17.$$

Therefore, our result improves results [9,10].

4. Conclusions

In this article, we study the oscillatory behavior of a class of nonlinear fourth-order differential equations and establish sufficient conditions for oscillation of a fourth-order differential equation by using Riccati transformation. Furthermore, in future work, we get some Hille and Nehari type and Philos type oscillation criteria of (1).

Author Contributions: O.B.: Writing original draft, and writing review and editing. M.P.: Formal analysis, writing review and editing, funding and supervision. All authors have read and agreed to the published version of the manuscript.

Acknowledgments: The authors thank the reviewers for for their useful comments, which led to the improvement of the content of the paper.

References

1. Hale, J.K. *Theory of Functional Differential Equations*; Springer: New York, NY, USA, 1977.
2. Agarwal, R.; Grace, S.; O'Regan, D. *Oscillation Theory for Difference and Functional Differential Equations*; Kluwer Academic Publishers: Dordrecht, The Netherlands, 2000.
3. Alzabut, J.; Tunc, C. Existence of Periodic solutions for a type of Rayleigh equation with state-dependent delay. *Electron. J. Differ. Equ.* **2012**, *2012*, 1–8. [CrossRef]
4. Philos, C.G. A new criterion for the oscillatory and asymptotic behavior of delay differential equations. *Bull. Acad. Pol. Sci. Sér. Sci. Math.* **1981**, *39*, 61–64.
5. Philos, C.G. On the existence of non-oscillatory solutions tending to zero at ∞ for differential equations with positive delays. *Arch. Math.* **1981**, *36*, 168–178. [CrossRef]
6. Bazighifan, O.; Cesarano, C. Some New Oscillation Criteria for Second-Order Neutral Differential Equations with Delayed Arguments. *Mathematics* **2019**, *7*, 619. [CrossRef]
7. Cesarano, C.; Bazighifan, O. Qualitative behavior of solutions of second order differential equations. *Symmetry* **2019**, *11*, 777. [CrossRef]
8. Agarwal, R.P.; Zhang, C.; Li, T. Some remarks on oscillation of second order neutral differential equations. *Appl. Math. Compt.* **2016**, *274*, 178–181. [CrossRef]
9. Agarwal, R.P.; Grace, S.R.; O'Regan, D. Oscillation criteria for certain nth order differential equations with deviating arguments. *J. Math. Appl. Anal.* **2001**, *262*, 601–622. [CrossRef]
10. Baculikova, B.; Dzurina, J.; Graef, J.R. On the oscillation of higher-order delay differential equations. *J. Math.* **2012**, *187*, 387–400. [CrossRef]
11. Grace, S.R. Oscillation theorems for nth-order differential equations with deviating arguments. *J. Math. Appl. Anal.* **1984**, *101*, 268–296. [CrossRef]
12. Xu, Z.; Xia, Y. Integral averaging technique and oscillation of certain even order delay differential equations. *J. Math. Appl. Anal.* **2004**, *292*, 238–246. [CrossRef]
13. Bazighifan, O.; Elabbasy, E.M.; Moaaz, O. Oscillation of higher-order differential equations with distributed delay. *J. Inequal. Appl.* **2019**, *55*, 1–9. [CrossRef]
14. Zhang, C.; Agarwal, R.P.; Bohner, M.; Li, T. New results for oscillatory behavior of even-order half-linear delay differential equations. *Appl. Math. Lett.* **2013**, *26*, 179–183. [CrossRef]
15. Zhang, C.; Li, T.; Sun, B.; Thandapani, E. On the oscillation of higher-order half-linear delay differential equations. *Appl. Math. Lett.* **2011**, *24*, 1618–1621. [CrossRef]
16. Moaaz, O.; Park, C.; Muhib, A.; Bazighifan, O. Oscillation criteria for a class of even-order neutral delay differential equations. *J. Appl. Math. Comput.* **2020**, *2020*, 1–11. [CrossRef]

17. Moaaz, O.; Jan, A.; Omar, B. A New Approach in the Study of Oscillation Criteria of Even-Order Neutral Differential Equations. *Mathematics* **2020**, *8*, 197. [CrossRef]
18. Bazighifan, O.; Cesarano, C. A Philos-Type Oscillation Criteria for Fourth-Order Neutral Differential Equations. *Symmetry* **2020**, *12*, 379. [CrossRef]
19. Chatzarakis, G.E.; Elabbasy, E.M.; Bazighifan, O. An oscillation criterion in 4th-order neutral differential equations with a continuously distributed delay. *Adv. Differ. Equ.* **2019**, *336*, 1–9.
20. Cesarano, C.; Pinelas, S.; Al-Showaikh, F.; Bazighifan, O. Asymptotic Properties of Solutions of Fourth-Order Delay Differential Equations. *Symmetry* **2019**, *11*, 628. [CrossRef]
21. Moaaz, O.; Elabbasy, E.M.; Bazighifan, O. On the asymptotic behavior of fourth-order functional differential equations. *Adv. Differ. Equ.* **2017**, *2017*, 261. [CrossRef]
22. Cesarano, C.; Bazighifan, O. Oscillation of fourth-order functional differential equations with distributed delay. *Axioms* **2019**, *8*, 61. [CrossRef]
23. Moaaz, O.; El-Nabulsi, R.; Bazighifan, O. Oscillatory behavior of fourth-order differential equations with neutral delay. *Symmetry* **2020**, *12*, 371. [CrossRef]
24. Parhi, N.; Tripathy, A. On oscillatory fourth order linear neutral differential equations-I. *Math. Slovaca* **2004**, *54*, 389–410.
25. Zhang, C.; Li, T.; Saker, S. Oscillation of fourth-order delay differential equations. *J. Math. Sci.* **2014**, *201*, 296–308. [CrossRef]

Two Stochastic Differential Equations for Modeling Oscillabolastic-Type Behavior

Antonio Barrera [1,†], Patricia Román-Román [2,3,†] and Francisco Torres-Ruiz [2,3,*,†]

1 Departamento de Análisis Matemático, Estadística e Investigación Operativa y Matemática Aplicada, Facultad de Ciencias, Universidad de Málaga, Bulevar Louis Pasteur, 31, 29010 Málaga, Spain; antonio.barrera@uma.es
2 Departamento de Estadística e Investigación Operativa, Facultad de Ciencias, Universidad de Granada, Avenida Fuente Nueva s/n, 18071 Granada, Spain; proman@ugr.es
3 Instituto de Matemáticas de la Universidad de Granada (IEMath-GR), Calle Ventanilla 11, 18001 Granada, Spain
* Correspondence: fdeasis@ugr.es.
† These authors contributed equally to this work.

Abstract: Stochastic models based on deterministic ones play an important role in the description of growth phenomena. In particular, models showing oscillatory behavior are suitable for modeling phenomena in several application areas, among which the field of biomedicine stands out. The oscillabolastic growth curve is an example of such oscillatory models. In this work, two stochastic models based on diffusion processes related to the oscillabolastic curve are proposed. Each of them is the solution of a stochastic differential equation obtained by modifying, in a different way, the original ordinary differential equation giving rise to the curve. After obtaining the distributions of the processes, the problem of estimating the parameters is analyzed by means of the maximum likelihood method. Due to the parametric structure of the processes, the resulting systems of equations are quite complex and require numerical methods for their resolution. The problem of obtaining initial solutions is addressed and a strategy is established for this purpose. Finally, a simulation study is carried out.

Keywords: diffusion processes; growth model; oscillabolastic curve; stochastic differential equations

1. Introduction

Ordinary differential equations are one of the most frequently used mathematical tools for modeling and describing dynamical systems in general and growth phenomena in particular. In such models a quantitative variable evolves with time according to certain parameters and deterministic functions, and for their properties they are commonly applied to several fields of research.

Studying such models requires, on the one hand, understanding the evolving mechanisms of dynamical systems to properly explain their behavior. On the other, we must be able to include in them external information different from that supplied by the variables under study. In doing so we modify their behavior and exert external control on the evolution of the phenomenon under consideration.

However, in many situations, the resulting models do not adequately reproduce the observed phenomenon. The main reason is that deterministic models do not take into account certain disturbances that can greatly influence the final behavior. These disturbances may have multiple and various origins, which are not always quantifiable or may even be unknown. As stated by Li et al. [1], *"noises are ubiquitous in both nature and human society, such as fire, earthquake, climate warming, financial crisis, etc."*. Many applications require the modeling of external elements whose nature can be considered stochastic, either because of the impossibility of making measurements or because said measurements present a high degree of uncertainty. It is, therefore, necessary to modify the deterministic model to introduce a random element.

Faced with that need, the design of models is obtained from the ordinary differential equation for the deterministic model, in which a random element is introduced, usually a function of a white noise. The result are stochastic models. Such modification requires considering stochastic differential equations (SDE) whose solutions are, under some conditions, diffusion processes. Not only does the use of these stochastic models provide a more realistic explanation of the variables under study, but it also allows us to study other important aspects (like inference or certain time variables that represent some features of the phenomenon).

As previously mentioned, ordinary differential equations are modified to give rise to a certain curve by introducing in them a white noise function. The choice of such function (usually motivated by specific characteristics of the studied phenomenon) can lead to a variety of stochastic processes, although they are related to the same deterministic curve. In addition, that choice can determine our specific approach to the stochastic solution process. In fact, even if the SDE has a solution (and sometimes it is not an explicit one), the resulting diffusion process is difficult to handle because the associated Kolmogorov equations have not yet been solved. For example, Feller pioneered this methodology by applying it to the case of the logistic equation and introducing an additional error term that was directly proportional to the square root of the variable. However, the backward and forward equations associated with the diffusion process solution of the SDE have never been solved (see Schurz [2] for other possible choices of the error term and associated diffusion processes). Furthermore, and regarding the logistic case, in the few cases where the transition density has been calculated (see Tuckwell [3]), it was impossible to find an explicit expression for certain important characteristics such as the mean of the process. This has led to theoretical works that analyze the existence of a solution for the SDE, study its stability properties (see Sun and Wang [4]), or propose simulation techniques (see Hu and Wang [5]). These considerations can be extended to other types of situations beyond the logistic case.

In light of the above, efforts have been made to address SDEs whose solutions are more manageable stochastic processes and also reproduce a certain pattern of observed behavior, i.e., their mean follows a given growth curve. This allows certain issues to be directly addressed, including inference or the study of temporal variables (such as first-passage-times), without having to resort to approximations such as those derived from the discretization of the SDE. The modification of certain ordinary differential equations by adding a multiplicative noise has given rise to processes whose mean is a growth curve. In this sense it is worth highlighting the role of the lognormal process with exogenous factors. As a matter of fact, concrete choices of such exogenous factors lead to processes by which we may study patterns of behavior modeled by a wide range of growth curves (see Román-Román et al. [6,7] and references therein). These references are related to the one-dimensional case, although there are extensions to multidimensionality such as the one proposed by Rupšys in [8], where a 4-variate Bertalanffy-type SDE is considered.

Other lines of action within the scope of SDEs should be mentioned. For example, the inclusion of delays in the modeling of certain phenomena (see Rupšys [9], Longtin [10], Sakthivel et al. [11]), the introduction of fractional Brownian motion and fractional calculus (Dung [12]), or the consideration of both methodologies at once (Moghaddam et al. [13]).

We must keep in mind that the ultimate objective of studying models such as these is modeling behavior patterns in real situations. Beyond the widely discussed classic growth curves (logistic, Gompertz, Bertalanffy, Richards, etc.), other interesting families of curves have been considered in recent times. Among them, the hyperbolastic functions of type I, II and III, introduced by Tabatabai et al. [14], stand out. The main feature of this type of curves is the introduction of hyperbolic functions into known models, thus revealing mobile inflection points and increasing the capability of the models to fit real data. Some works have proved the usefulness of such curves in the description of dynamical phenomena. For example, Eby et al. [15] used them to study the growth of the solid Ehrlich carcinoma under particular treatments, obtaining a more accurate representation than those yielded by other classic curves such as Gompertz or Weibull. Tabatabai et al. [16] used the hyperbolastic type III model to describe the behavior of embryonic stem cells. Later, other models derived from hyperbolastic curves have been introduced, specifically the oscillabolastic and the T-model (Tabatabai et al. [17,18]). Focusing on the first one, the oscillabolastic curve was introduced as an appropriate model to describe dynamic phenomena that present oscillatory behaviors. The introduction of parameter-dependent trigonometric functions allows for oscillatory dynamics to be incorporated into the model and regulates both the amplitude and frequency of such oscillations. A progressive reduction in amplitude guarantees the asymptotic stability of the oscillatory behavior. In that paper the authors applied the oscillating model to two data sets: the first dealing with the oscillabolastic modeling of Ehrlich ascites tumor cells, and the second with the mean signal intensity of Hes1 gene expression in response to serum stimulation.

The main objective of this article is to present two diffusion processes associated with the oscillabolastic curve. Each will be obtained from a differential equation verified by the curve, and include an additive and a multiplicative error term, respectively. The rest of the paper is organized as follows: in Section 2 the oscillabolastic deterministic model is introduced and studied from a classical as well as an alternative differential formulation. The adaptation of these two formulations to stochastic models takes place in Section 3. To provide practical instances of application to the models presented, Section 4 addresses the maximum likelihood estimation of the parameters of each model. The complexity of the models and their high number of parameters make the resolution of the likelihood equations somewhat difficult. The use of numerical methods will thus require the prior estimation of initial solutions for the system, which is why a methodology for its calculation is suggested at the end of that section. Section 5 is a simulation study for both models that includes the generation of data for each process and the application of our proposed estimation methodology.

2. The Oscillabolastic Curve

The original formulation of the oscillabolastic curve is given by the function $u : \mathbb{R} \to \mathbb{R}$,

$$u(t) = \kappa + \alpha \sin(\beta t)/t + \gamma \operatorname{arcsinh}(\theta t)/t, \tag{1}$$

where $\kappa, \alpha, \beta, \gamma$ and θ are real parameters. Clearly, the function $u(t)$ is defined for all t and $\lim_{t\to\infty} u(t) = \kappa$. As easily seen, for $t \to 0^+$ one has $u(t) = \kappa + \alpha\beta + \gamma\theta$, so one can define $u(t)$ also at $t = 0$. However, in the context of growth phenomena, t is a time index and therefore we consider the curve defined in $[t_0, +\infty)$, $t_0 \geq 0$.

To study the behavior of (1), we start by observing that its expression is determined by two terms, say $u(t) = u_1(t) + u_2(t)$, where $u_1(t) = \kappa + \alpha \sin(\beta t)/t$ and $u_2(t) = \gamma \operatorname{arcsinh}(\theta t)/t$. It can therefore be seen that the behavior of the curve is governed mainly by $u_1(t)$. This function summarizes the oscillatory trend of the model, as well as its evolution towards κ, which we consider to be the population limit in a growth model (and not the maximum sustainable capacity that is typical of monotonous and positive growth models). Parameter α is responsible for the amplitude of the oscillations, which become larger as $|\alpha|$ grows. Furthermore, their frequency is directly proportional to $|\beta|$.

The flexibility of the oscillabolastic curve is clearly increased thanks to the influence of $u_2(t)$. In this case, the hyperbolic arcsine allows $u_1(t)$ to be deformed, thus making the oscillatory pattern evolve along a curved path. This in turn has an impact on the sequence of local extremes of $u(t)$ (local minima and maxima at each oscillation), since one of the functions ($u(t)$ or $u_1(t)$) will present a wider range at initial values than the other, which will remain closer to constant value κ. In any case, both will eventually converge to κ. Figure 1 shows the two parts of an oscillabolastic curve for $\kappa = 1, \alpha = -1, \beta = 1, \gamma = -1$ and $\theta = 1$, as well as the resulting curve $u(t)$. It can be seen how $u_2(t)$ influences the final curve by modifying the behavior of $u_1(t)$. This interaction is indicated by a dashed line representing translated function $u_2(t) + \nu$, where ν is the average difference between $u(t)$ and $u_2(t)$.

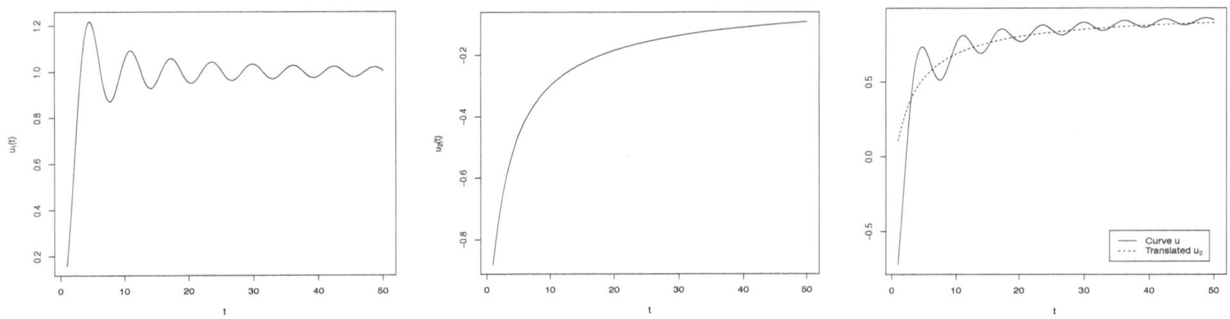

(a) (b) (c)

Figure 1. $u_1(t)$ curve (**a**), $u_2(t)$ curve (**b**), and oscillabolastic $u(t)$ curve together with $u_2(t)$ (dashed line) moved to indicate the trend (**c**). The translation of $u_2(t)$ is the average between the difference $u(t) - u_2(t)$ for every t. Parameters: $\kappa = 1, \alpha = -1, \beta = 1, \gamma = -1, \theta = 1, t_0 = 1$.

Another fundamental aspect of $u_2(t)$ is its ability to make the final curve smoother, and to partially compensate the amplitude of oscillations from $u_1(t)$. In this sense, high values of γ/α cause a deformation of the oscillations, increasing the difference between the increasing and decreasing phases.

In any case, the behavior of the oscillabolastic curve $u(t)$ is characterized by four parameters. It is not easy to establish a clear pattern of influence for each without taking into account the others, as well as the time interval in which the curve is evaluated.

An important characteristic of the oscillabolastic curve is the presence of multiple inflections, a logical consequence of the oscillations of the model. Equation $d^2u(t)/dt^2 = 0$ leads to the following equation, verified for any inflection time t_*,

$$2\gamma \operatorname{arcsinh}(\theta t_*) - \frac{\gamma\theta t_*(2 + 3\theta^2 t_*^2)}{(1+\theta^2 t_*^2)^{3/2}} = 2\alpha\beta t_* \cos(\beta t_*) + \alpha \sin(\beta t_*)(\beta^2 t_*^2 - 2). \tag{2}$$

From an analytical approach, the function $u(t)$ satisfies, under the initial condition $u(t_0) = u_0 \in \mathbb{R}$, the linear ordinary differential equation

$$\frac{d}{dt}u(t) = (g_\delta(t) - u(t))\, t^{-1}, \tag{3}$$

for $t \geq t_0 > 0$, being

$$g_\delta(t) = \kappa + \alpha\beta\cos(\beta t) + \gamma\,\theta\left(1 + \theta^2 t^2\right)^{-1/2} \tag{4}$$

a real function defined in the same domain as $u(t)$ and where δ denotes the parametric vector $\delta = (\kappa, \alpha, \beta, \gamma, \theta)^T$.

On the other hand, it is possible to reformulate the oscillabolastic curve as

$$u(t) = \kappa\left(1 + \alpha' \sin(\beta t)/t + \gamma' \operatorname{arcsinh}(\theta t)/t\right),$$

where the relationship between the parameters of both formulations would be given by $\alpha' = \alpha/\kappa$ and $\gamma' = \gamma/\kappa$.

In this case, it can be verified that the reformulated curve is the solution to the linear ordinary differential equation

$$\frac{d}{dt}u(t) = h_{\delta'}(t)\, u(t), \qquad t \geq t_0 > 0, \tag{5}$$

with initial condition $u(t_0) = u_0$, and where function $h_{\delta'}(t)$ is expressed as

$$h_{\delta'}(t) = \frac{\alpha'\left(\beta t\cos(\beta t) - \sin(\beta t)\right) + \gamma'\left(\frac{\theta t}{\sqrt{1+\theta^2 t^2}} - \operatorname{arcsinh}(\theta t)\right)}{t^2 + \alpha' t\sin(\beta t) + \gamma' t\operatorname{arcsinh}(\theta t)}, \qquad t \geq t_0 > 0, \tag{6}$$

being $\delta' = (\alpha', \beta, \gamma', \theta)^T$.

Both differential equations, (3) and (5), define deterministic models of evolution for the oscillabolastic curve. The most notable difference between them is their structure: in the first case we have a non-homogeneous linear differential equation with non-constant coefficients, while in the second it is reduced to a multiplicative model.

Given the potential applications of both models, their stochastic extensions will be addressed in the following sections. The main idea is to make the mean of the resulting stochastic process able to describe oscillabolastic-type behaviors, which opens the door for observed patterns of this type to be reproduced.

3. Oscillabolastic-Type Diffusion Processes

This section introduces stochastic extensions for the deterministic oscillabolastic model in light of the two differential equations described in the previous section. The starting points will be ordinary differential Equations (3) and (5). Although both are linear differential equations, the diffusion processes to which they give rise have different characteristics. The former is a Gaussian process (of the non-homogeneous Ornstein–Uhlenbeck type), while the second is related to the non-homogeneous geometric Brownian motion, or lognormal process with exogenous factors.

3.1. *Oscillabolastic Gaussian Diffusion Process*

This process is obtained by adding a random component to (3). Specifically, a white noise with variance σ^2 is included in the model, where $\sigma > 0$ is a parameter that marks the random influence on the model. Writing the resulting equation in the traditional form of stochastic differential equations, we have

$$dX^G(t) = \left(g_\delta(t) - X^G(t)\right)t^{-1}dt + \sigma\, dW(t), \tag{7}$$

where $\{W(t); t \geq t_0\}$ is the standard Wiener process in $\mathbb{R}$, which we consider independent from the initial variable $X^G(t_0)$, whereas g_δ is given by (4).

Equation (7) is a linear SDE verifying the conditions of existence and uniqueness of solution. In fact, functions $t \mapsto -t^{-1}$ and $t \mapsto g_\delta(t)t^{-1}$ are measurable and bounded in every subinterval $[t_0, T]$, with $T > t_0$, of $[t_0, +\infty)$, which guarantees the existence and uniqueness of a global solution (i.e., defined in $[t_0, +\infty)$) for each initial condition $X_0^G = X^G(t_0)$.

This solution is a stochastic diffusion process $\{X^G(t); t \geq t_0\}$, adapted to the natural filtration generated by the Borel sets in one dimension and characterized by drift $(g_\delta(t) - x)/t$ and constant infinitesimal variance σ^2. The explicit formulation of the process is given in terms of the solution of ordinary differential equation $dx(t)/dt = -x(t)/t$, leading to

$$X^G(t) = \left(t_0\, X_0^G + \int_{t_0}^{t} g_\delta(s)ds + \sigma\int_{t_0}^{t} s\, dW(s)\right)t^{-1}.$$

Initial condition X_0^G is fundamental to establishing the nature of the process. Indeed, if X_0^G is a constant (degenerate) or a Gaussian random variable, then the solution of (7) is a Gaussian diffusion process (from now on we will consider both scenarios). In addition, in the case where X_0^G is constant, the independence of the increments of the process is also guaranteed.

On the other hand, condition $\mathrm{E}\left[|X_0^G|^p\right] < \infty$ guarantees the existence of the p-th order moment for $X^G(t)$. Specifically, the expressions of the mean and covariance functions are respectively

$$m_\delta(t) = \mathrm{E}[X^G(t)] = \left(t_0\mathrm{E}[X_0^G] + \kappa\Delta_0 t + \alpha\Delta_0\sin(\beta t) + \gamma\Delta_0\,\mathrm{arcsinh}(\theta t)\right)t^{-1},$$
$$C(t,s) = \mathrm{Cov}(X^G(t), X^G(s)) = (\sigma_0 t_0)^2/(st) + \sigma^2\left(t^3 \wedge s^3 - t_0^3\right)/(3st),$$

for $s,t \in [t_0, +\infty)$, being $\sigma_0^2 = \mathrm{Var}[X_0^G]$. Here Δ_0 denotes increasing operator $\Delta_0 f(t) = f(t) - f(t_0)$ for a function f, whereas $t \wedge s = Min(s,t)$.

Regarding the mean function $m_\delta(t)$, note that it can be expressed as

$$m_\delta(t) = u(t) + \frac{t_0}{t}\left(\mathrm{E}[X_0^G] - u_0\right),$$

where $u_0 = u(t_0)$ is the initial value of the oscillabolastic curve. Also note that when constructing the stochastic model, the main goal is for the mean function of the process to coincide with that of a prescribed oscillabolastic model, i.e., $m_\delta(t) = u(t)$ for all t. Regarding the last expression, this condition holds when $\mathrm{E}[X_0^G] = u_0$. However, the asymptotic tendency of the process (the behavior of its mean at the limit) is similar to that of the oscillabolastic deterministic model. Indeed, the difference between the mean and the curve, for each time instant, decreases as t tends to ∞, i.e.,

$$\lim_{t\to\infty} |m_\delta(t) - u(t)| = 0.$$

With regard to the transition density function, which is required to implement the inference procedures that will be described in the next section, for any time instants $t > s \geq t_0$ the conditioned variable $X^G(t)|X^G(s) = y$ is normal, with mean and variance given respectively by expressions

$$\mu_\delta(s,t) = m_\delta(t) + \sigma(t)\,\rho(s,t)\,(y - m_\delta(s))/\sigma(s),$$
$$\nu(s,t)^2 = \left(1 - \rho(s,t)^2\right)\sigma(t)^2,$$

where $\sigma(t) = C(t,t)^{1/2}$ is the standard deviation of $X^G(t)$ (resp. for s) and $\rho(s,t) = C(s,t)/\left(\sigma(s)\sigma(t)\right)$ is the correlation function. Both expressions can also be written in the form

$$\mu_\delta(s,t) = \frac{1}{t}\left(sy + \int_s^t g_\delta(r)dr\right)$$

and

$$\nu(s,t)^2 = \frac{\sigma^2}{3}\left(\frac{t^3 - s^3}{t^2}\right),$$

from which the probability transition density function is given by

$$f(x,t|y,s) = \left(2\pi\frac{\sigma^2}{3}\left(\frac{t^3 - s^3}{t^2}\right)\right)^{-\frac{1}{2}} \exp\left(-\frac{\left(x - \frac{s}{t}y - \frac{1}{t}\int_s^t g_\delta(r)dr\right)^2}{2\frac{\sigma^2}{3}\left(\frac{t^3-s^3}{t^2}\right)}\right). \tag{8}$$

3.2. Oscillabolastic Diffusion Process Derived from the Non-Homogeneous Lognormal Diffusion Process

From Equation (5), it is possible to obtain a stochastic model whose mean function coincides with the original deterministic model, i.e., the oscillabolastic curve. As a matter of fact, this differential equation represents a Malthusian model with a time-dependent fertility $h_{\delta'}(t)$ given by (6).

The introduction of a white noise $\varrho(t)$ with variance σ^2 in this fertility component, accomplished by replacing $h_{\delta'}(t)$ with $h_{\delta'}(t) + \varrho(t)$, leads to the SDE

$$dX^L(t) = h_{\delta'}(t)X^L(t)\,dt + \sigma X^L(t)\,dW(t), \qquad X^L(t_0) = X_0^L, \tag{9}$$

where $W(t)$ is the standard Wiener process, independent from X_0^L, $t \geq t_0$. Please note that the reformulation of the original curve leads to a model with one parameter less than the Gaussian model.

In this case, the existence and uniqueness of the solution is ensured by the continuity of the $h_{\delta'}(t)$ function. Thus, the solution of (9) is a stochastic diffusion process $\{X^L(t); t \geq t_0\}$ characterized by drift $h_{\delta'}(t)x$ and infinitesimal variance $\sigma^2 x^2$. Furthermore, a closed-form expression for the solution is given, for $t \geq t_0$, by

$$X^L(t) = X_0^L \exp\left(H_\xi(t_0,t) + \sigma(W(t) - W(t_0))\right),$$

where

$$H_\xi(s,t) = \int_s^t h_{\delta'}(u)du - \frac{\sigma^2}{2}(t-s) = \log\frac{Q_{\delta'}(t)}{Q_{\delta'}(s)} - \frac{\sigma^2}{2}(t-s), \qquad s < t,\ \xi = \left(\delta'^T, \sigma^2\right)^T,$$

being

$$Q_{\delta'}(t) = 1 + \alpha' \sin(\beta t)/t + \gamma' \operatorname{arcsinh}(\theta t)/t.$$

This process is a particular case of the non-homogeneous lognormal diffusion, or lognormal diffusion process with exogenous factors. The inclusion of the $h_{\delta'}(t)$ function in the drift justifies the terminology: the infinitesimal mean of the process can be affected by the inclusion of time-dependent external factors.

As regards the distribution of the process, if X_0^L is distributed according to a lognormal distribution $\Lambda_1\left[\mu_0; \sigma_0^2\right]$ or is a degenerate random variable ($P[X_0^L = x_0] = 1$), all the finite-dimensional distributions of the process are lognormal. Concretely, $\forall n \in \mathbb{N}$ and $t_1 < \cdots < t_n$, vector $(X^L(t_1), \ldots, X^L(t_n))^T$ is distributed according to a n-dimensional lognormal distribution $\Lambda_n[\boldsymbol{\varepsilon}, \boldsymbol{\Sigma}]$, where the components of the vector $\boldsymbol{\varepsilon}$ are $\varepsilon_i = \mu_0 + H_\xi(t_0, t_i)$, $i = 1, \ldots, n$, being $\sigma_{ij} = \sigma_0^2 + \sigma^2(\min(t_i, t_j) - t_0)$, $i, j = 1, \ldots, n$, those of the matrix $\boldsymbol{\Sigma}$.

As far as the transition probability is concerned, it is obtained from that of $(X^L(s), X^L(t))^T$, $s < t$. Concretely,

$$X^L(t) \mid X^L(s) = y \rightsquigarrow \Lambda_1\left(\log y + H_\xi(s,t), \sigma^2(t-s)\right), \qquad s < t,$$

whose density function is

$$f(x,t|y,s) = \frac{1}{x\sqrt{2\pi\sigma^2(t-s)}} \exp\left(-\frac{\log\frac{x}{y} - H_\xi(s,t)}{2\sigma^2(t-s)}\right). \tag{10}$$

Once the distribution of the process has been established, we may calculate a variety of its associated characteristics including the mean and conditioned mean functions, whose expressions are

$$m_{\delta'}(t) = E[X_0^L]\,\frac{Q_{\delta'}(t)}{Q_{\delta'}(t_0)}$$

and

$$m_{\delta'}(t|t_0) = E[X^L(t)|X^L(t_0) = x_0] = x_0 \exp\left(H_\xi(t_0,t) + \frac{\sigma^2}{2}\right) = x_0\,\frac{Q_{\delta'}(t)}{Q_{\delta'}(t_0)},$$

respectively. Both functions are oscillabolastic curves of the type introduced in the previous section. Other characteristics of the process can be obtained from the results included in Román et al. [6].

4. Estimation of the Parameters

The main useful application of the diffusion processes described in the previous section is the description of dynamic phenomena subject to random influences, in particular growth phenomena. For that reason, the estimation of the parameters included in the models is especially relevant. Among the procedures available we will employ the maximum likelihood estimation method. The estimation of the parameters involves solving a system of equations, which will require using adequate numerical techniques.

Let us consider d sample paths observed at time instants t_{ij}, $(i = 1, \ldots, d,\ j = 1, \ldots, n_i)$. Please note that neither the sample sizes nor the times of observation have to be the same, although we will suppose that the first time of observation is common for all sample paths, i.e., $t_{i1} = t_0$, $i = 1, \ldots, d$. Let $\mathbf{X}_i^T$ be the vector containing the random variables of the i-th sample-path, i.e., $\mathbf{X}_i = (X(t_{i1}), \ldots, X(t_{i,n_i}))^T$, $i = 1, \ldots, d$, and denote $\mathbf{X} = \left(\mathbf{X}_1^T | \cdots | \mathbf{X}_d^T\right)^T$.

For simplicity, we consider a degenerate initial distribution, i.e., $\mathrm{P}[X_0 = x_0] = 1$. The reason is that the parameters of the distribution in the non-degenerate case (normal or lognormal depending on the process) are usually considered functionally independent from the parameters involved in the transition distribution, so the estimation of these last parameters is not influenced by the estimation of the others. In any case, if X_0 were not degenerate its parameters could not be estimated from the values of the sample paths at the first time instant of observation.

In what follows, and for the sake of simplicity in the notation, we will not differentiate between the processes when noting the observed values. Obviously, the context determines the meaning in each case.

4.1. The Gaussian Case

Let us consider a fixed value $\mathbf{x}$ of $\mathbf{X}$. From (8) the likelihood function can be obtained, whose logarithm is

$$\log \mathrm{L}_{\mathbf{x}}(\xi) = -\frac{n}{2}\log\frac{2\pi}{3} - \frac{n}{2}\log\sigma^2 - \frac{1}{2}\sum_{i=1}^{d}\sum_{j=2}^{n_i}\log\frac{\Delta t_{ij}^3}{t_{ij}^2} - \frac{3}{2\sigma^2}\sum_{i=1}^{d}\sum_{j=2}^{n_i}\frac{1}{\Delta t_{ij}^3}\left(\Delta\left(t_{ij}x_{ij}\right) - G_{ij}^{\delta}\right)^2,$$

where $n = \sum_{i=1}^{d}(n_i - 1)$ and $G_{ij}^{\delta} := \int_{t_{ij-1}}^{t_{ij}} g_\delta(s)ds$. Now operator Δ (applied to a generic function f) denotes the increments $\Delta f(t_{ij}) = f(t_{ij}) - f(t_{ij-1})$.

To obtain the maximum likelihood estimates, we must calculate the derivatives of $\log \mathrm{L}_{\mathbf{x}}(\xi)$ with respect to the parameters. In this sense, the derivatives with respect to the components of δ are

$$\partial_\tau \log \mathrm{L}_{\mathbf{x}}(\xi) = \frac{3}{\sigma^2}\sum_{i=1}^{d}\sum_{j=2}^{n_i}\frac{1}{\Delta t_{ij}^3}\left(\Delta\left(t_{ij}x_{ij}\right) - G_{ij}^{\delta}\right)\partial_\tau G_{ij}^{\delta}, \qquad \tau \in \delta \tag{11}$$

where

$$\partial_\tau G_{ij}^{\delta} = \begin{cases} \Delta t_{ij} & \text{if } \tau = \kappa, \\ \Delta\sin(\beta t_{ij}) & \text{if } \tau = \alpha, \\ \alpha\,\Delta\left(t_{ij}\cos(\beta t_{ij})\right) & \text{if } \tau = \beta, \\ \Delta\,\mathrm{arcsinh}(\theta t_{ij}) & \text{if } \tau = \gamma, \\ \gamma\,\Delta\left(t_{ij}^2\Big/\sqrt{1+\theta^2 t_{ij}^2}\right) & \text{if } \tau = \theta. \end{cases}$$

On the other hand, the partial derivative with respect to σ^2 is

$$\partial_{\sigma^2} \log \mathrm{L}_{\mathbf{x}}(\xi) = -\frac{n}{2\sigma^2} + \frac{3}{2\left(\sigma^2\right)^2} \sum_{i=1}^{d} \sum_{j=2}^{n_i} \frac{1}{\Delta t_{ij}^3} \left(\Delta\left(t_{ij}x_{ij}\right) - G_{ij}^{\delta}\right)^2. \tag{12}$$

Finally, by using (11) and (12), and equating to zero the partial derivative of the loglikelihood, one gets:

$$\begin{aligned}
\sum_{i=1}^{d}\sum_{j=2}^{n_i} \frac{1}{\Delta t_{ij}^3} \left(\Delta\left(t_{ij}x_{ij}\right) - k\Delta t_{ij} - \alpha\Delta\sin(\beta t_{ij}) - \gamma\Delta\,\mathrm{arcsinh}(\theta t_{ij})\right)\Delta t_{ij} &= 0,\\
\sum_{i=1}^{d}\sum_{j=2}^{n_i} \frac{1}{\Delta t_{ij}^3} \left(\Delta\left(t_{ij}x_{ij}\right) - k\Delta t_{ij} - \alpha\Delta\sin(\beta t_{ij}) - \gamma\Delta\,\mathrm{arcsinh}(\theta t_{ij})\right)\Delta\sin(\beta t_{ij}) &= 0,\\
\sum_{i=1}^{d}\sum_{j=2}^{n_i} \frac{1}{\Delta t_{ij}^3} \left(\Delta\left(t_{ij}x_{ij}\right) - k\Delta t_{ij} - \alpha\Delta\sin(\beta t_{ij}) - \gamma\Delta\,\mathrm{arcsinh}(\theta t_{ij})\right)\alpha\Delta\left(t_{ij}\cos(\beta t_{ij})\right) &= 0,\\
\sum_{i=1}^{d}\sum_{j=2}^{n_i} \frac{1}{\Delta t_{ij}^3} \left(\Delta\left(t_{ij}x_{ij}\right) - k\Delta t_{ij} - \alpha\Delta\sin(\beta t_{ij}) - \gamma\Delta\,\mathrm{arcsinh}(\theta t_{ij})\right)\Delta\,\mathrm{arcsinh}(\theta t_{ij}) &= 0,\\
\sum_{i=1}^{d}\sum_{j=2}^{n_i} \frac{1}{\Delta t_{ij}^3} \left(\Delta\left(t_{ij}x_{ij}\right) - k\Delta t_{ij} - \alpha\Delta\sin(\beta t_{ij}) - \gamma\Delta\,\mathrm{arcsinh}(\theta t_{ij})\right)\gamma\Delta\left(t_{ij}^2 \Big/ \sqrt{1+\theta^2 t_{ij}^2}\right) &= 0,\\
\frac{3}{n\sigma^2}\sum_{i=1}^{d}\sum_{j=2}^{n_i} \frac{1}{\Delta t_{ij}^3} \left(\Delta\left(t_{ij}x_{ij}\right) - k\Delta t_{ij} - \alpha\Delta\sin(\beta t_{ij}) - \gamma\Delta\,\mathrm{arcsinh}(\theta t_{ij})\right)^2 - 1 &= 0.
\end{aligned} \tag{13}$$

The resolution of this system will be determined by the application of numerical techniques. Such procedures will require a set of initial solutions. This problem, common to the two models presented, will be addressed in Section 4.3.

4.2. Lognormal Case

Regarding oscillabolastic process $X^L(t)$, given by SDE (9), the maximum likelihood estimation can be addressed by means of the procedure described in [6]. This approach has already been used succesfully in the inferential treatment of the Gompertz multisigmoidal diffusion process [7], as well as in the generalization of the classic Weibull model, in particular for the hyperbolastic type III diffusion process [19].

In this case, and considering probability transition density function (10), the logarithm of the likelihood function is given by

$$\log \mathrm{L}_{\mathbf{x}}(\xi) = -\frac{n}{2}\log(2\pi) - \frac{n}{2}\log\sigma^2 - \frac{Z_1 + \Phi_\xi - 2\Gamma_\xi}{2\sigma^2},$$

where

$$Z_1 = \sum_{i=1}^{d}\sum_{j=2}^{n_i} \frac{1}{\Delta t_{ij}} \log\frac{x_{ij}}{x_{ij-1}}, \quad \Phi_\xi = \sum_{i=1}^{d}\sum_{j=2}^{n_i} \frac{1}{\Delta t_{ij}}\left(m_{ij}^{\xi}\right)^2, \quad \Gamma_\xi = \sum_{i=1}^{d}\sum_{j=2}^{n_i} \frac{m_{ij}^{\xi}}{\Delta t_{ij}} \log\frac{x_{ij}}{x_{ij-1}}$$

being $m_{ij}^{\xi} = H_\xi(t_{ij-1}, t_{ij})$.

The estimate of vector ξ (see [6] for details) follows by solving the system of equations

$$\Psi_{\delta'} - \Omega_\xi = 0, \tag{14}$$

$$Z_1 + \Phi_\xi - 2\Gamma_\xi - \sigma^2 Z_2 + \sigma^2 Y_\xi = n\sigma^2, \tag{15}$$

where
$$\Omega_\xi = \frac{1}{2}\frac{\partial \Phi_\xi}{\partial \delta'^T}, \quad \Psi_{\delta'} = \frac{\partial \Gamma_\xi}{\partial \delta'^T}, \quad Y_\xi = -\frac{\partial \Phi_\xi}{\partial \sigma^2}, \quad Z_2 = -2\frac{\partial \Gamma_\xi}{\partial \sigma^2}.$$

These functions are obtained from the partial derivatives of the m_{ij}^ξ functions:

$$_\tau\omega_{\delta'} = \frac{\partial m_{ij}^\xi}{\partial \tau} = \frac{1}{Q_{\delta'}(t_{ij})}\left.\frac{\partial Q_{\delta'}(t)}{\partial \tau}\right|_{t=t_{ij}} - \frac{1}{Q_{\delta'}(t_{ij-1})}\left.\frac{\partial Q_{\delta'}(t)}{\partial \tau}\right|_{t=t_{ij-1}}, \qquad \tau \in \delta',$$
$$\frac{\partial m_{ij}^\xi}{\partial \sigma^2} = -\frac{1}{2}\Delta t_{ij},$$

where
$$\frac{\partial Q_{\delta'}}{\partial \tau} = \begin{cases} \sin(\beta t)/t & \text{if } \tau = \alpha', \\ \alpha' \cos(\beta t) & \text{if } \tau = \beta, \\ \operatorname{arcsinh}(\theta t)/t & \text{if } \tau = \gamma', \\ \gamma'/\sqrt{1+\theta^2 t^2} & \text{if } \tau = \theta. \end{cases}$$

From these last expressions, and from those of Ω_ξ and $\Psi_{\delta'}$, the subsystem of Equation (14) can be written in the form
$$_\tau\Xi_{\delta'} + \frac{\sigma^2}{2}\, {_\tau\overline{\Omega}_{\delta'}} = 0, \ \tau \in \delta', \tag{16}$$

with
$$_\tau\Xi_{\delta'} = \sum_{i=1}^{d}\sum_{j=2}^{n_i} \frac{\log\left(x_{i,j}/x_{i,j-1}\right) - \log\left(Q_{\delta'}(t_{ij})/Q_{\delta'}(t_{i,j-1})\right)}{\Delta t_{ij}}\, {_\tau\omega_{\delta'}^{i,j,j-1}}, \quad {_\tau\overline{\Omega}_{\delta'}} = \sum_{i=1}^{d}\sum_{j=2}^{n_i} {_\tau\omega_{\delta'}^{i,j,j-1}}, \ \tau \in \delta'.$$

On the other hand, and after some calculus, Equation (15) transforms into
$$\sigma^2\left[n + \sigma^2 Z_3/4\right] + 2X_1^{\delta'} - X_2^{\delta'} - Z_1 = 0, \tag{17}$$

being
$$X_1^{\delta'} = \sum_{i=1}^{d}\sum_{j=2}^{n_i} \frac{\left(\log\left(Q_{\delta'}(t_{ij})/Q_{\delta'}(t_{i,j-1})\right)\right)^2}{\Delta t_{ij}}, \quad X_2^{\delta'} = \sum_{i=1}^{d}\sum_{j=2}^{n_i} \frac{\log\left(x_{i,j}/x_{i,j-1}\right)\, \log\left(Q_{\delta'}(t_{ij})/Q_{\delta'}(t_{i,j-1})\right)}{\Delta t_{ij}},$$

and $Z_3 = \sum_{i=1}^{d}\sum_{j=2}^{n_i} \Delta t_{ij}$.

As in the previous case, system of Equations (16) and (17) cannot be solved explicitly, and it is, therefore, necessary to use numerical methods such as Newton-Raphson's, for which an initial solution is required. In the next subsection we present a strategy to provide such a solution based on the information provided by the sample data, and applicable to the two resulting systems of equations.

4.3. Initial Solutions

As already mentioned, the systems of likelihood equations obtained in the preceding subsections have a high degree of complexity due to the high number of parameters involved (although in the lognormal case this number is reduced) and to the presence of random data. Their resolution, therefore, will require the use of numerical techniques. For such procedures to be applied a prior choice must be made regarding initial solutions.

This section proposes a strategy based on the information provided by the sample data. The key to the procedure focuses on the sample mean function, which is calculated from the values of the sample paths at each time instant of observation t_i. Since the procedure needs to evaluate this function over the entire observation interval, a spline function is later fitted to the mean values of the observed sample paths. Furthermore, the procedure is applicable to the two processes introduced earlier. The goal is to make the theoretical mean functions of both processes approximate the oscillabolastic curve (1), although in the Gaussian case this approach is asymptotic. We expose the strategy for the Gaussian case. For the lognormal-type process it is deduced by simply taking into account the relationship between the parameters which has been described in Section 2.

The following is the suggested procedure for each parameter:

- κ: This is the limit value of the model, and represents its asymptotic behavior when $t \to +\infty$. However, due to oscillations, this value is not the maximum (which usually occurs in monotonous positive growth models, giving rise to the *carrying capacity* of the system under consideration). Since the curve approaches κ, a possible option to obtain an initial value κ_0 is to take the last value of the sample mean function. However, this depends largely on the length of the time interval, as well as on the amplitude of the oscillations and/or of the curvature of the model. These effects can lead to estimates of κ far from its true value. For greater accuracy, an alternative approach would be to take the average value between the sequences of local extreme values (maxima and minima) of the sample mean function.
- $\alpha, \beta, \gamma, \theta$: The calculation of initial values for these parameters is quite complex. The starting point is to have an approximation for one of the instants of time (t_*) in which an inflection is observed in the mean of the process. To obtain it, a maximum of the derivative of the sample mean function is numerically approached.

  On the other hand, it can be observed that (1) is a linear function in α and γ. Therefore, considering t_0 and t_*, and the value κ_0 previously calculated, it is possible to obtain an expression of both parameters as a function of β and θ from the system of equations

$$\begin{aligned}\kappa_0 &= u_0 - \alpha\sin(\beta t_0)/t_0 - \gamma\,\text{arcsinh}(\theta t_0)/t_0,\\ \kappa_0 &= u_* - \alpha\sin(\beta t_*)/t_* - \gamma\,\text{arcsinh}(\theta t_*)/t_*.\end{aligned} \tag{18}$$

  Please note that u_0 and u_* can be obtained from the spline function previously adjusted to the mean of the sample paths. We shall denote by $\alpha_{\beta,\theta}$ and $\gamma_{\beta,\theta}$ the resulting values. Now the problem has been reduced to finding initial values for β and θ. To this end we look for pairs of values (β, θ) within a two-dimensional bounded region satisfying the inflection condition (2) for a predefined error threshold, say ϵ, i.e., it must hold

$$\left|\frac{\gamma_{\beta,\theta}\theta t_*(2+3\theta^2 t_*^2)}{(1+\theta^2 t_*^2)^{3/2}} - 2\gamma_{\beta,\theta}\,\text{arcsinh}(\theta t_*) + 2\alpha_{\beta,\theta}\beta t_*\cos(\beta t_*) + \alpha_{\beta,\theta}\sin(\beta t_*)(\beta^2 t_*^2 - 2)\right| < \epsilon,$$

  being t_* the inflection time instant previously obtained. We suggest using values of ϵ between 0.0001 and 0.1, depending on the order of magnitude of the sample data. Initial values β_0 and θ_0 will be the mean of the resulting values for each parameter, respectively. Finally, the initial values for α and γ are obtained from $\alpha_{\beta_0,\theta_0}$ and $\gamma_{\beta_0,\theta_0}$, respectively.

- σ^2: The initial value for σ^2 is obtained from the estimation of the variance of the one-dimensional distributions of the processes. For instance, in the case of the Gaussian oscillabolastic process, the variance is (please remember that we have considered a degenerate initial distribution)

$$Var[X(t)] = \sigma^2\frac{t^3 - t_0^3}{3t^2},$$

so, if we denote by σ_i^2 the sample variance at t_i, a linear regression model of the σ_i^2 values versus t_i $(i = 1, \ldots, n)$ is considered in order to obtain the estimated value for σ^2. It should be noted that the sample variance can sometimes present a rather volatile behavior. Therefore, in practice it is advisable to procure a smoothed version of said function before obtaining values for σ_i^2.

For the other oscillabolastic-type diffusion process, for which (see Section 3.2) $X(t) \rightsquigarrow \Lambda_1[\varepsilon, \varsigma]$ being $\varsigma = \sigma^2(t - t_0)$, we can avoid using the sample variance function. In point of fact, it is well-known that for a lognormal distribution $\Lambda_1[\varepsilon, \varsigma]$ the quotient between the arithmetic mean and the geometric one provides an estimation of ς. This leads, for each t_i, and when the initial distribution is degenerate, to an estimate of $\sigma^2(t_i - t_0)$ in the distribution of $X(t)$, say, $\sigma_i^2 = 2\log(m_i/m_i^g)$, $i = 1, \ldots,$ where m_i and m_i^g are, respectively, the values of the mean and the geometric sample mean of the sample paths at t_i. Again, the initial value for σ^2 is obtained by performing a simple linear regression of σ_i^2 values against t_i.

Please note that if X_0 is not a degenerate random variable, σ_i^2 provides an estimate of $\sigma_0^2 + \sigma^2(t_i - t_0)$ for each t_i. In that case, σ_0^2 can be previously estimated from the values of the sample paths at t_0 and then its values introduced in the regression function.

5. Simulation Study

In this section, a simulation study illustrates the behavior of the oscillabolastic process, as well as the application of the proposed methodology for estimating the parameters of the processes under consideration. One of its most interesting aspects is how initial solutions are obtained for the subsequent estimation of the parameters of the process, to which end we have applied the methodology proposed in the previous subsection.

5.1. Simulation in the Gaussian Case

Data from the oscillabolastic Gaussian process are obtained by simulating 20 sample paths in the time interval $[1, 10]$, with degenerate initial distribution $x_0 = 1$. As for the values of the parameters, we have started by considering $\kappa = 2.5, \alpha = -2, \beta = 1.5, \gamma = 1$ and $\theta = 0.5$. The parameter corresponding to the infinitesimal variance is, in this case, $\sigma^2 = 0.0025$. Simulated paths are shown in Figure 2.

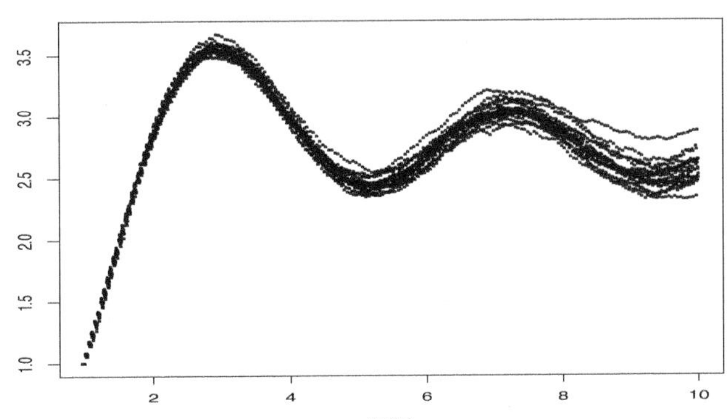

Figure 2. Simulated sample paths for the oscillabolastic Gaussian process. Parameters: $\kappa = 2.5$, $\alpha = -2, \beta = 1.5, \gamma = 1, \theta = 0.5$ and $\sigma^2 = 0.0025$.

Following the method described in Section 4.1, the system of maximum likelihood Equation (13) is solved by numerical procedures. To do this, initial solutions to the parameters are obtained by applying the methodology described in Section 4.3. The initial value for κ is selected by taking the last value of the observed mean, in this case $\kappa_0 = 2.58$. One inflection point is calculated from (2) after fitting a spline function to the observed mean values, resulting in $t_* = 1.38$, and being $u_* = 1.71$ the value of the curve at this point. With this information, initial values β_0 and θ_0 for β and θ, respectively, are found by holding the inflection condition with an error threshold $\epsilon = 5 \times 10^{-4}$. The search region is determined by the square $[0.1, 3] \times [0.1, 1]$. In fact, the number of oscillations of the observed data suggests that β must lie between 0.1 and 3 (and might even be 2). On the other hand, the absence of a pronounced curvature implies low values of θ. Finally, the initial values are $\beta_0 = 1.5$ and $\theta_0 = 0.3$. From these values, and using the expressions $\alpha_{\beta_0,\theta_0}$ and $\gamma_{\beta_0,\theta_0}$ obtained from (18), we find the corresponding initial values for α and β, concretely $\alpha_0 = -1.97$ and $\gamma_0 = 1.1$. Finally, the initial value for σ^2 is obtained from the estimation of the slope of the regression line of observed variability against time, resulting in $\sigma_0^2 = 0.0029$.

Once the initial values of the parameters have been determined, they are used to solve the system of equations. Detailed results for estimated values of each parameter are shown in Table 1. Please note that in this case, initial values coincide with estimated ones, mainly due to low variability of the sample paths.

Table 1. Results for simulated Gaussian oscillabolastic diffusion.

	κ	α	β	γ	θ	σ^2
Estimated values	2.584	−1.973	1.500	1.1	0.300	0.0029

To evaluate the quality of the estimation of the process, we have considered the absolute relative errors between the sample mean of the simulated process and the estimated one, i.e.,

$$\text{RAE} = \frac{1}{n} \sum_{j=1}^{n} \frac{|m_j - \hat{m}_j|}{m_j},$$

where m_j and $\hat{m}_j$ are the values of the sample mean function and the ones estimated at t_j, $j = 1, \ldots, n$.

In this case, the value of this error is RAE $= 0.024$, which indicates a very good fit between the observed and estimated mean functions from the values found for the parameters. This can be visualized in Figure 3, in which both mean functions are plotted.

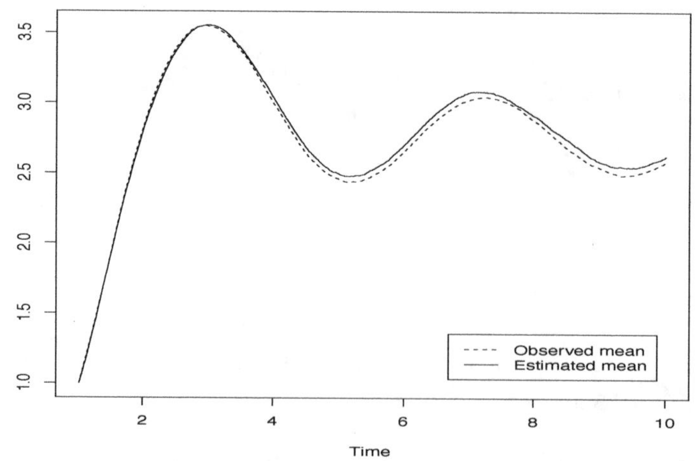

Figure 3. Simulated and estimated mean functions. Gaussian oscillabolastic process.

A new simulation study has been carried out in order to check the process estimation for oscillatory behavior, for example when the parameter β that influences the frequency of the oscillations is varied, and to accommodate for the variability of the sample paths. The simulation pattern is the same as the one considered in the previous case. Parameter β takes values 2, 3, and 4. As previously mentioned, the number of oscillations in the data may determinate the range of values of β. Also, the curvature of the trend followed by the oscillations may be a good indicator for the range of parameter θ. Initial values coming from this bounded region lead to the estimation of α and γ.

Final results for different combinations of parameters are shown in Table 2. A graphical comparison between simulated and estimated mean functions, for different parameters, is shown in Figure 4. It is observed that despite the initial difference between the mean function and the original oscillabolastic curve, the Gaussian diffusion process performs well with the proposed methodology for the parameter estimation. Please note that the last two results are the same but for different regions of β_0. Results are the same up to the sign of α and β, due to the odd condition of the sine function.

Also, three simulations has been carried out for different values of the infinitesimal variance, $\sigma^2 = 0.0001, 0.0025$ and 0.01, with parameters $\kappa = 2.3, \alpha = -2, \beta = 2, \gamma = 1$, and $\theta = 0.5$ from $t_0 = 1$ to 10, starting at $x_0 = 1$. Please note that final performance is very good although, as expected, RAE increases with variability. Results are shown in Table 3 and Figure 5.

Table 2. Results for simulated Gaussian diffusion with different parameters.

$\beta = 2$	κ	α	β	γ	θ	σ^2
Original	1.837	-2	2	1	0.500	0.0025
Initial	1.850	-1.792	1.564	0.812	0.561	0.0029
Estimated	1.855	-1.790	1.565	0.813	0.561	0.0029
		$\beta_0 \in [1,3]$	Inflection $t_* = 2.97$			RAE = 0.011
$\beta = 3$	κ	α	β	γ	θ	σ^2
Original	0.801	-2	3	1	0.500	0.0025
Initial	0.980	-1.942	3	0.611	0.550	0.0013
Estimated	0.977	-1.950	3	0.620	0.550	0.0014
		$\beta_0 \in [2,5]$	Inflection $t_* = 1.98$			RAE = 0.07
$\beta = 4$	κ	α	β	γ	θ	σ^2
Original	-1.495	-2	4	1	0.500	0.0025
Initial	-1.464	-2.012	4.000	0.836	0.551	0.0020
Estimated	-1.463	-2.012	4	0.838	0.550	0.0021
		$\beta_0 \in [3,5]$	Inflection $t_* = 5.48$			RAE = 0.02

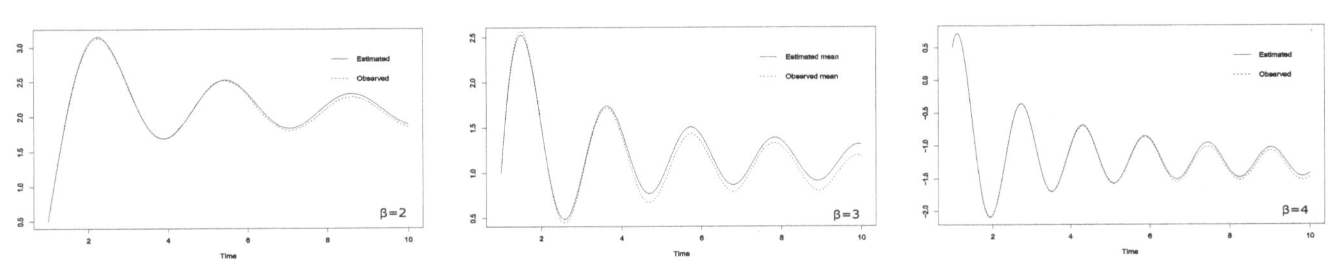

Figure 4. Simulated and estimated mean functions for variations of parameters of Table 2.

Table 3. Results for simulated Gaussian diffusion for different values of σ^2.

	κ	α	β	γ	θ	σ^2
$\sigma^2 = \mathbf{0.0001}$						
Original	2.337	−2	2	1	0.500	0.0001
Initial	2.392	−1.864	2	1.032	0.300	0.0001
Estimated	2.392	−1.869	2	1.038	0.300	0.0001
						RAE = 0.013
$\sigma^2 = \mathbf{0.0025}$						
Original	2.337	−2	2	1	0.500	0.0025
Initial	2.369	−1.865	2	1.330	0.250	0.0020
Estimated	2.369	−1.868	2	1.332	0.250	0.0022
						RAE = 0.013
$\sigma^2 = \mathbf{0.01}$						
Original	2.337	−2	2	1	0.500	0.01
Initial	2.479	−1.867	2	0.741	0.300	0.012
Estimated	2.483	−2.059	2.002	0.806	0.304	0.029
						RAE = 0.016

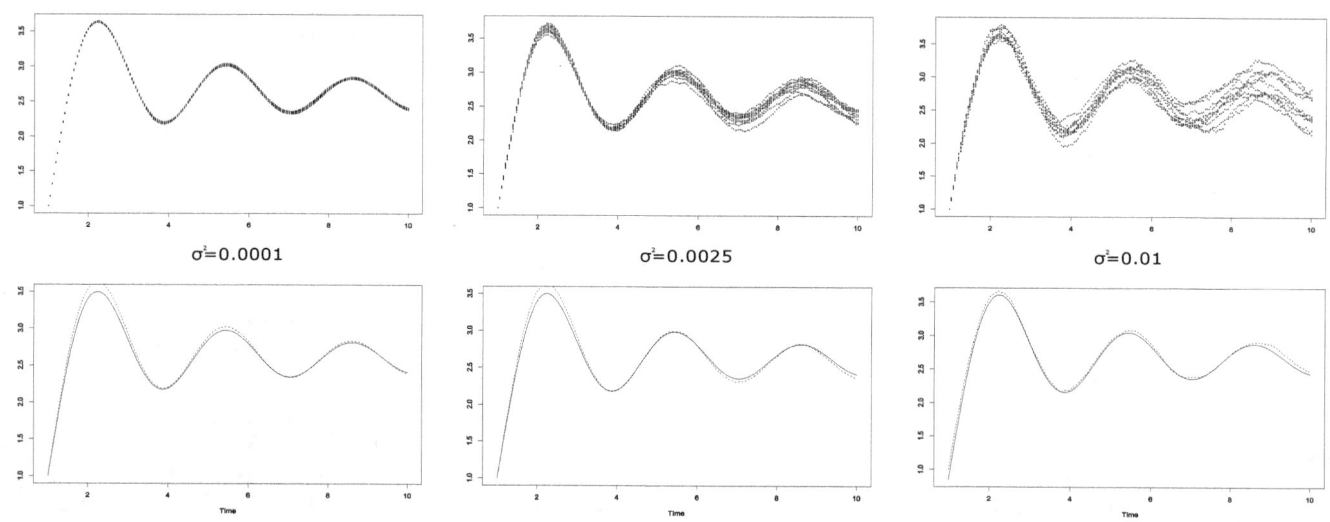

Figure 5. Sample paths (top) and simulated vs. estimated mean functions (bottom) for $\sigma^2 = 0.0001, 0.0025, 0.01$, (see Table 3).

5.2. Simulation in the Lognormal Case

The simulation scheme applied to the lognormal oscillabolastic diffusion process is similar to the one used earlier. In this case, the initial value of the degenerate distribution is $x_0 = 0.5$. Original values for the parameters are $\alpha' = -1, \beta = 2, \gamma' = 1, \theta = 0.5$, and $\sigma^2 = 10^{-3}$. Figure 6 displays the simulated paths (note that in the lognormal case, and in contrast to the Gaussian one, variability grows faster due to its dependence on the state).

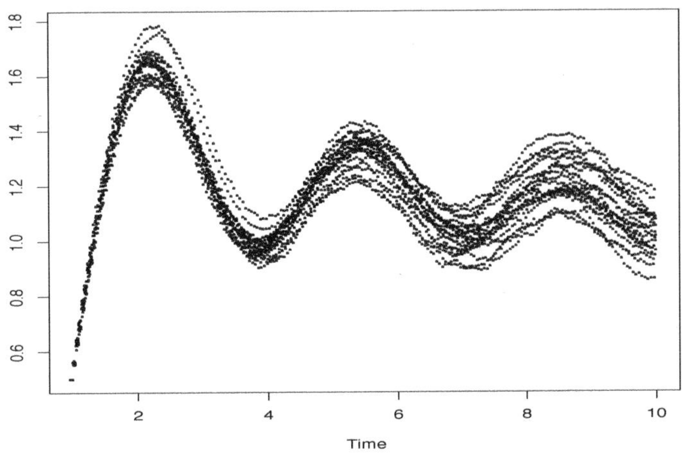

Figure 6. Simulated sample paths for the oscillabolastic lognormal-type diffusion process. Parameters: $\alpha' = -1, \beta = 2, \gamma' = 1, \theta = 0.5$ and $\sigma^2 = 10^{-3}$.

The initial values for the parameters follow from the methodology used in the lognormal case. The search region for β_0 and θ_0 is the same as in the Gaussian simulation example above. The number of oscillations (in relation to the time interval) and the curvature of the observations suggest that this may be a good choice. An inflection condition with $t_* = 2.97$, $u_* = 1.32$ and error $\epsilon = 0.1$ results in initial values $\beta_0 = 2$ and $\theta_0 = 0.55$, respectively. Because of increasing variability, very low error thresholds are not recommended, as they would decrease performance.

Initial values of α' and γ' have been calculated following the same methodology set out in the previous example, resulting in $\alpha'_0 = -0.85$ and $\gamma'_0 = 0.48$. Finally, the initial value of infinitesimal variance σ^2 follows from the regression analysis of the values obtained from the arithmetic and geometric mean of the observed mean values. In this case, $\sigma_0^2 = 0.001$ is obtained.

Using these initial values in the numerical resolution of the maximum likelihood system of Equations (16) and (17) leads to $\alpha' = -0.948, \beta = 2, \gamma' = 0.747, \theta = 0.564$, and $\sigma^2 = 0.001$. Table 4 summarizes the results.

Table 4. Results for simulated lognormal oscillabolastic diffusion.

	α'	β	γ'	θ	σ^2
Estimated values	−0.948	2	0.747	0.564	0.001

Figure 7 shows the simulated and estimated mean functions. Note how the adjustment provided by the mean function is good and is supported by the value of the RAE, which is RAE = 0.007.

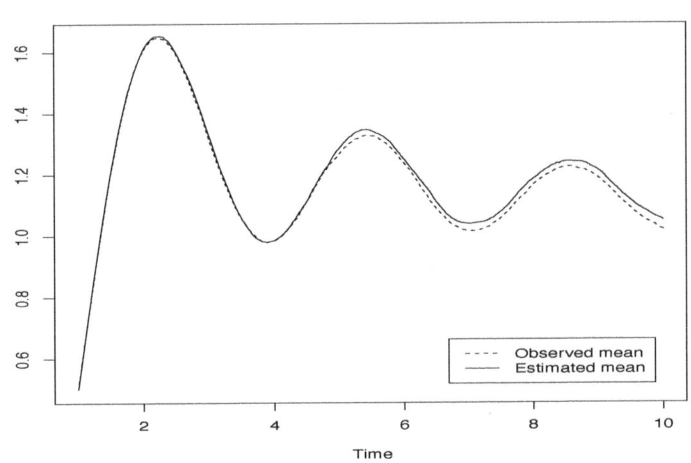

Figure 7. Simulated and estimated mean functions. Oscillabolastic lognormal-type diffusion process.

As for the Gaussian oscillabolastic diffusion process, another simulation study has been carried out by varying, in this case, parameters β (which affects the frequency of the oscillations) and σ^2 (responsible for the variability of the sample paths). Following the same methodology as in the Gaussian case, the ranges of β and θ can be directly proposed from the observations.

In this case, paths are simulated for values of $\beta = 2, 4, 6$. Final estimated values, as well as the search range for initial values of β are shown in Table 5. Furthermore, Figure 8 shows simulated and estimated mean functions for some combinations of parameters. RAE for every one of them is also shown in the table. These values suggest that the model and the proposed estimation methodology perform very well.

On the other hand, Table 6 and Figure 9 also show good performance for different values of variability, in particular $\sigma^2 = 0.0001, 0.0025, 0.01$ for parameters $\alpha' = -0.5.\beta = 2, \gamma' = 1, \theta = 0.5$ and $x_0 = 1$ for $t \in [t_0, 10]$ with $t_0 = 1$.

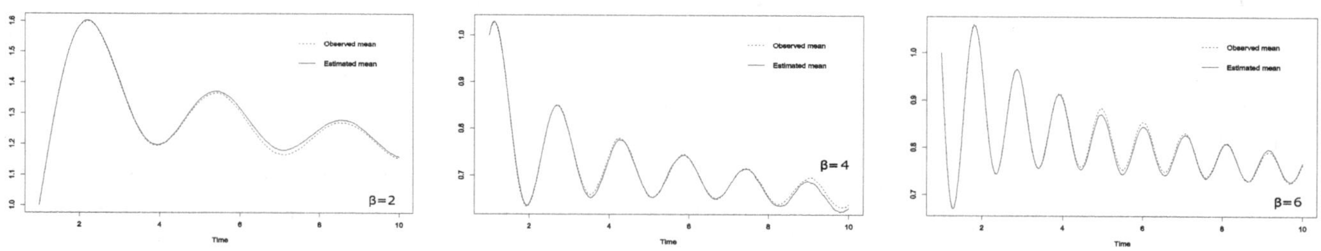

Figure 8. Simulated and estimated mean functions for variations of parameters of Table 5.

Table 5. Results for simulated lognormal diffusion with different parameters.

$\beta = 2$					
	α'	β	γ'	θ	σ^2
Original	−0.500	2	1	0.500	0.0001
Initial	−0.372	2.047	0.353	0.526	0.00002
Estimated	−0.492	1.999	0.813	0.585	0.0001
	$\beta_0 \in [2, 5]$	Inflection $t_* = 1.03$		RAE = 0.001	
$\beta = 4$					
	α'	β	γ'	θ	σ^2
Original	−0.500	4	1	0.500	0.0001
Initial	−0.314	4.067	0.196	0.600	0.0002
Estimated	−0.477	3.999	0.694	0.637	0.0002
	$\beta_0 \in [2, 5]$	Inflection $t_* = 3.90$		RAE = 0.002	
$\beta = 6$					
	α'	β	γ'	θ	σ^2
Original	−0.500	6	1	0.500	0.0001
Initial	−0.329	6	0.274	0.550	0.0002
Estimated	−0.474	6.000	0.705	0.618	0.0002
	$\beta_0 \in [4, 7]$	Inflection $t_* = 0.99$		RAE = 0.004	

Table 6. Results for simulated lognormal diffusion for different values of σ^2.

$\sigma^2 = \mathbf{0.0001}$					
	α'	β	γ'	θ	σ^2
Original	−0.500	2	1	0.500	0.0001
Initial	−0.516	1.789	0.557	0.665	0.00003
Estimated	−0.489	2.001	0.915	0.520	0.0001
					RAE = 0.003
$\sigma^2 = \mathbf{0.0025}$					
	α'	β	γ'	θ	σ^2
Original	−0.500	2	1	0.500	0.002
Initial	−0.498	1.809	0.451	0.667	0.0004
Estimated	−0.498	1.809	0.451	0.667	0.0004
					RAE = 0.004
$\sigma^2 = \mathbf{0.01}$					
	α'	β	γ'	θ	σ^2
Original	−0.500	2	1	0.500	0.010
Initial	−0.506	1.794	0.410	0.674	0.007
Estimated	−0.466	1.979	0.360	0.866	0.006
					RAE = 0.012

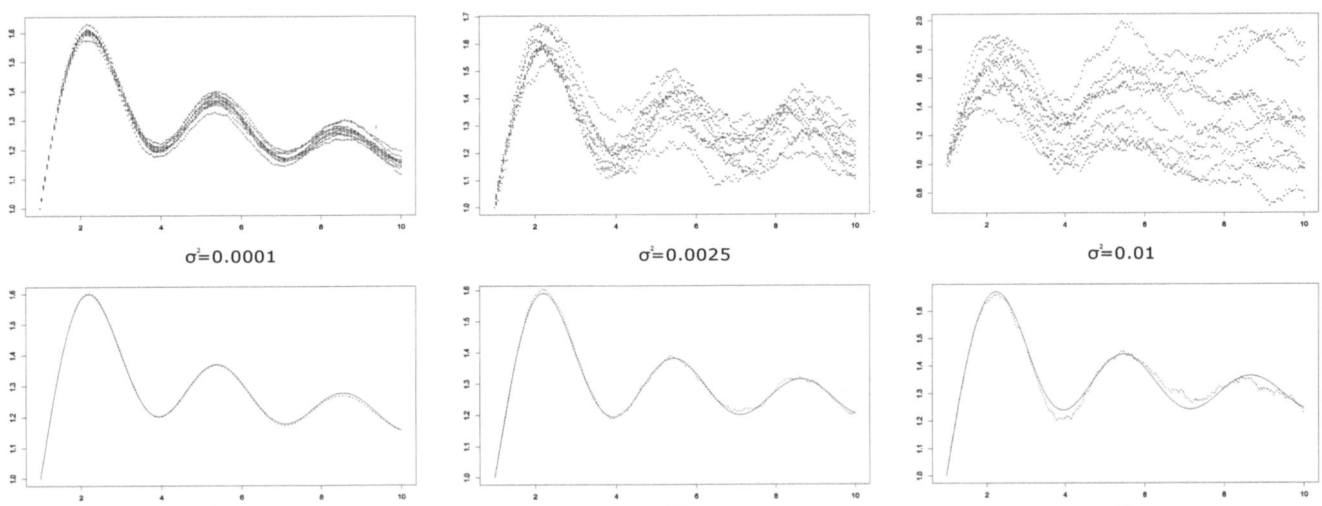

Figure 9. Sample paths (top) and simulated vs. estimated mean functions (bottom) for $\sigma^2 = 0.0001, 0.0025, 0.01$, (see Table 6).

6. Conclusions

Classic growth curves are generally described as arising from ordinary differential equations, which can be studied by themselves or as part of more sophisticated models. In the present paper, we have focused on the study of the oscillabolastic curve, which is particularly interesting for its ability to describe dynamic oscillatory phenomena.

Stochastic models are obtained from deterministic ones by including in them certain random fluctuations that can significantly influence the evolution of dynamic phenomena. This work considers two oscillabolastic stochastic models that retain the properties of the original curve and can be applied to practical situations. Two strategies have been proposed for managing the original differential equation and producing two different stochastic models. One of them is a Gaussian process while the other is a derivative of the lognormal process with exogenous factors.

One of the main problems derived from such models is the development of inference strategies. The estimation of the parameters by maximum likelihood implies the numerical resolution of two systems of equations. We have proposed a methodology to obtain initial solutions for the parameters based on the main traits of the model, namely the presence of an oscillabolastic trend (in the lognormal case) or an asymptotic oscillabolastic behavior (in the Gaussian case). To apply these procedures, a simulation study has been carried out taking into account a variety of error measures. This simulation study confirms the validity of the methodology proposed.

Author Contributions: The three authors have participated equally in the development of this work, either in the theoretical developments or in the applied aspects. The paper was also written and reviewed cooperatively. All authors have read and agreed to the published version of the manuscript.

References

1. Li, W.; Wang, K.; Su, H. Optimal harvesting policy for stochastic logistic population model. *Appl. Math. Comput.* **2011**, *218*, 157–162. [CrossRef]
2. Schurz, H. Modeling, analysis and discretization of stochastic logistic equations. *Int. J. Numer. Anal. Model.* **2007**, *4*, 178–197.
3. Tuckwell, H.C. A study of some diffusion models of population growth. *Theor. Popul. Biol.* **1974**, *5*, 345–357. [CrossRef]
4. Sun, X.; Wang, Y. Stability analysis of a stochastic logistic model with nonlinear diffusion term. *Appl. Math. Model.* **2008**, *32*, 2067–2075. [CrossRef]
5. Hu, G.; Wang, K. The estimation of probability distribution of SDE by only one sample trajectory. *Comput. Math. Appl.* **2011**, *62*, 1798–1806. [CrossRef]
6. Román-Román, P.; Serrano-Pérez, J.J.; Torres-Ruiz, F. Some Notes about inference for the lognormal diffusion process with exogenous factors. *Mathematics* **2018**, *6*, 85. [CrossRef]
7. Román-Román, P.; Serrano-Pérez, J.J.; Torres-Ruiz, F. A Note on Estimation of Multi-Sigmoidal Gompertz Functions with Random Noise. *Mathematics* **2019**, *7*, 541. [CrossRef]
8. Rupšys, P. Understanding the Evolution of Tree Size Diversity within the Multivariate Nonsymmetrical Diffusion Process and Information Measures. *Mathematics* **2019**, *7*, 761. [CrossRef]
9. Rupšys, P. Stationary densities and parameter estimation for delayed stochastic logistic growth laws with application in biomedical studies. *WSEAS Trans. Biol. Biomed.* **2019**, *5*, 117–132.
10. Longtin, A. Stochastic Delay-Differential Equations. In *Complex Time-Delay Systems. Understanding Complex Systems*; Atay, F.M., Ed.; Springer: Berlin, Germany, 2010; pp. 177–195.
11. Sakthivel, R.; Revathi, P.; Ren, Y.; Shen, G. Retarded stochastic differential equations with infinite delay driven by Rosenblatt process. *Stoch. Anal. Appl.* **2018**, *36*, 304–323. [CrossRef]
12. Dung, N.T. Fractional stochastics differential equations with applications to finance. *J. Math. Anal. Appl.* **2013**, *397*, 334–348. [CrossRef]
13. Moghaddam, B.P.; Mendes Lopes, A.; Tenreiro Machado, J.A.; Mostaghim, Z.S. Computational scheme for solving nonlinear fractional stochastic differential equations with delay. *Stoch. Anal. Appl.* **2019**, *37*, 893–908. [CrossRef]
14. Tabatatai, M.; Williams, D.K.; Bursac, Z. Hyperbolastic growth models: theory and application. *Theor. Biol. Med. Model.* **2005**, *2*, 1–13. [CrossRef]
15. Eby, W.M.; Tabatabai, M.A.; Bursac, Z. Hyperbolastic modeling of tumor growth with a combined treatment of iodoacetate and dimethylsulphoxide. *BMC Cancer* **2010**, *10*, 509. [CrossRef] [PubMed]
16. Tabatabai, M.A.; Bursac, Z.; Eby, W.M.; Singh, K.P. Mathematical modeling of stem cell proliferation. *Med. Biol. Eng. Comput.* **2011**, *49*, 253–262. [CrossRef] [PubMed]
17. Tabatabai, M.A.; Eby, W.M.; Bursac, Z. Oscillabolastic model, a new model for oscillatory dynamics, applied to the analysis of Hes1 gene expression and Ehrlich ascites tumor growth. *J. Biomed. Inform.* **2012**, *45*, 401–407. [CrossRef] [PubMed]
18. Tabatabai, M.A.; Eby, W.M.; Singh, K.P.; Bae, S. T model of growth and its application in systems of tumor-immune dynamics. *Math. Biosci. Eng.* **2013**, *10*, 925–938. [CrossRef] [PubMed]
19. Barrera, A.; Román-Román, P.; Torres-Ruiz, F. Hyperbolastic type-III diffusion process: Obtaining from the generalized Weibull diffusion process. *Math. Biosci. Eng.* **2019**, *17*, 814–833. [CrossRef] [PubMed]

4

Oscillation Criteria of Higher-Order Neutral Differential Equations with Several Deviating Arguments

Osama Moaaz [1,†], **Ioannis Dassios** [2,*,†] **and Omar Bazighifan** [3,4,†]

1 Department of Mathematics, Faculty of Science, Mansoura University, Mansoura 35516, Egypt; o_moaaz@mans.edu.eg
2 AMPSAS, University College Dublin, D4 Dublin, Ireland
3 Department of Mathematics, Faculty of Science, Hadhramout University, Hadhramout 50512, Yemen; o.bazighifan@gmail.com
4 Department of Mathematics, Faculty of Education, Seiyun University, Hadhramout 50512, Yemen
* Correspondence: ioannis.dassios@ucd.ie
† These authors contributed equally to this work.

Abstract: This work is concerned with the oscillatory behavior of solutions of even-order neutral differential equations. By using the technique of Riccati transformation and comparison principles with the second-order differential equations, we obtain a new Philos-type criterion. Our results extend and improve some known results in the literature. An example is given to illustrate our main results.

Keywords: even-order differential equations; neutral delay; oscillation

---

1. Introduction

In this article, we investigate the asymptotic behavior of solutions of even-order neutral differential equation of the form

$$\left(b(t)\left(z^{(n-1)}(t)\right)^{\gamma}\right)' + \sum_{i=1}^{k} q_i(t)\,u^{\gamma}(\delta_i(t)) = 0, \tag{1}$$

where $t \geq t_0$, $n \geq 4$ is an even natural number, $k \geq 1$ is an integer and $z(t) := u(t) + p(t)\,u(\sigma(t))$. Throughout this paper, we assume the following conditions to hold:

(P_1) γ is a quotient of odd positive integers;
(P_2) $b \in C[t_0, \infty)$, $b(t) > 0$, $b'(t) \geq 0$;
(P_3) $\sigma \in C^1[t_0, \infty)$, $\delta_i \in C[t_0, \infty)$, $\sigma'(t) > 0$, $\delta(t) \leq \delta_i(t)$, $\sigma(t) \leq t$ and $\lim_{t\to\infty} \sigma(t) = \lim_{t\to\infty} \delta_i(t) = \infty$, $i = 1, 2, ..., k$;
(P_4) $p, q_i \in C[t_0, \infty)$, $q_i(t) > 0$, $0 \leq p(t) < p_0 < \infty$ and

$$\int_{t_0}^{\infty} b^{-1/\gamma}(s)\,\mathrm{d}s = \infty \tag{2}$$

Definition 1. *The function $u \in C^3[t_u, \infty)$, $t_u \geq t_0$, is called a solution of (1), if $b(t)\left(z^{(n-1)}(t)\right)^{\gamma} \in C^1[t_u, \infty)$, and $u(t)$ satisfies (1) on $[t_u, \infty)$. Moreover, a solution of (1) is called oscillatory if it has arbitrarily large zeros on $[t_u, \infty)$, and otherwise is called to be nonoscillatory.*

Definition 2. *Let*

$$D = \{(t,s) \in \mathbb{R}^2 : t \geq s \geq t_0\} \text{ and } D_0 = \{(t,s) \in \mathbb{R}^2 : t > s \geq t_0\}.$$

A kernel function $H_i \in p(D,\mathbb{R})$ is said to belong to the function class $\Im$, written by $H \in \Im$, if, for $i = 1,2$,

(i) *$H_i(t,s) = 0$ for $t \geq t_0$, $H_i(t,s) > 0$, $(t,s) \in D_0$;*
(ii) *$H_i(t,s)$ has a continuous and nonpositive partial derivative $\partial H_i/\partial s$ on D_0 and there exist functions $\sigma, \vartheta \in C^1([t_0,\infty),(0,\infty))$ and $h_i \in C(D_0,\mathbb{R})$ such that*

$$\frac{\partial}{\partial s}H_1(t,s) + \frac{\theta'(s)}{\theta(s)}H_1(t,s) = h_1(t,s)H_1^{\gamma/(\gamma+1)}(t,s) \tag{3}$$

and

$$\frac{\partial}{\partial s}H_2(t,s) + \frac{\upsilon'(s)}{\upsilon(s)}H_2(t,s) = h_2(t,s)\sqrt{H_2(t,s)}. \tag{4}$$

The oscillation theory of differential equations with deviating arguments was initiated in a pioneering paper [1] of Fite, which appeared in the first quarter of the twentieth century. Delay equations play an important role in applications of real life. One area of active research in recent times is to study the sufficient criteria for oscillation of differential equations, see [1–11], and oscillation of neutral differential equations has become an important area of research, see [12–30]. Having in mind such applications, for instance, in electrical engineering, we cite models that describe electrical power systems, see [18]. Neutral differential equations also have wide applications in applied mathematics [31,32], physics [33], ecology [34] and engineering [35].

In the following, we show some previous results in the literature related to this paper: Moaaz et al. [23] proved that if there exist positive functions η, $\zeta \in C^1([t_0,\infty),\mathbb{R})$ such that the differential equations

$$\psi'(t) + \left(\frac{\mu\left(\delta^{-1}(\eta(t))\right)^{n-1}}{(n-1)!r^{1/\alpha}\left(\delta^{-1}(\eta(t))\right)}\right)^{\alpha} q(t)P_n^{\alpha}(\sigma(t))\psi\left(\delta^{-1}(\eta(t))\right) = 0$$

and

$$\phi'(t) + \delta^{-1}(\zeta(t))R_{n-3}(t)\phi\left(\delta^{-1}(\zeta(t))\right) = 0$$

are oscillatory, then (1) is oscillatory.

Zafer [29] proved that the even-order differential equation

$$z^{(n)}(t) + q(t)x(\sigma(t)) = 0 \tag{5}$$

is oscillatory if

$$\liminf_{t\to\infty}\int_{\sigma(t)}^{t} Q(s)\,ds > \frac{(n-1)2^{(n-1)(n-2)}}{e}, \tag{6}$$

or

$$\limsup_{t\to\infty}\int_{\sigma(t)}^{t} Q(s)\,ds > (n-1)2^{(n-1)(n-2)}, \quad \sigma'(t) \geq 0.$$

where $Q(t) := \sigma^{n-1}(t)(1-p(\sigma(t)))q(t)$.

Zhang and Yan [30] proved that (5) is oscillatory if either

$$\liminf_{t\to\infty}\int_{\sigma(t)}^{t} Q(s)\,ds > \frac{(n-1)!}{e}, \tag{7}$$

or

$$\limsup_{t\to\infty}\int_{\sigma(t)}^{t} Q(s)\,\mathrm{d}s > (n-1)!,\ \ \sigma(t)\geq 0.$$

It's easy to note that $(n-1)! < (n-1)\,2^{(n-1)(n-2)}$ for $n > 3$, and hence results in [30] improved results of Zafer in [29].

Xing et al. [28] proved that (1) is oscillatory if

$$\left(\delta^{-1}(t)\right)' \geq \delta_0 > 0,\ \sigma'(t)\geq\sigma_0>0,\ \sigma^{-1}(\delta(t)) < t$$

and

$$\liminf_{t\to\infty}\int_{\sigma^{-1}(\delta(t))}^{t}\frac{\widehat{q}(s)}{b(s)}\left(s^{n-1}\right)^{\gamma}\mathrm{d}s > \left(\frac{1}{\delta_0}+\frac{p_0^{\gamma}}{\delta_0\sigma_0}\right)\frac{((n-1)!)^{\gamma}}{\mathrm{e}}, \tag{8}$$

where $\widehat{q}(t) := \min\left\{q\left(\delta^{-1}(t)\right), q\left(\delta^{-1}(\sigma(t))\right)\right\}$.

Hence, [28] improved the results in [29,30].

In our paper, by carefully observing and employing some inequalities of different type, we provide a new criterion for oscillation of differential Equation (1). Here, we provide different criteria for oscillation, which can cover a larger area of different models of fourth order differential equations. We introduce a Riccati substitution and comparison principles with the second-order differential equations to obtain a new Philos-type criteria. Finally, we apply the main results to one example.

2. Some Auxiliary Lemmas

We shall employ the following lemmas:

Lemma 1 ([5]). *Let β be a ratio of two odd numbers, $V > 0$ and U are constants. Then*

$$Uu - Vu^{(\beta+1)/\beta} \leq \frac{\beta^{\beta}}{(\beta+1)^{\beta+1}}\frac{U^{\beta+1}}{V^{\beta}}.$$

Lemma 2 ([6]). *If the function u satisfies $u^{(i)}(t) > 0$, $i = 0,1,...,n$, and $u^{(n+1)}(t) < 0$, then*

$$\frac{u(t)}{t^{n}/n!} \geq \frac{u'(t)}{t^{n-1}/(n-1)!}.$$

Lemma 3 ([4]). *The equation*

$$\left(b(t)\left(u'(t)^{\gamma}\right)\right)' + q(t)\,u^{\gamma}(t) = 0, \tag{9}$$

where $b\in C[t_0,\infty)$, $b(t) > 0$ and $q(t) > 0$, is non-oscillatory if and only if there exist a $t\geq t_0$ and a function $v\in C^1[\ell,\infty)$ such that

$$v'(t) + \frac{\gamma}{b^{1/\gamma}(t)}v^{1+1/\gamma}(t) + q(t) \leq 0,$$

for $t\geq t_0$.

Lemma 4 ([2], Lemma 2.2.3). *Let $u\in C^n([t_0,\infty),(0,\infty))$. Assume that $u^{(n)}(t)$ is of fixed sign and not identically zero on $[t_0,\infty)$ and that there exists a $t_1\geq t_0$ such that $u^{(n-1)}(t)\,u^{(n)}(t)\leq 0$ for all $t\geq t_1$. If $\lim_{t\to\infty}u(t)\neq 0$, then for every $\mu\in(0,1)$ there exists $t_{\mu}\geq t_1$ such that*

$$u(t) \geq \frac{\mu}{(n-1)!}t^{n-1}\left|u^{(n-1)}(t)\right| \ \text{for}\ t\geq t_{\mu}.$$

3. Main Results

In this section, we give the main results of the article. Here, we define the next notation:

$$\begin{aligned}
P_k(t) &= \frac{1}{p(\sigma^{-1}(t))}\left(1-\frac{(\sigma^{-1}(\sigma^{-1}(t)))^{k-1}}{(\sigma^{-1}(t))^{k-1}\,p(\sigma^{-1}(\sigma^{-1}(t)))}\right), \quad \text{for } k=2,n,\\
R_0(t) &= \left(\frac{1}{b(t)}\int_t^\infty \sum_{i=1}^{k} q_i(s)\,P_2^\gamma(\delta_i(s))\,ds\right)^{1/\gamma},\\
\Theta(t) &= \gamma\frac{\mu_1}{(n-2)!}\left(\frac{b(t)}{b(\sigma^{-1}(\delta_i(t)))}\right)^{1/\gamma}\frac{(\sigma^{-1}(\delta_i(t)))'(\delta_i(t))'(\sigma^{-1}(\delta_i(t)))^{n-2}}{(b\theta)^{1/\gamma}(t)},\\
\tilde{\Theta}(t) &= \frac{h_1^{\gamma+1}(t,s)\,H_1^\gamma(t,s)}{(\gamma+1)^{\gamma+1}}\frac{((n-2)!)^\gamma\, b(\sigma^{-1}(\delta_i(t)))\,\theta(t)}{\left(\mu_1(\sigma^{-1}(\delta_i(t)))'(\delta_i(t))'(\sigma^{-1}(\delta_i(t)))^{n-2}\right)^\gamma}
\end{aligned}$$

and

$$R_m(t)=\int_t^\infty R_{m-1}(s)\,ds,\quad m=1,2,\ldots,n-3.$$

Lemma 5 ([8], Lemma 1.2). *Assume that u is an eventually positive solution of (1). Then, there exist two possible cases:*

$(\mathbf{S}_1)$ $z(t)>0,\ z'(t)>0,\ z''(t)>0,\ z^{(n-1)}(t)>0,\ z^{(n)}(t)<0,$

$(\mathbf{S}_2)$ $z(t)>0,\ z^{(j)}(t)>0,\ z^{(j+1)}(t)<0$ *for all odd integer* $j\in\{1,3,\ldots,n-3\},\ z^{(n-1)}(t)>0,\ z^{(n)}(t)<0,$

for $t\ge t_1$, where $t_1\ge t_0$ is sufficiently large.

Lemma 6. *Let u be an eventually positive solution of (1) and*

$$\left(\sigma^{-1}\left(\sigma^{-1}(t)\right)\right)^{n-1}<\left(\sigma^{-1}(t)\right)^{n-1}p\left(\sigma^{-1}\left(\sigma^{-1}(t)\right)\right). \tag{10}$$

Then

$$u(t)\ge\frac{z(\sigma^{-1}(t))}{p(\sigma^{-1}(t))}-\frac{1}{p(\sigma^{-1}(t))}\frac{z(\sigma^{-1}(\sigma^{-1}(t)))}{p(\sigma^{-1}(\sigma^{-1}(t)))}. \tag{11}$$

Proof. Let u be an eventually positive solution of (1) on $[t_0,\infty)$. From the definition of $z(t)$, we see that

$$p(t)\,u(\sigma(t))=z(t)-u(t)$$

and so

$$p\left(\sigma^{-1}(t)\right)u(t)=z\left(\sigma^{-1}(t)\right)-z\left(\sigma^{-1}(t)\right).$$

Repeating the same process, we obtain

$$u(t)=\frac{1}{p(\sigma^{-1}(t))}\left(z\left(\sigma^{-1}(t)\right)-\left(\frac{z(\sigma^{-1}(\sigma^{-1}(t)))}{p(\sigma^{-1}(\sigma^{-1}(t)))}-\frac{u(\sigma^{-1}(\sigma^{-1}(t)))}{p(\sigma^{-1}(\sigma^{-1}(t)))}\right)\right),$$

which yields

$$u(t)\ge\frac{z(\sigma^{-1}(t))}{p(\sigma^{-1}(t))}-\frac{1}{p(\sigma^{-1}(t))}\frac{z(\sigma^{-1}(\sigma^{-1}(t)))}{p(\sigma^{-1}(\sigma^{-1}(t)))}.$$

Thus, (11) holds. This completes the proof. □

Lemma 7. *Assume that u is an eventually positive solution of (1) and*

$$\left(b(t)\left(z^{(n-1)}(t)\right)^{\gamma}\right)' \leq -z^{\gamma}\left(\sigma^{-1}(\delta(t))\right)\sum_{i=1}^{k} q_i(t) P_n^{\gamma}(\delta_i(t)), \text{ if } z \text{ satisfies } (\mathbf{S}_1) \tag{12}$$

and

$$z''(t) + R_{n-3}(t) z\left(\sigma^{-1}(\delta(t))\right) \leq 0, \text{ if } z \text{ satisfies } (\mathbf{S}_2). \tag{13}$$

Proof. Let u be an eventually positive solution of (1) on $[t_0, \infty)$. It follows from Lemma 5 that there exist two possible cases $(\mathbf{S}_1)$ and $(\mathbf{S}_2)$.

Suppose that Case $(\mathbf{S}_1)$ holds. From Lemma 2, we obtain $z(t) \geq \frac{1}{(n-1)} t z'(t)$ and hence the function $t^{1-n} z(t)$ is nonincreasing, which with the fact that $\sigma(t) \leq t$ gives

$$\left(\sigma^{-1}(t)\right)^{n-1} z\left(\sigma^{-1}\left(\sigma^{-1}(t)\right)\right) \leq \left(\sigma^{-1}\left(\sigma^{-1}(t)\right)\right)^{n-1} z\left(\sigma^{-1}(t)\right). \tag{14}$$

Combining (11) and (14), we conclude that

$$\begin{aligned} u(t) &\geq \frac{1}{p(\sigma^{-1}(t))}\left(1 - \frac{(\sigma^{-1}(\sigma^{-1}(t)))^{n-1}}{(\sigma^{-1}(t))^{n-1} p(\sigma^{-1}(\sigma^{-1}(t)))}\right) z\left(\sigma^{-1}(t)\right) \\ &= P_n(t) z\left(\sigma^{-1}(t)\right). \end{aligned} \tag{15}$$

From (1) and (15), we obtain

$$\begin{aligned} \left(b(t)\left(z^{(n-1)}(t)\right)^{\gamma}\right)' &\leq -\sum_{i=1}^{k} q_i(t) P_n^{\gamma}(\delta_i(t)) z^{\gamma}\left(\sigma^{-1}(\delta_i(t))\right) \\ &\leq -z^{\gamma}\left(\sigma^{-1}(\delta(t))\right)\sum_{i=1}^{k} q_i(t) P_n^{\gamma}(\delta_i(t)). \end{aligned}$$

Thus, (12) holds.

Suppose that Case $(\mathbf{S}_2)$ holds. From Lemma 2, we find

$$z(t) \geq t z'(t) \tag{16}$$

and thus the function $t^{-1} z(t)$ is nonincreasing, eventually. Since $\sigma^{-1}(t) \leq \sigma^{-1}(\sigma^{-1}(t))$, we obtain

$$\sigma^{-1}(t) z\left(\sigma^{-1}\left(\sigma^{-1}(t)\right)\right) \leq \sigma^{-1}\left(\sigma^{-1}(t)\right) z\left(\sigma^{-1}(t)\right). \tag{17}$$

Combining (11) and (17), we find

$$\begin{aligned} u(t) &\geq \frac{1}{p(\sigma^{-1}(t))}\left(1 - \frac{(\sigma^{-1}(\sigma^{-1}(t)))}{(\sigma^{-1}(t)) p(\sigma^{-1}(\sigma^{-1}(t)))}\right) z\left(\sigma^{-1}(t)\right) \\ &= P_2(t) z\left(\sigma^{-1}(t)\right), \end{aligned}$$

which with (1) yields

$$\left(b(t)\left(z^{(n-1)}(t)\right)^{\gamma}\right)' + \sum_{i=1}^{k} q_i(t) P_2^{\gamma}(\delta_i(t)) z^{\gamma}\left(\sigma^{-1}(\delta_i(t))\right) \leq 0. \tag{18}$$

Integrating the (18) from t to ∞, we obtain

$$z^{(n-1)}(t) \geq b_0(t)\, z\left(\sigma^{-1}(\delta(t))\right).$$

Integrating this inequality from t to ∞ a total of $n-3$ times, we obtain

$$z''(t) + R_{n-3}(t)\, z\left(\sigma^{-1}(\delta(t))\right) \leq 0.$$

Thus, (13) holds. This completes the proof. □

Theorem 1. *Let (2) and (10) hold. If there exist positive functions* $\theta, v \in C^1([t_0,\infty),\mathbb{R})$ *such that*

$$\limsup_{t\to\infty} \frac{1}{H_1(t,t_1)} \int_{t_1}^{t} \left(H_1(t,s)\,\psi(s) - \tilde{\Theta}(s)\right) ds = \infty \tag{19}$$

and

$$\limsup_{t\to\infty} \frac{1}{H_2(t,t_1)} \int_{t_1}^{t} \left(H_2(t,s)\,\psi^*(s) - \frac{v(s)\,h_2^2(t,s)}{4}\right) ds = \infty, \tag{20}$$

where

$$\psi(s) = \theta(t) \sum_{i=1}^{k} q_i(t)\, P_n^{\gamma}(\delta_i(t)), \quad \psi^*(s) = v(t)\, b_{n-3}(t) \left(\frac{\sigma^{-1}(\delta(t))}{t}\right)$$

and

$$\tilde{\Theta}(s) = \frac{h_1^{\gamma+1}(t,s)\, H_1^{\gamma}(t,s)}{(\gamma+1)^{\gamma+1}} \frac{((n-2)!)^{\gamma}\, b\left(\sigma^{-1}(\delta(t))\right)\theta(t)}{\left(\mu_1 \left(\sigma^{-1}(\delta(t))\right)' (\delta(t))' \left(\sigma^{-1}(\delta(t))\right)^{n-2}\right)^{\gamma}},$$

then (1) is oscillatory.

Proof. Let u be a non-oscillatory solution of (1) on $[t_0,\infty)$. Without loss of generality, we can assume that u is eventually positive. It follows from Lemma 5 that there exist two possible cases $(\mathbf{S}_1)$ and $(\mathbf{S}_2)$.

Let $(\mathbf{S}_1)$ hold. From Lemma 7, we arrive at (12). Next, we define a function ξ by

$$\xi(t) := \theta(t) \frac{b(t)\left(z^{(n-1)}(t)\right)^{\gamma}}{z^{\gamma}\left(\sigma^{-1}(\delta(t))\right)} > 0.$$

Differentiating and using (12), we obtain

$$\begin{aligned} \xi'(t) \leq\ & \frac{\theta'(t)}{\theta(t)}\xi(t) - \theta(t) \sum_{i=1}^{k} q_i(t)\, P_n^{\gamma}(\delta_i(t)) \\ & -\gamma\theta(t) \frac{b(t)\left(z^{(n-1)}(t)\right)^{\gamma} \left(\sigma^{-1}(\delta(t))\right)' (\delta(t))'\, z_u'\left(\sigma^{-1}(\delta(t))\right)}{z_u^{\gamma+1}\left(\sigma^{-1}(\delta(t))\right)}. \end{aligned} \tag{21}$$

Recalling that $b(t)\left(z^{(n-1)}(t)\right)^{\gamma}$ is decreasing, we get

$$b\left(\sigma^{-1}(\delta(t))\right)\left(z^{(n-1)}\left(\sigma^{-1}(\delta(t))\right)\right)^{\gamma} \geq b(t)\left(z^{(n-1)}(t)\right)^{\gamma}.$$

This yields

$$\left(z^{(n-1)}\left(\sigma^{-1}(\delta(t))\right)\right)^{\gamma} \geq \frac{b(t)}{b\left(\sigma^{-1}(\delta(t))\right)} \left(z^{(n-1)}(t)\right)^{\gamma}. \tag{22}$$

It follows from Lemma 4 that

$$z'\left(\sigma^{-1}\left(\delta\left(t\right)\right)\right) \geq \frac{\mu_1}{(n-2)!}\left(\sigma^{-1}\left(\delta\left(t\right)\right)\right)^{n-2} z^{(n-1)}\left(\sigma^{-1}\left(\delta\left(t\right)\right)\right), \tag{23}$$

for all $\mu_1 \in (0,1)$ and every sufficiently large t. Thus, by (21), (22) and (23), we get

$$\begin{aligned} \xi'\left(t\right) &\leq \frac{\theta'\left(t\right)}{\theta\left(t\right)}\xi\left(t\right) - \theta\left(t\right)\sum_{i=1}^{k} q_i\left(t\right) P_n^{\gamma}\left(\delta_i\left(t\right)\right) \\ &-\gamma\theta\left(t\right)\frac{\mu_1}{(n-2)!}\left(\frac{b\left(t\right)}{b\left(\sigma^{-1}\left(\delta\left(t\right)\right)\right)}\right)^{1/\gamma}\frac{b\left(t\right)\left(z^{(n-1)}\left(t\right)\right)^{\gamma+1}\left(\sigma^{-1}\left(\delta\left(t\right)\right)\right)'\left(\delta\left(t\right)\right)'\left(\sigma^{-1}\left(\delta\left(t\right)\right)\right)^{n-2}}{z^{\gamma+1}\left(\sigma^{-1}\left(\delta\left(t\right)\right)\right)}. \end{aligned}$$

Hence,

$$\begin{aligned} \xi'\left(t\right) &\leq \frac{\theta'\left(t\right)}{\theta\left(t\right)}\xi\left(t\right) - \theta\left(t\right)\sum_{i=1}^{k} q_i\left(t\right) P_n^{\gamma}\left(\delta_i\left(t\right)\right) \\ &-\Theta\left(t\right)\xi^{\frac{\gamma+1}{\gamma}}\left(t\right). \end{aligned} \tag{24}$$

Multiplying (24) by $H_1\left(t,s\right)$ and integrating the resulting inequality from t_1 to t; we find that

$$\begin{aligned} \int_{t_1}^{t} H_1\left(t,s\right)\psi\left(s\right) ds &\leq \xi\left(t_1\right) H_1\left(t,t_1\right) + \int_{t_1}^{t}\left(\frac{\partial}{\partial s} H_1\left(t,s\right) + \frac{\theta'\left(s\right)}{\theta\left(s\right)} H_1\left(t,s\right)\right)\xi\left(s\right) ds \\ &- \int_{t_1}^{t}\Theta\left(s\right) H_1\left(t,s\right)\xi^{\frac{\gamma+1}{\gamma}}\left(s\right) ds. \end{aligned}$$

From (3), we get

$$\begin{aligned} \int_{t_1}^{t} H_1\left(t,s\right)\psi\left(s\right) ds &\leq \xi\left(t_1\right) H_1\left(t,t_1\right) + \int_{t_1}^{t} h_1\left(t,s\right) H_1^{\gamma/(\gamma+1)}\left(t,s\right)\xi\left(s\right) ds \\ &- \int_{t_1}^{t}\Theta\left(s\right) H_1\left(t,s\right)\xi^{\frac{\gamma+1}{\gamma}}\left(s\right) ds. \end{aligned} \tag{25}$$

Using Lemma 1 with $V = \Theta\left(s\right) H_1\left(t,s\right)$, $U = h_1\left(t,s\right) H_1^{\gamma/(\gamma+1)}\left(t,s\right)$ and $u = \xi\left(s\right)$, we get

$$\begin{aligned} & h_1\left(t,s\right) H_1^{\gamma/(\gamma+1)}\left(t,s\right)\xi\left(s\right) - \Theta\left(s\right) H_1\left(t,s\right)\xi^{\frac{\gamma+1}{\gamma}}\left(s\right) \\ \leq\ & \frac{h_1^{\gamma+1}\left(t,s\right) H_1^{\gamma}\left(t,s\right)}{(\gamma+1)^{\gamma+1}}\frac{\left((n-2)!\right)^{\gamma} b\left(\sigma^{-1}\left(\delta\left(t\right)\right)\right)\theta\left(t\right)}{\left(\mu_1\left(\sigma^{-1}\left(\delta\left(t\right)\right)\right)'\left(\delta\left(t\right)\right)'\left(\sigma^{-1}\left(\delta\left(t\right)\right)\right)^{n-2}\right)^{\gamma}}, \end{aligned}$$

which with (25) gives

$$\frac{1}{H_1\left(t,t_1\right)}\int_{t_1}^{t}\left(H_1\left(t,s\right)\psi\left(s\right) - \tilde{\Theta}\left(s\right)\right) ds \leq \xi\left(t_1\right),$$

which contradicts (19).

On the other hand, let $(\mathbf{S}_2)$ hold. Using Lemma 7, we get that (13) holds. Now, we define

$$\varphi\left(t\right) = v\left(t\right)\frac{z'\left(t\right)}{z\left(t\right)}. \tag{26}$$

Then $\varphi(t) > 0$ for $t \geq t_1$. By differentiating φ and using (13), we find

$$\begin{aligned}\varphi'(t) &= \frac{v'(t)}{v(t)}\varphi(t) + v(t)\frac{z''(t)}{z(t)} - v(t)\left(\frac{z'(t)}{z(t)}\right)^2 \\ &\leq \frac{v'(t)}{v(t)}\varphi(t) - v(t)\,b_{n-3}(t)\frac{z\left(\sigma^{-1}(\delta(t))\right)}{z(t)} - \frac{1}{v(t)}\varphi^2(t). \end{aligned} \tag{27}$$

By using Lemma 2, we find that

$$z(t) \geq tz'(t). \tag{28}$$

From (28), we get that

$$z\left(\sigma^{-1}(\delta(t))\right) \geq \frac{\sigma^{-1}(\delta(t))}{t}z(t). \tag{29}$$

Thus, from (27) and (29), we obtain

$$\varphi'(t) \leq \frac{v'(t)}{v(t)}\varphi(t) - v(t)\,R_{n-3}(t)\left(\frac{\sigma^{-1}(\delta(t))}{t}\right) - \frac{1}{v(t)}\varphi^2(t). \tag{30}$$

Multiplying (30) by $H_2(t,s)$ and integrating the resulting from t_1 to t, we obtain

$$\begin{aligned}\int_{t_1}^{t} H_2(t,s)\,\psi^*(s)\,ds &\leq \varphi(t_1)\,H_2(t,t_1) \\ &\quad + \int_{t_1}^{t}\left(\frac{\partial}{\partial s}H_2(t,s) + \frac{v'(s)}{v(s)}H_2(t,s)\right)\varphi(s)\,ds \\ &\quad - \int_{t_1}^{t}\frac{1}{v(s)}H_2(t,s)\,\varphi^2(s)\,ds.\end{aligned}$$

Thus,

$$\begin{aligned}\int_{t_1}^{t} H_2(t,s)\,\psi^*(s)\,ds &\leq \varphi(t_1)\,H_2(t,t_1) + \int_{t_1}^{t} h_2(t,s)\sqrt{H_2(t,s)}\varphi(s)\,ds \\ &\quad - \int_{t_1}^{t}\frac{1}{v(s)}H_2(t,s)\,\varphi^2(s)\,ds \\ &\leq \varphi(t_1)\,H_2(t,t_1) + \int_{t_1}^{t}\frac{v(s)\,h_2^2(t,s)}{4}ds\end{aligned}$$

and so

$$\frac{1}{H_2(t,t_1)}\int_{t_1}^{t}\left(H_2(t,s)\,\psi^*(s) - \frac{v(s)\,h_2^2(t,s)}{4}\right)ds \leq \varphi(t_1),$$

which contradicts (20). This completes the proof. □

In the next theorem, we establish new oscillation results for (1) by using the theory of comparison with a second order differential equation.

Theorem 2. *Assume that the equation*

$$y''(t) + y(t)\sum_{i=1}^{k} q_i(t)\,P_n^{\gamma}(\delta_i(t)) = 0 \tag{31}$$

and

$$\left[b(t)\left(y'(t)\right)^{\gamma}\right]' + R_{n-3}(t)\left(\frac{\sigma^{-1}(\delta(t))}{t}\right)y^{\gamma}(t) = 0, \tag{32}$$

are oscillatory, then every solution of (1) is oscillatory.

Proof. Suppose to the contrary that (1) has a eventually positive solution u and by virtue of Lemma 3. From Theorem 1, we set $\theta(t) = 1$ in (24), then we get

$$\xi'(\ell) + \Theta(t)\xi^{\frac{\gamma+1}{\gamma}} + \sum_{i=1}^{k} q_i(t) P_n^{\gamma}(\delta_i(t)) \leq 0.$$

Thus, we can see that Equation (31) is nonoscillatory, which is a contradiction. If we now set $v(t) = 1$ in (30), then we obtain

$$\varphi'(t) + R_{n-3}(t)\left(\frac{\sigma^{-1}(\delta(t))}{t}\right) + \varphi^2(t) \leq 0.$$

Hence, Equation (32) is nonoscillatory, which is a contradiction.
Theorem 2 is proved. □

Corollary 1. *If conditions (19) and (20) in Theorem 1 are replaced by the following conditions:*

$$\limsup_{t\to\infty} \frac{1}{H_1(t,t_1)} \int_{t_1}^{t} H_1(t,s)\psi(s)\,ds = \infty$$

and

$$\limsup_{t\to\infty} \frac{1}{H_1(t,t_1)} \int_{t_1}^{t} \tilde{\Theta}(s)\,ds < \infty.$$

Moreover,

$$\limsup_{t\to\infty} \frac{1}{H_2(t,t_1)} \int_{t_1}^{t} H_2(t,s)\psi^*(s)\,ds = \infty$$

and

$$\limsup_{t\to\infty} \frac{1}{H_2(t,t_1)} \int_{t_1}^{t} v(s)h_2^2(t,s)\,ds < \infty,$$

then (1) is oscillatory.

Corollary 2. *Let (10) holds. If there exist positive functions $v, \theta \in^1 ([t_0,\infty), \mathbb{R})$ such that*

$$\int_{t_0}^{\infty}\left(\theta(s)\sum_{i=1}^{k} q_i(s) P_n^{\gamma}(\delta_i(s)) - \varpi(s)\right) ds = \infty \tag{33}$$

and

$$\int_{t_0}^{\infty}\left(P_1 v(s)\int_{t}^{\infty}\left(\frac{1}{r(\varrho)}\int_{\varrho}^{\infty}\sum_{i=1}^{k} q_i(s)\left(\frac{\tau^{-1}(\sigma(s))}{s}\right)^{\alpha} ds\right)^{1/\alpha} d\varrho - \pi(s)\right) ds = \infty, \tag{34}$$

where

$$\varpi(t) := \frac{(n-2)!^{\alpha}}{(\alpha+1)^{\alpha+1}} \frac{r(\tau^{-1}(\sigma(t)))(\theta'(t))^{\alpha+1}}{\left(\mu_1\theta(t)(\tau^{-1}(\sigma(t)))'(\tau^{-1}(\sigma(t)))^{n-2}\right)^{\alpha}}$$

and

$$\pi(t) := \frac{(v'(s))^2}{4v(s)},$$

then (1) is oscillatory.

Example 1. *Consider the equation*

$$\left(x(t) + 16x\left(\frac{1}{2}t\right)\right)^{(4)} + \frac{q_0}{t^4}x\left(\frac{1}{3}t\right) = 0, \ t \geq 1, \tag{35}$$

where $q_0 > 0$. *We note that* $r(t) = 1$, $p(t) = 16$, $\tau(t) = t/2$, $\sigma(t) = t/3$ *and* $q(t) = q_0/t^4$.
Thus, we have

$$P_1(t) = \frac{1}{32}, \; P_2(t) = \frac{7}{128}.$$

Now, we obtain

$$\int_{t_0}^{\infty} \left(\theta(s) \sum_{i=1}^{k} q_i(s) P_n^{\gamma}(\delta_i(s)) - \varpi(s) \right) \mathrm{d}s = \infty$$

and

$$\begin{aligned} & \int_{t_0}^{\infty} \left(P_1 v(s) \int_t^{\infty} \left(\frac{1}{r(\varrho)} \int_{\varrho}^{\infty} \sum_{i=1}^{k} q_i(s) \left(\frac{\tau^{-1}(\sigma(s))}{s} \right)^{\alpha} \mathrm{d}s \right)^{1/\alpha} \mathrm{d}\varrho - \pi(t) \right) \mathrm{d}s \\ = & \int_{t_0}^{\infty} \left(\frac{7q_0}{1152} - \frac{1}{4} \right) \mathrm{d}s, \\ = & \infty, \quad \text{if } q_0 > 41.14. \end{aligned}$$

Thus, by using Corollary 2, Equation (35) is oscillatory if $q_0 > 41.14$.

4. Conclusions

The aim of this article was to provide a study of asymptotic nature for a class of even-order neutral delay differential equations. We used a generalized Riccati substitution and the integral averaging technique to ensure that every solution of the studied equation is oscillatory. The results presented here complement some of the known results reported in the literature.

A further extension of this article is to use our results to study a class of systems of higher order neutral differential equations as well as of fractional order. For all these there is already some research in progress.

Author Contributions: O.M. and O.B.: Writing–original draft, and writing–review & editing. I.D.: Formal analysis, writing–review & editing, funding and supervision. All authors have read and agreed to the published version of the manuscript.

Acknowledgments: The authors thank the referees for their careful reading and valuable comments.

References

1. Fite, W.B. Concerning the zeros of the solutions of certain differential equations. *Trans. Am. Math. Soc.* **1918**, *19*, 341–352. [CrossRef]
2. Agarwal, R.; Grace, S.; O'Regan, D. *Oscillation Theory for Difference and Functional Differential Equations*; Kluwer Acad. Publ.: Dordrecht, The Netherlands, 2000.
3. Hale, J.K. *Theory of Functional Differential Equations*; Springer: New York, NY, USA, 1977.
4. Bazighifan, O.; Elabbasy, E.M.; Moaaz, O. Oscillation of higher-order differential equations with distributed delay. *J. Inequal. Appl.* **2019**, *55*, 1–9. [CrossRef]
5. Elabbasy, E.M.; Cesarano, C.; Bazighifan, O.; Moaaz, O. Asymptotic and oscillatory behavior of solutions of a class of higher order differential equation. *Symmetry* **2019**, *11*, 1434. [CrossRef]
6. Kiguradze, I.T.; Chanturiya, T.A. *Asymptotic Properties of Solutions of Nonautonomous Ordinary Differential Equations*; Kluwer Acad. Publ.: Dordrecht, The Netherlands, 1993.
7. Kitamura, Y.; Kusano, T. Oscillation of first-order nonlinear differential equations with deviating arguments. *Proc. Am. Math. Soc.* **1980**, *78*, 64–68. [CrossRef]

8. Philos, C.G. A new criterion for the oscillatory and asymptotic behavior of delay differential equations. *Bull. Acad. Pol. Sci. Sér. Sci. Math.* **1981**, *39*, 61–64.
9. Philos, C.G. On the existence of non-oscillatory solutions tending to zero at ∞ for differential equations with positive delays. *Arch. Math.* **1981**, *36*, 168–178. [CrossRef]
10. Ladas, G.; Lakshmikantham, V.; Papadakis, L.S. *Oscillations of Higher-Order Retarded Differential Equations Generated By the Retarded Arguments, Delay and Functional Differential Equations and their Applications*; Academic Press: New York, NY, USA, 1972; pp. 219–231.
11. Koplatadze, R.G.; Chanturija, T.A. Oscillating and monotone solutions of first-order differential equations with deviating argument. *Differ. Uravn.* **1982**, *18*, 1463–1465. (In Russian)
12. Agarwal, R.P.; Bohner, M.; Li, T.; Zhang, C. A new approach in the study of oscillatory behavior of even-order neutral delay diferential equations. *Appl. Math. Comput.* **2013**, *225*, 787–794.
13. Baculikova, B.; Dzurina, J. Oscillation theorems for second-order nonlinear neutral differential equations. *Comput. Math. Appl.* **2011**, *62*, 4472–4478. [CrossRef]
14. Baculikova, B.; Dzurina, J.; Li, T. Oscillation results for even-order quasi linear neutral functional differential equations. *Electron. J. Difer. Equ.* **2011**, *2011*, 1–9.
15. Baculikova, B.; Dzurina, J. Oscillation theorems for higher order neutral diferential equations. *Appl. Math. Comput.* **2012**, *219*, 3769–3778.
16. Bazighifan, O.; Cesarano, C. Some New Oscillation Criteria for Second-Order Neutral Differential Equations with Delayed Arguments. *Mathematics* **2019**, *7*, 619. [CrossRef]
17. Chatzarakis, G.E.; Elabbasy, E.M.; Bazighifan, O. An oscillation criterion in 4th-order neutral differential equations with a continuously distributed delay. *Adv. Differ. Equ.* **2019**, *336*, 1–9.
18. Liu, M.; Dassios, I.; Milano, F. On the Stability Analysis of Systems of Neutral Delay Differential Equations. *Circuits Syst. Signal Process.* **2019**, 38, 1639–1653. [CrossRef]
19. Elabbasy, E.M.; Hassan, T.S.; Moaaz, O. Oscillation behavior of second-order nonlinear neutral differential equations with deviating arguments. *Opusc. Math.* **2012**, *32*, 719–730. [CrossRef]
20. Li, T.; Han, Z.; Zhao, P.; Sun, S. Oscillation of even-order neutral delay differential equations. *Adv. Differ. Equ.* **2010**, *2010*, 184180 . [CrossRef]
21. Moaaz, O. New criteria for oscillation of nonlinear neutral differential equations. *Adv. Differ. Equ.* **2019**, *2019*, 484. [CrossRef]
22. Moaaz, O.; Elabbasy, E.M.; Muhib, A. Oscillation criteria for even-order neutral differential equations with distributed deviating arguments. *Adv. Differ. Equ.* **2019**, *297*, 1–10. [CrossRef]
23. Moaaz, O.; Awrejcewicz, J.; Bazighifan, O. A New Approach in the Study of Oscillation Criteria of Even-Order Neutral Differential Equations. *Mathematics* **2020**, *2020*, 179. [CrossRef]
24. Moaaz, O.; El-Nabulsi, R.; Bazighifan, O. Oscillatory behavior of fourth-order differential equations with neutral delay. *Symmetry* **2020**, *12*, 371. [CrossRef]
25. Moaaz, O.; Elabbasy, E.M.; Shaaban, E. Oscillation criteria for a class of third order damped differential equations. *Arab. J. Math. Sci.* **2018**, *24*, 16–30. [CrossRef]
26. Parhi, N.; Tripathy, A. On oscillatory fourth order linear neutral differential equations-I. *Math. Slovaca* **2004**, *54*, 389–410.
27. Moaaz, O.; Elabbasy, E.M.; Bazighifan, O. On the asymptotic behavior of fourth-order functional differential equations. *Adv. Differ. Equ.* **2017**, *2017*, 261. [CrossRef]
28. Xing, G.; Li, T.; Zhang, C. Oscillation of higher-order quasi linear neutral differential equations. *Adv. Differ. Equ.* **2011**, *2011*, 45. [CrossRef]
29. Zafer, A. Oscillation criteria for even order neutral differential equations. *Appl. Math. Lett.* **1998**, *11*, 21–25. [CrossRef]
30. Zhang, Q.; Yan, J. Oscillation behavior of even order neutral differential equations with variable coefficients. *Appl. Math. Lett.* **2006**, *19*, 1202–1206. [CrossRef]
31. Braverman, E.; Karpuz, B. On oscillation of differential and difference equations with non-monotone delays. *Appl. Math. Comput.* **2011**, *218*, 3880–3887. [CrossRef]
32. Pao, B.M.; Liu, C.; Yin, G. *Topics in Stochastic Analysis and Nonparametric Estimation*; Science Business Media: New York, NY, USA, 2008.

33. Kolmanovskii, V.B.; Nosov, V.R. *Stability and Periodic Modes of Control Systems with Aftereffect*; Nauka: Moscow, Russia, 1981.
34. Gopalsamy, K.; Zhang, B. On a neutral delay-logistic equation. *Dyn. Stab. Syst.* **1987**, *2*, 183. [CrossRef]
35. Bellen, N.G.A.; Ruehli, A. Methods for linear systems of circuit delay differential equations of neutral type. *IEEE Trans. Circuits Syst. I Fundam. Theory Appl.* **1999**, *1*, 212. [CrossRef]

5

Improved Oscillation Results for Functional Nonlinear Dynamic Equations of Second Order

Taher S. Hassan [1,2,*,†] **Yuangong Sun** [3,*,†] **and Amir Abdel Menaem** [4,†]

1 Department of Mathematics, Faculty of Science, University of Ha'il, Ha'il 2440, Saudi Arabia
2 Department of Mathematics, Faculty of Science, Mansoura University, Mansoura 35516, Egypt
3 School of Mathematical Sciences, University of Jinan, Jinan 250022, China
4 Department of Automated Electrical Systems, Ural Power Engineering Institute, Ural Federal University, 620002 Yekaterinburg, Russia; abdel.menaem@urfu.ru
* Correspondence: tshassan@mans.edu.eg (T.S.H.); ss_sunyg@ujn.edu.cn (Y.S.)
† These authors contributed equally to this work.

Abstract: In this paper, the functional dynamic equation of second order is studied on an arbitrary time scale under milder restrictions without the assumed conditions in the recent literature. The Nehari, Hille, and Ohriska type oscillation criteria of the equation are investigated. The presented results confirm that the study of the equation in this formula is superior to other previous studies. Some examples are addressed to demonstrate the finding.

Keywords: time scales; functional dynamic equations; second order; oscillation criteria

---

1. Introduction

In order to combine continuous and discrete analysis, the theory of dynamic equations on time scales was proposed by Stefan Hilger in [1]. There are different types of time scales applied in many applications (see [2]). The cases when the time scale $\mathbb{T}$ as an arbitrary closed subset is equal to the reals or to the integers represent the classical theories of differential and of difference equations. The theory of dynamic equations includes the classical theories for the differential equations and difference equations cases and other cases in between these classical cases. That is, we are eligible to consider the q-difference equations when $\mathbb{T} = q^{\mathbb{N}_0} := \{q^k : k \in \mathbb{N}_0 \text{ for } q > 1\}$ which has significant applications in quantum theory (see [3]) and different types of time scales like $\mathbb{T} = h\mathbb{N}$, $\mathbb{T} = \mathbb{N}^2$ and $\mathbb{T} = \mathbb{T}_n$ (the set of the harmonic numbers) can also be applied. For more details of time scales calculus, see [2,4,5]. The study of nonlinear dynamic equations is considered in this work because these equations arise in various real-world problems like the turbulent flow of a polytrophic gas in a porous medium, non-Newtonian fluid theory, and in the study of $p-$Laplace equations. Therefore, we are interested in the oscillatory behavior of the nonlinear functional dynamic equation of second order with deviating arguments

$$\left[a(\zeta)\varphi_\gamma\left(z^\Delta(\zeta)\right)\right]^\Delta + q(\zeta)\varphi_\beta\left(z(\eta(\zeta))\right) = 0 \tag{1}$$

on an above-unbounded time scale $\mathbb{T}$, where $\varphi_\alpha(u) := |u|^\alpha \text{sgn} u$, $\alpha > 0$; a and q are positive rd-continuous functions on $\mathbb{T}$ such that

$$\int^\infty \frac{\Delta \varkappa}{a^{\frac{1}{\gamma}}(\varkappa)} = \infty; \tag{2}$$

and $\eta : \mathbb{T} \to \mathbb{T}$ is a rd-continuous function such that $\lim_{\zeta\to\infty} \eta(\zeta) = \infty$.

By a solution of Equation (1) we mean a nontrivial real-valued function $z \in C^1_{rd}[\zeta_z, \infty)_{\mathbb{T}}$ for some $\zeta_z \geq \zeta_0$ with $\zeta_0 \in \mathbb{T}$ such that $z^\Delta, a(\zeta)\varphi_\gamma\left(z^\Delta(\zeta)\right) \in C^1_{rd}[\zeta_z, \infty)_{\mathbb{T}}$ and $z(\zeta)$ satisfies Equation (1) on $[\zeta_z, \infty)_{\mathbb{T}}$, where C_{rd} is the space of right-dense continuous functions. It should be mentioned that in a particular case when $\mathbb{T} = \mathbb{R}$ then

$$\sigma(\zeta) = \zeta,\ \mu(\zeta) = 0,\ g^\Delta(\zeta) = g'(\zeta),\ \int_a^b g(\zeta)\Delta\zeta = \int_a^b g(\zeta)d\zeta,$$

and (1) turns as the nonlinear functional differential equation

$$\left[a(\zeta)\varphi_\gamma\left(z'(\zeta)\right)\right]' + q(\zeta)\varphi_\beta\left(z(\eta(\zeta))\right) = 0. \tag{3}$$

The oscillation properties of Equation (3) and special cases were investigated by Nehari [6], Fite [7], Hille [8], Wong [9], Erbe [10], and Ohriska [11] as follows: The oscillatory behavior of the linear differential equation of second order

$$z''(\zeta) + q(\zeta)z(\zeta) = 0, \tag{4}$$

is investigated in Nehari [6] and showed that if

$$\liminf_{\zeta\to\infty} \frac{1}{\zeta}\int_{\zeta_0}^{\zeta} \varpropto^2 q(\varpropto)d\varpropto > \frac{1}{4}, \tag{5}$$

then all solutions of (4) are oscillatory. Fite [7] proved that if

$$\int_{\zeta_0}^{\infty} q(\varpropto)d\varpropto = \infty, \tag{6}$$

then all solutions of Equation (4) are oscillatory. Hille [8] developed the condition (6) and illustrated that if

$$\liminf_{\zeta\to\infty} \zeta\int_{\zeta}^{\infty} q(\varpropto)d\varpropto > \frac{1}{4}, \tag{7}$$

then all solutions of Equation (4) are oscillatory. For the delay differential equation

$$z''(\zeta) + q(\zeta)z(\eta(\zeta)) = 0, \tag{8}$$

the Hille-type condition (7) is generalized by Wong [9], where $\eta(\zeta) \geq \gamma\zeta$ with $0 < \gamma < 1$, and showed that if

$$\liminf_{\zeta\to\infty} \zeta\int_{\zeta}^{\infty} q(\varpropto)d\varpropto > \frac{1}{4\gamma}, \tag{9}$$

then all solutions of (8) are oscillatory. Erbe [10] enhanced the condition (9) and examined that if

$$\liminf_{\zeta\to\infty} \zeta\int_{\zeta}^{\infty} q(\varpropto)\frac{\eta(\varpropto)}{\varpropto}d\varpropto > \frac{1}{4}, \tag{10}$$

then all solutions of (8) are oscillatory where $\eta(\zeta) \leq \zeta$. Ohriska [11] proved that, if

$$\limsup_{\zeta\to\infty} \zeta\int_{\zeta}^{\infty} q(\varpropto)\frac{\eta(\varpropto)}{\varpropto}d\varpropto > 1, \tag{11}$$

then all solutions of (8) are oscillatory.

When $\mathbb{T} = \mathbb{Z}$, then

$$\sigma(\zeta) = \zeta + 1,\ \mu(\zeta) = 1,\ g^\Delta(\zeta) = \Delta g(\zeta),\ \int_a^b g(\zeta)\Delta\zeta = \sum_{\zeta=a}^{b-1} g(\zeta),$$

and (1) turns as the nonlinear functional difference equation

$$\Delta\left[a(\zeta)\varphi_{\gamma}\left(\Delta z(\zeta)\right)\right]+q(\zeta)\varphi_{\beta}\left(z(\eta(\zeta))\right)=0. \tag{12}$$

The oscillation of Equation (12) when $a(\zeta)=1$, $\eta(\zeta)=\zeta$, and $\gamma=\beta$ is the quotient of odd positive integers was elaborated by Thandapani et al. [12] in which $q(\zeta)$ is a positive sequence and showed that every solution of (12) is oscillatory, if

$$\sum_{k=k_0}^{\infty} q(k)=\infty.$$

We will examine that our results not only unite some of the known oscillation results for differential and difference equations but they also can be applied on other cases in which the oscillatory behavior of solutions for these equations on various types of time scales was not known. Note that, if $\mathbb{T}=h\mathbb{Z}$, $h>0$, then

$$\sigma(\zeta)=\zeta+h,\ \mu(\zeta)=h,\ z^{\Delta}(\zeta)=\Delta_h z(\zeta)=\frac{z(\zeta+h)-z(\zeta)}{h},$$

$$\int_a^b g(\zeta)\Delta\zeta=\sum_{k=0}^{\frac{b-a-h}{h}} g(a+kh)h,$$

and (1) turns as the nonlinear functional difference equation

$$\Delta_h\left[a(\zeta)\varphi_{\gamma}\left(\Delta_h z(\zeta)\right)\right]+q(\zeta)\varphi_{\beta}\left(z(\eta(\zeta))\right)=0. \tag{13}$$

If

$$\mathbb{T}=q^{\mathbb{N}_0}=\{\zeta:\zeta=q^k,\ k\in\mathbb{N}_0,\ q>1\},$$

then

$$\sigma(\zeta)=q\,\zeta,\ \mu(\zeta)=(q-1)\zeta,\ z^{\Delta}(\zeta)=\Delta_q z(\zeta)=(z(q\,\zeta)-z(\zeta))/(q-1)\,\zeta,$$

$$\int_{\zeta_0}^{\infty} g(\zeta)\Delta\zeta=\sum_{k=n_0}^{\infty} g(q^k)\mu(q^k),$$

where $t_0=q^{n_0}$, and (1) turns as the second order $q-$nonlinear difference equation

$$\Delta_q\left[a(\zeta)\varphi_{\gamma}\left(\Delta_q z(\zeta)\right)\right]+q(\zeta)\varphi_{\beta}\left(z(\eta(\zeta))\right)=0. \tag{14}$$

If

$$\mathbb{T}=\mathbb{N}_0^2:=\{n^2:n\in\mathbb{N}_0\},$$

then

$$\sigma(\zeta)=(\sqrt{\zeta}+1)^2,\ \mu(\zeta)=1+2\sqrt{\zeta},\ \Delta_N z(\zeta)=\frac{z((\sqrt{\zeta}+1)^2)-z(\zeta)}{1+2\sqrt{\zeta}},$$

and (1) turns as the second order nonlinear difference equation

$$\Delta_N\left[a(\zeta)\varphi_{\gamma}\left(\Delta_N z(\zeta)\right)\right]+q(\zeta)\varphi_{\beta}\left(z(\eta(\zeta))\right)=0. \tag{15}$$

If $\mathbb{T}=\{H_n:n\in\mathbb{N}_0\}$ where H_n is the harmonic numbers defined by

$$H_0=0,\ H_n=\sum_{k=1}^{n}\frac{1}{k},\ n\in\mathbb{N},$$

then

$$\sigma(H_n) = H_{n+1},\ \mu(H_n) = \frac{1}{n+1},\ z^{\Delta}(t) = \Delta_{H_n} z(H_n) = (n+1)\Delta z(H_n),$$

and (1) turns as the second order nonlinear harmonic difference equation

$$\Delta_{H_n}\left[a(H_n)\varphi_{\gamma}\left(\Delta_{H_n} z(H_n)\right)\right] + q(H_n)\varphi_{\beta}\left(z(\eta(H_n))\right) = 0. \tag{16}$$

For dynamic equations, Erbe et al. in [13,14] expanded the Hille and Nehari oscillation criteria to the half-linear delay dynamic equation of second order

$$(a(\zeta)(z^{\Delta}(\zeta))^{\gamma})^{\Delta} + q(\zeta)z^{\gamma}(\eta(\zeta)) = 0, \tag{17}$$

where γ is a quotient of odd positive integers,

$$\eta(\zeta) \leq \zeta,\ a^{\Delta}(\zeta) \geq 0,\ \int_{\zeta_0}^{\infty} \eta^{\gamma}(\zeta)q(\zeta)\Delta\zeta = \infty. \tag{18}$$

The authors showed that if either of the following conditions holds

$$\liminf_{\zeta\to\infty} \zeta^{\gamma}\int_{\sigma(\zeta)}^{\infty} q(\propto)\left(\frac{\eta(\propto)}{\sigma(\propto)}\right)^{\gamma}\Delta\propto > \frac{\gamma^{\gamma}}{l^{\gamma^2}(\gamma+1)^{\gamma+1}}, \tag{19}$$

or

$$\liminf_{\zeta\to\infty} \zeta^{\gamma}\int_{\sigma(\zeta)}^{\infty} q(\propto)\left(\frac{\eta(\propto)}{\sigma(\propto)}\right)^{\gamma}\Delta\propto + \liminf_{\zeta\to\infty}\frac{1}{\zeta}\int_{\zeta_0}^{\zeta}\propto^{\gamma+1}q(\propto)\left(\frac{\eta(\propto)}{\sigma(\propto)}\right)^{\gamma}\Delta\propto > \frac{1}{l^{\gamma(\gamma+1)}},$$

where $l := \liminf_{\zeta\to\infty}\dfrac{\zeta}{\sigma(\zeta)}$, then all solutions of (17) are oscillatory. We refer the reader to related results [15–35] and the references cited therein.

A natural question now is: Do the oscillation criteria (5), (6), (7) and (11) for the differential equations of second order by Nehari, Fite, Hille and Ohriska extend to the nonlinear dynamic equation of second order (1) without the restrictive condition (18) in both cases $\eta(\zeta) \leq \zeta$ and $\eta(\zeta) \geq \zeta$, and when $\beta \geq \gamma$ and $\beta \leq \gamma$.

The aim of this paper is to propose an obvious answer to the above question. We will establish Nehari, Hille and Ohriska type oscillation criteria for (1) without imposing the restrictive condition (18), which generalize and improve the aforementioned results in the literature.

2. Oscillation Criteria of (1) when $\beta \geq \gamma$

In the subsequent results, we will use the subsequent notations

$$A(\zeta) := \int_{\zeta_0}^{\zeta}\frac{\Delta\propto}{a^{\frac{1}{\gamma}}(\propto)} \quad \text{and} \quad l := \liminf_{\zeta\to\infty}\frac{A(\zeta)}{A(\sigma(\zeta))} \leq 1,$$

and

$$\phi(\zeta) := \begin{cases} 1, & \eta(\zeta) \geq \zeta, \\ \left(\dfrac{A(\eta(\zeta))}{A(\zeta)}\right)^{\beta}, & \eta(\zeta) \leq \zeta. \end{cases}$$

Furthermore, $l > 0$ is assuming in the next results.

First, we derive Nehari type to the nonlinear dynamic equation of second order (1).

Theorem 1. *Let* (2) *holds, and*

$$\begin{aligned}
&\liminf_{\zeta\to\infty} \frac{1}{A(\zeta)} \int_T^\zeta A^{\gamma+1}(\varkappa)\phi(\varkappa)q(\varkappa)\Delta\varkappa > \frac{1}{l^{\gamma(\gamma+1)}}\left(1-\frac{l^\gamma}{\gamma l^\gamma+1}\right), \quad 0<\gamma\le 1,\\
&\liminf_{\zeta\to\infty} \frac{1}{A(\zeta)} \int_T^\zeta A^{\gamma+1}(\varkappa)\phi(\varkappa)q(\varkappa)\Delta\varkappa > \frac{\gamma}{l^{\gamma(\gamma+1)}(\gamma+l^\gamma)}, \quad \gamma\ge 1,
\end{aligned} \tag{20}$$

for enough large $T\in[\zeta_0,\infty)_{\mathbb{T}}$. Then all solutions of Equation (1) *are oscillatory.*

Proof. Assume $z(t)$ is a nonoscillatory solution of Equation (1) on $[\zeta_0,\infty)_{\mathbb{T}}$. Thus, without loss of generality, let $z(\zeta)>0$ and $z(\eta(\zeta))>0$ on $[\zeta_0,\infty)_{\mathbb{T}}$. Since $q\in C_{rd}([\zeta_0,\infty)_{\mathbb{T}},\mathbb{R}^+)$ and then

$$\left[a(\zeta)\varphi_\gamma\left(z^\Delta(\zeta)\right)\right]^\Delta<0 \quad \text{for } \zeta\ge\zeta_0.$$

Hence $z^\Delta(\zeta)>0$, otherwise, it leads to a contradiction. Define

$$w(\zeta):=\frac{a(\zeta)\varphi_\gamma\left(z^\Delta(\zeta)\right)}{z^\gamma(\zeta)}.$$

Using the product and quotient rules, we reach

$$\begin{aligned}
w^\Delta(\zeta) &= \left(\frac{a(\zeta)\varphi_\gamma\left(z^\Delta(\zeta)\right)}{z^\gamma(\zeta)}\right)^\Delta\\
&= \frac{1}{z^\gamma(\zeta)}\left[a(\zeta)\varphi_\gamma\left(z^\Delta(\zeta)\right)\right]^\Delta\\
&\quad +\left(\frac{1}{z^\gamma(\zeta)}\right)^\Delta\left[a(\zeta)\varphi_\gamma\left(z^\Delta(\zeta)\right)\right]^\sigma\\
&= \frac{\left[a(\zeta)\varphi_\gamma\left(z^\Delta(\zeta)\right)\right]^\Delta}{z^\gamma(\zeta)} - \frac{\left(z^\gamma(\zeta)\right)^\Delta}{z^\gamma(\zeta)z^\gamma(\sigma(\zeta))}\left[a(\zeta)\varphi_\gamma\left(z^\Delta(\zeta)\right)\right]^\sigma.
\end{aligned} \tag{21}$$

From (1) and the definition of $w(\zeta)$, we have

$$w^\Delta(\zeta)=-\left(\frac{z(\eta(\zeta))}{z(\zeta)}\right)^\beta z^{\beta-\gamma}(\zeta)\,q(\zeta)-\frac{\left(z^\gamma(\zeta)\right)^\Delta}{z^\gamma(\zeta)}w(\sigma(\zeta)).$$

Since $z^\Delta>0$, then $z(\zeta)\ge z(\zeta_0)$ for $\zeta\ge\zeta_0$ and so

$$z^{\beta-\gamma}(\zeta)\ge z^{\beta-\gamma}(\zeta_0)=:k>0 \quad \text{for } \zeta\ge\zeta_0.$$

Therefore,

$$w^\Delta(\zeta)\le -k\left(\frac{z(\eta(\zeta))}{z(\zeta)}\right)^\beta q(\zeta)-\frac{\left(z^\gamma(\zeta)\right)^\Delta}{z^\gamma(\zeta)}w(\sigma(\zeta)).$$

Let $\zeta\in[\zeta_0,\infty)_{\mathbb{T}}$ be fixed. If $\eta(\zeta)\ge\zeta$, then $z(\eta(\zeta))\ge z(\zeta)$ by the fact that $z^\Delta>0$. Now the case $\eta(\zeta)\le\zeta$ is considered. Since $\left(a\,\varphi_\gamma\left(z^\Delta\right)\right)^\Delta<0$ on $[\zeta_0,\infty)_{\mathbb{T}}$, we achieve

$$\begin{aligned}
z(\zeta) &\ge z(\zeta)-z(\zeta_1)=\int_{\zeta_0}^\zeta z^\Delta(\varkappa)\Delta\varkappa\\
&\ge a^{\frac{1}{\gamma}}(\zeta)z^\Delta(\zeta)\int_{\zeta_0}^\zeta \frac{\Delta\varkappa}{a^{\frac{1}{\gamma}}(\varkappa)}\\
&= a^{\frac{1}{\gamma}}(\zeta)z^\Delta(\zeta)A(\zeta).
\end{aligned}$$

Therefore

$$\begin{aligned}\left[\frac{z(\zeta)}{A(\zeta)}\right]^{\Delta} &= \frac{A(\zeta)z^{\Delta}(\zeta)-z(\zeta)a^{-\frac{1}{\gamma}}(\zeta)}{A(\zeta)A^{\sigma}(\zeta)}\\ &= \frac{a^{-\frac{1}{\gamma}}(\zeta)}{A(\zeta)A^{\sigma}(\zeta)}\left(a^{\frac{1}{\gamma}}(\zeta)z^{\Delta}(\zeta)A(\zeta)-z(\zeta)\right)\\ &\leq 0, \quad \zeta\in(\zeta_0,\infty)_{\mathbb{T}}.\end{aligned}$$

So there exists a $\zeta_1 \in (\zeta_0,\infty)_{\mathbb{T}}$ such that $\eta(\zeta) \in (\zeta_0,\infty)_{\mathbb{T}}$ for $\zeta \geq \zeta_1$ and so

$$\frac{z(\eta(\zeta))}{z(\zeta)} \geq \frac{A(\eta(\zeta))}{A(\zeta)} \quad \text{for } \zeta\in[\zeta_1,\infty)_{\mathbb{T}}.$$

In both cases and from the definition of $\phi(\zeta)$ we have that

$$\left(\frac{z(\eta(\zeta))}{z(\zeta)}\right)^{\beta} \geq \phi(\zeta), \tag{22}$$

and so

$$w^{\Delta}(\zeta) \leq -k\,\phi(\zeta)q(\zeta) - \frac{(z^{\gamma}(\zeta))^{\Delta}}{z^{\gamma}(\zeta)}w(\sigma(\zeta)), \quad \zeta\in[\zeta_1,\infty)_{\mathbb{T}}. \tag{23}$$

Then by using the Pötzsche chain rule ([2], Theorem 1.90), we get that

$$\begin{aligned}(z^{\gamma}(\zeta))^{\Delta} &= \gamma\left(\int_0^1 \left[z(\zeta)+h\mu(\zeta)z^{\Delta}(\zeta)\right]^{\gamma-1}dh\right)z^{\Delta}(\zeta)\\ &= \gamma\left(\int_0^1 [(1-h)\,z(\zeta)+hz(\sigma(\zeta))]^{\gamma-1}dh\right)z^{\Delta}(\zeta)\\ &\geq \begin{cases}\gamma z^{\gamma-1}(\sigma(\zeta))\,z^{\Delta}(\zeta), & 0<\gamma\leq 1,\\ \gamma z^{\gamma-1}(\zeta)z^{\Delta}(\zeta), & \gamma\geq 1.\end{cases}\end{aligned}$$

If $0<\gamma\leq 1$, then

$$w^{\Delta}(\zeta) < -k\,\phi(\zeta)q(\zeta) - \gamma\frac{z^{\Delta}(\zeta)}{z(\sigma(\zeta))}\left(\frac{z(\sigma(\zeta))}{z(\zeta)}\right)^{\gamma}w(\sigma(\zeta));$$

and if $\gamma \geq 1$, then

$$w^{\Delta}(\zeta) \leq -k\,\phi(\zeta)q(\zeta) - \gamma\frac{z^{\Delta}(\zeta)}{z(\sigma(\zeta))}\frac{z(\sigma(\zeta))}{z(\zeta)}w(\sigma(\zeta)).$$

Note that $z^{\Delta}>0$ and $\left(a\,\varphi_{\gamma}(z^{\Delta})\right)^{\Delta}<0$ on $[\zeta_1,\infty)_{\mathbb{T}}$, we see for $\gamma>0$,

$$\begin{aligned}w^{\Delta}(\zeta) &\leq -k\,\phi(\zeta)q(\zeta) - \gamma\frac{z^{\Delta}(\zeta)}{z(\sigma(\zeta))}w(\sigma(\zeta))\\ &\leq -k\,\phi(\zeta)q(\zeta) - \gamma a^{-\frac{1}{\gamma}}(\zeta)w^{1+\frac{1}{\gamma}}(\sigma(\zeta)), \quad \zeta\in[\zeta_1,\infty)_{\mathbb{T}}.\end{aligned} \tag{24}$$

Multiplying both sides of (24) by $A^{\gamma+1}(\zeta)$ and integrating from ζ_2 to $\zeta\in[\zeta_2,\infty)_{\mathbb{T}}$, we get

$$\begin{aligned}\int_{\zeta_2}^{\zeta} A^{\gamma+1}(\varkappa)w^{\Delta}(\varkappa)\Delta\varkappa &\leq -k\int_{\zeta_2}^{\zeta} A^{\gamma+1}(\varkappa)\phi(\varkappa)q(\varkappa)\Delta\varkappa\\ &\quad -\gamma\int_{\zeta_2}^{\zeta} a^{-\frac{1}{\gamma}}(\varkappa)\left(A^{\gamma}(\varkappa)\,w(\sigma(\varkappa))\right)^{\frac{\gamma+1}{\gamma}}\Delta\varkappa.\end{aligned}$$

By integration by parts, we have

$$\begin{aligned} A^{\gamma+1}(\zeta)w(\zeta) &\leq A^{\gamma+1}(\zeta_2)w(\zeta_2) + \int_{\zeta_2}^{\zeta} \left(A^{\gamma+1}(\varkappa)\right)^{\Delta} w(\sigma(\varkappa))\,\Delta\varkappa \\ &\quad -k\int_{\zeta_2}^{\zeta} A^{\gamma+1}(\varkappa)\phi(\varkappa)q(\varkappa)\Delta\varkappa \\ &\quad -\gamma\int_{\zeta_2}^{\zeta} a^{-\frac{1}{\gamma}}(\varkappa)\left(A^{\gamma}(\varkappa)\,w(\sigma(\varkappa))\right)^{\frac{\gamma+1}{\gamma}}\Delta\varkappa. \end{aligned}$$

Using the Pötzsche chain rule, we arrive

$$\begin{aligned} \left(A^{\gamma+1}(\varkappa)\right)^{\Delta} &= (\gamma+1)\int_0^1 [A(\varkappa)+h\mu(\varkappa)A^{\Delta}(\varkappa)]^{\gamma}dh\,\frac{1}{a^{1/\gamma}(\varkappa)} \\ &= (\gamma+1)\int_0^1 [(1-h)\,A(\varkappa)+hA\,(\sigma(\varkappa))]^{\gamma}dh\,\frac{1}{a^{1/\gamma}(\varkappa)} \\ &\leq (\gamma+1)\,\frac{A^{\gamma}\,(\sigma(\varkappa))}{a^{1/\gamma}(\varkappa)}. \end{aligned} \tag{25}$$

Hence

$$\begin{aligned} A^{\gamma+1}(\zeta)w(\zeta) &\leq A^{\gamma+1}(\zeta_2)w(\zeta_2) - \int_{\zeta_2}^{\zeta} A^{\gamma+1}(\varkappa)\phi(\varkappa)q(\varkappa)\Delta\varkappa \\ &\quad +(\gamma+1)\int_{\zeta_2}^{\zeta}\frac{1}{a^{1/\gamma}(\varkappa)}\left[\frac{A\,(\sigma(\varkappa))}{A(\varkappa)}\right]^{\gamma} A^{\gamma}(\varkappa)w\,(\sigma(\varkappa))\,\Delta\varkappa \\ &\quad -\gamma\int_{\zeta_2}^{\zeta}\frac{1}{a^{1/\gamma}(\varkappa)}\left(A^{\gamma}(\varkappa)\,w\,(\sigma(\varkappa))\right)^{\frac{\gamma+1}{\gamma}}\Delta\varkappa. \end{aligned}$$

It follows that $w^{\Delta}(\zeta)\leq 0$ on $[\zeta_1,\infty)_{\mathbb{T}}$. Let $\varepsilon>0$, then we choose $\zeta_2\in[\zeta_1,\infty)_{\mathbb{T}}$, enough large, so for $\zeta\in[\zeta_2,\infty)_{\mathbb{T}}$,

$$A^{\gamma}(\zeta)\,w\,(\sigma(\zeta))\geq a_*-\varepsilon, \tag{26}$$

and

$$\frac{A(\zeta)}{A\,(\sigma(\zeta))}\geq l-\varepsilon, \tag{27}$$

where a_* is defined by

$$a_* := \liminf_{\zeta\to\infty} A^{\gamma}(\zeta)w\,(\sigma(\zeta))\leq 1. \tag{28}$$

By (27), we then get that

$$A^{\gamma+1}(\zeta)w(\zeta)\leq A^{\gamma+1}(\zeta_2)w(\zeta_2)-k\int_{\zeta_2}^{\zeta}A^{\gamma+1}(\varkappa)\phi(\varkappa)q(\varkappa)\Delta\varkappa$$

$$+\int_{\zeta_2}^{\zeta}\frac{1}{a^{1/\gamma}(\varkappa)}\left[\frac{\gamma+1}{(l-\varepsilon)^{\gamma}}A^{\gamma}(\varkappa)\,w\,(\sigma(\varkappa))-\gamma\left(A^{\gamma}(\varkappa)\,w\,(\sigma(\varkappa))\right)^{\frac{\gamma+1}{\gamma}}\right]\Delta\varkappa.$$

Using the inequality

$$Yu-Xu^{\frac{\gamma+1}{\gamma}}\leq\frac{\gamma^{\gamma}}{(\gamma+1)^{\gamma+1}}\frac{Y^{\gamma+1}}{X^{\gamma}} \tag{29}$$

with $X=\gamma$, $Y=\dfrac{\gamma+1}{(l-\varepsilon)^{\gamma}}$ and $u=A^{\gamma}(\varkappa)\,w\,(\sigma(\varkappa))$, we get

$$A^{\gamma+1}(\zeta)w(\zeta) \leq A^{\gamma+1}(\zeta_2)w(\zeta_2) - k\int_{\zeta_2}^{\zeta} A^{\gamma+1}(\varkappa)\phi(\varkappa)q(\varkappa)\Delta\varkappa + \frac{1}{(l-\varepsilon)^{\gamma(\gamma+1)}}\left[A(\zeta)-A(\zeta_2)\right].$$

Dividing both sides by $A(\zeta)$, we obtain

$$A^{\gamma}(\zeta)w(\zeta) \leq \frac{A^{\gamma+1}(\zeta_2)w(\zeta_2)}{A(\zeta)} - \frac{k}{A(\zeta)}\int_{\zeta_2}^{\zeta} A^{\gamma+1}(\varkappa)\phi(\varkappa)q(\varkappa)\Delta\varkappa + \frac{1}{(l-\varepsilon)^{\gamma(\gamma+1)}}\left[1-\frac{A(\zeta_2)}{A(\zeta)}\right].$$

Since $w^{\sigma}(\zeta) \leq w(\zeta)$ we get

$$A^{\gamma}(\zeta)w(\sigma(\zeta)) \leq \frac{A^{\gamma+1}(\zeta_2)w(\zeta_2)}{A(\zeta)} - \frac{k}{A(\zeta)}\int_{\zeta_2}^{\zeta} A^{\gamma+1}(\varkappa)\phi(\varkappa)q(\varkappa)\Delta\varkappa + \frac{1}{(l-\varepsilon)^{\gamma(\gamma+1)}}\left[1-\frac{A(\zeta_2)}{A(\zeta)}\right].$$

Taking the lim sup of both sides as $\zeta \to \infty$ we get

$$A_* \leq -\liminf_{\zeta\to\infty}\frac{k}{A(\zeta)}\int_{\zeta_2}^{\zeta} A^{\gamma+1}(\varkappa)\phi(\varkappa)q(\varkappa)\Delta\varkappa + \frac{1}{(l-\varepsilon)^{\gamma(\gamma+1)}}.$$

where

$$A_* := \limsup_{\zeta\to\infty} A^{\gamma}(\zeta)w(\sigma(\zeta)).$$

Since $k,\ \varepsilon > 0$ are arbitrary constants, we obtain

$$A_* \leq -\liminf_{\zeta\to\infty}\frac{1}{A(\zeta)}\int_{\zeta_2}^{\zeta} A^{\gamma+1}(\varkappa)\phi(\varkappa)q(\varkappa)\Delta\varkappa + \frac{1}{l^{\gamma(\gamma+1)}}. \tag{30}$$

Now, multiplying both sides of (24) by $A^{\gamma+1}(\zeta)$, we get

$$\begin{aligned} A^{\gamma+1}(\zeta)\,w^{\Delta}(\zeta) &\leq -k\,A^{\gamma+1}(\zeta)\,\phi(\zeta)q(\zeta) - \gamma a^{-1/\gamma}(\zeta)A^{\gamma+1}(\zeta)\,w^{1+\frac{1}{\gamma}}(\sigma(\zeta)) \\ &= -A^{\gamma+1}(\zeta)\,\phi(\zeta)q(\zeta) \\ &\quad -\gamma a^{-1/\gamma}(\zeta)A^{\gamma}(\zeta)\,w(\sigma(\zeta))\;A(\zeta)\,w^{\frac{1}{\gamma}}(\sigma(\zeta)). \end{aligned}$$

Using (26) gives

$$A^{\gamma+1}(\zeta)\,w^{\Delta}(\zeta) \leq -k\,A^{\gamma+1}(\zeta)\,\phi(\zeta)q(\zeta) - \vartheta a^{-1/\gamma}(\zeta), \quad \zeta\in[\zeta_2,\infty)_{\mathbb{T}}, \tag{31}$$

where $\vartheta = \gamma(a_* - \varepsilon)^{1+\frac{1}{\gamma}}$. Integrating the inequality (31) from ζ_2 to $\zeta \in [\zeta_2,\infty)_{\mathbb{T}}$, we get

$$\int_{\zeta_2}^{\zeta} A^{\gamma+1}(\varkappa)\,w^{\Delta}(\varkappa)\Delta\varkappa \leq -k\int_{\zeta_2}^{\zeta} A^{\gamma+1}(\varkappa)\,\phi(\varkappa)q(\varkappa)\Delta\varkappa - \vartheta\int_{\zeta_2}^{\zeta} a^{-1/\gamma}(\varkappa)\Delta\varkappa.$$

Using integrating by parts, we get

$$\begin{aligned} A^{\gamma+1}(\zeta)\, w(\zeta) &\leq A^{\gamma+1}(\zeta_2)\, w^{\Delta}(\zeta_2) + \int_{\zeta_2}^{\zeta} \left[A^{\gamma+1}(\backsim)\right]^{\Delta} w(\sigma(\backsim))\, \Delta\backsim \\ &\quad -k \int_{\zeta_2}^{\zeta} A^{\gamma+1}(\backsim)\, \phi(\backsim) q(\backsim) \Delta\backsim - \vartheta\left[A(\zeta) - A(\zeta_2)\right]. \end{aligned} \tag{32}$$

We consider the forthcoming two cases:

(I) When $0 < \gamma \leq 1$. Using the product rule, we have

$$\left[A^{\gamma+1}(\backsim)\right]^{\Delta} = \left[A^{\gamma}(\backsim)\, A(\backsim)\right]^{\Delta} = \left[A^{\gamma}(\backsim)\right]^{\Delta} A(\backsim) + A^{\gamma}(\sigma(\backsim))\, A^{\Delta}(\backsim).$$

Again use the Pötzsche chain rule, we get

$$\begin{aligned} \left(A^{\gamma}(\backsim)\right)^{\Delta} &= \gamma\left(\int_0^1 \left[A(\backsim) + h\mu(\backsim) A^{\Delta}(\backsim)\right]^{\gamma-1} \mathrm{d}h\right) A^{\Delta}(\backsim) \\ &= \gamma\left(\int_0^1 \left[(1-h)\, A(\backsim) + hA(\sigma(\backsim))\right]^{\gamma-1} \mathrm{d}h\right) A^{\Delta}(\backsim) \\ &\leq \gamma A^{\gamma-1}(\backsim)\, A^{\Delta}(\backsim). \end{aligned}$$

Then

$$\left[A^{\gamma+1}(\backsim)\right]^{\Delta} \leq \left(\gamma A^{\gamma}(\backsim) + A^{\gamma}(\sigma(\backsim))\right) A^{\Delta}(\backsim).$$

and so

$$\begin{aligned} A^{\gamma+1}(\zeta)\, w(\zeta) &\leq A^{\gamma+1}(\zeta_2)\, w^{\Delta}(\zeta_2) \\ &\quad + \int_{\zeta_2}^{\zeta} \left(\gamma A^{\gamma}(\backsim) + A^{\gamma}(\sigma(\backsim))\right) A^{\Delta}(\backsim)\, w(\sigma(\backsim))\, \Delta\backsim \\ &\quad -k \int_{\zeta_2}^{\zeta} A^{\gamma+1}(\backsim)\, \phi(\backsim) q(\backsim) \Delta\backsim - \vartheta\left[A(\zeta) - A(\zeta_2)\right] \\ &= A^{\gamma+1}(\zeta_2)\, w^{\Delta}(\zeta_2) \\ &\quad + \int_{\zeta_2}^{\zeta} \left(\gamma + \left[\frac{A(\sigma(\backsim))}{A(\backsim)}\right]^{\gamma}\right) A^{\Delta}(\backsim)\ A^{\gamma}(\backsim)\, w(\sigma(\backsim))\, \Delta\backsim \\ &\quad -k \int_{\zeta_2}^{\zeta} A^{\gamma+1}(\backsim)\, \phi(\backsim) q(\backsim) \Delta\backsim) - \vartheta\left[A(\zeta) - A(\zeta_2)\right] \\ &\leq A^{\gamma+1}(\zeta_2)\, w^{\Delta}(\zeta_2) + \left[\gamma + \frac{1}{(l-\varepsilon)^{\gamma}}\right] (A_* + \varepsilon)\left[A(\zeta) - A(\zeta_2)\right] \\ &\quad -k \int_{\zeta_2}^{\zeta} A^{\gamma+1}(\backsim)\, \phi(\backsim) q(\backsim) \Delta\backsim) - \vartheta\left[A(\zeta) - A(\zeta_2)\right]. \end{aligned}$$

Dividing both sides by $A(\zeta)$, we have

$$\begin{aligned} A^{\gamma}(\zeta)\, w(\sigma(\zeta)) &\leq A^{\gamma}(\zeta)\, w(\zeta) \leq \frac{A^{\gamma+1}(\zeta_2)\, w^{\Delta}(\zeta_2)}{A(\zeta)} \\ &\quad + \left[\gamma + \frac{1}{(l-\varepsilon)^{\gamma}}\right] (A_* + \varepsilon)\left[1 - \frac{A(\zeta_2)}{A(\zeta)}\right] \\ &\quad - \frac{k}{A(\zeta)} \int_{\zeta_2}^{\zeta} A^{\gamma+1}(\backsim)\, \phi(\backsim) q(\backsim) \Delta\backsim - \vartheta\left[1 - \frac{A(\zeta_2)}{A(\zeta)}\right]. \end{aligned}$$

Taking the lim sup of both sides as $\zeta \to \infty$ and using (2), we get

$$A_* \leq \left[\gamma + \frac{1}{(l-\varepsilon)^\gamma}\right](A_* + \varepsilon) - \liminf_{\zeta\to\infty} \frac{k}{A(\zeta)} \int_{\zeta_2}^{\zeta} A^{\gamma+1}(\varkappa)\,\phi(\varkappa) q(\varkappa)\Delta\varkappa - \vartheta.$$

Since k and $\varepsilon > 0$ are arbitrary constants, we achieve the demanded inequality

$$\liminf_{\zeta\to\infty} \frac{1}{A(\zeta)} \int_{\zeta_2}^{\zeta} A^{\gamma+1}(\varkappa)\,\phi(\varkappa) q(\varkappa)\Delta\varkappa \leq A_*\left[\gamma - 1 + \frac{1}{l^\gamma}\right] - \gamma a_*^{1+\frac{1}{\gamma}}. \tag{33}$$

From (30) and (33), we obtain

$$\liminf_{\zeta\to\infty} \frac{1}{A(\zeta)} \int_{\zeta_2}^{\zeta} A^{\gamma+1}(\varkappa)\phi(\varkappa) q(\varkappa)\Delta\varkappa \leq \frac{1}{l^{\gamma(\gamma+1)}}\left(1 - \frac{l^\gamma}{\gamma l^\gamma + 1}\right),$$

which contradicts the condition (20) if $0 < \gamma \leq 1$.

(II) When $\gamma \geq 1$. Using the product rule, we have

$$\left[A^{\gamma+1}(\varkappa)\right]^\Delta = \left[A^\gamma(\varkappa)\, A(\varkappa)\right]^\Delta = \left[A^\gamma(\varkappa)\right]^\Delta A(\sigma(\varkappa)) + A^\gamma(\varkappa)\, A^\Delta(\varkappa).$$

Again by the Pötzsche chain rule we obtain

$$\begin{aligned}
(A^\gamma(\varkappa))^\Delta &= \gamma\left(\int_0^1 \left[A(\varkappa) + h\mu(\varkappa)A^\Delta(\varkappa)\right]^{\gamma-1} \mathrm{d}h\right) A^\Delta(\varkappa)\\
&= \gamma\left(\int_0^1 \left[(1-h)\, A(\varkappa) + hA(\sigma(\varkappa))\right]^{\gamma-1} \mathrm{d}h\right) A^\Delta(\varkappa)\\
&\leq \gamma A^{\gamma-1}(\sigma(\varkappa))\, A^\Delta(\varkappa).
\end{aligned}$$

Then

$$\left[A^{\gamma+1}(\varkappa)\right]^\Delta \leq \left(\gamma A^\gamma(\sigma(\varkappa)) + A^\gamma(\varkappa)\right) A^\Delta(\varkappa).$$

and so

$$\begin{aligned}
A^{\gamma+1}(\zeta)\, w(\zeta) &\leq A^{\gamma+1}(\zeta_2)\, w^\Delta(\zeta_2)\\
&\quad + \int_{\zeta_2}^{\zeta} \left(\gamma A^\gamma(\sigma(\varkappa)) + A^\gamma(\varkappa)\right) A^\Delta(\varkappa)\, w(\sigma(\varkappa))\,\Delta\varkappa\\
&\quad -k\int_{\zeta_2}^{\zeta} A^{\gamma+1}(\varkappa)\,\phi(\varkappa) q(\varkappa)\Delta\varkappa - \vartheta\left[A(\zeta) - A(\zeta_2)\right]\\
&= A^{\gamma+1}(\zeta_2)\, w^\Delta(\zeta_2)\\
&\quad + \int_{\zeta_2}^{\zeta} \left(\gamma\left[\frac{A(\sigma(\varkappa))}{A(\varkappa)}\right]^\gamma + 1\right) A^\Delta(\varkappa)\;\; A^\gamma(\varkappa)\, w(\sigma(\varkappa))\,\Delta\varkappa\\
&\quad -k\int_{\zeta_2}^{\zeta} A^{\gamma+1}(\varkappa)\,\phi(\varkappa) q(\varkappa)\Delta\varkappa) - \vartheta\left[A(\zeta) - A(\zeta_2)\right]\\
&\leq A^{\gamma+1}(\zeta_2)\, w^\Delta(\zeta_2) + \left(\frac{\gamma}{(l-\varepsilon)^\gamma} + 1\right)(A_* + \varepsilon)\left[A(\zeta) - A(\zeta_2)\right]\\
&\quad -k\int_{\zeta_2}^{\zeta} A^{\gamma+1}(\varkappa)\,\phi(\varkappa) q(\varkappa)\Delta\varkappa - \vartheta\left[A(\zeta) - A(\zeta_2)\right].
\end{aligned}$$

Dividing both sides by $A(\zeta)$, we have

$$A^{\gamma}(\zeta)\,w(\zeta) \le \frac{A^{\gamma+1}(\zeta_2)\,w^{\Delta}(\zeta_2)}{A(\zeta)} + \left(\frac{\gamma}{(l-\varepsilon)^{\gamma}}+1\right)(A_*+\varepsilon)\left[1-\frac{A(\zeta_2)}{A(\zeta)}\right] - \frac{k}{A(\zeta)}\int_{\zeta_2}^{\zeta} A^{\gamma+1}(\varkappa)\,\phi(\varkappa)q(\varkappa)\Delta\varkappa - \vartheta\left[1-\frac{A(\zeta_2)}{A(\zeta)}\right].$$

Taking the lim sup of both sides as $\zeta \to \infty$ and by (2), we obtain

$$A_* \le \left(\frac{\gamma}{(l-\varepsilon)^{\gamma}}+1\right)(A_*+\varepsilon) - \liminf_{\zeta\to\infty}\frac{k}{A(\zeta)}\int_{\zeta_2}^{\zeta} A^{\gamma+1}(\varkappa)\,\phi(\varkappa)q(\varkappa)\Delta\varkappa - \vartheta.$$

Since $k,\ \varepsilon > 0$ are arbitrary constants, we reach the demanded inequality

$$\liminf_{\zeta\to\infty}\frac{1}{A(\zeta)}\int_{\zeta_2}^{\zeta} A^{\gamma+1}(\varkappa)\,\phi(\varkappa)q(\varkappa)\Delta\varkappa \le \gamma\left(\frac{A_*}{l^{\gamma}} - a_*^{1+\frac{1}{\gamma}}\right). \tag{34}$$

From (30) and (34), we get

$$\liminf_{\zeta\to\infty}\frac{1}{A(\zeta)}\int_{\zeta_2}^{\zeta} A^{\gamma+1}(\varkappa)\,\phi(\varkappa)q(\varkappa)\Delta\varkappa \le \frac{\gamma}{l^{\gamma(\gamma+1)}(\gamma+l^{\gamma})},$$

which is in contrast to the condition (20) if $\gamma \ge 1$. The proof is accomplished. □

Theorem 2. *Let* (2) *holds, and*

$$\liminf_{\zeta\to\infty}\frac{1}{A(\zeta)}\int_{T}^{\zeta} A^{\gamma+1}(\varkappa)\phi(\varkappa)q(\varkappa)\Delta\varkappa > \frac{1}{l^{\gamma(\gamma+1)}}\left(1-\frac{l^{\gamma}}{\gamma+1}\right), \tag{35}$$

for enough large $T \in [\zeta_0,\infty)_{\mathbb{T}}$. *Then all solutions of Equation* (1) *are oscillatory.*

Proof. Assume z is a nonoscillatory solution of Equation (1) on $[\zeta_0,\infty)_{\mathbb{T}}$. Thus, without loss of generality, let $z(\zeta) > 0$ and $z(\eta(\zeta)) > 0$ on $[\zeta_0,\infty)_{\mathbb{T}}$. As shown in the proof of Theorem 1, we obtain

$$A^{\gamma+1}(\zeta)\,w(\zeta) \le A^{\gamma+1}(\zeta_2)\,w^{\Delta}(\zeta_2) + \int_{\zeta_2}^{\zeta}\left[A^{\gamma+1}(\varkappa)\right]^{\Delta} w(\sigma(\varkappa))\,\Delta\varkappa - k\int_{\zeta_2}^{\zeta} A^{\gamma+1}(\varkappa)\,\phi(\varkappa)q(\varkappa)\Delta\varkappa - \vartheta\left[A(\zeta)-A(\zeta_2)\right], \tag{36}$$

where $\vartheta = \gamma\,(a_*-\varepsilon)^{1+\frac{1}{\gamma}}$. In addition, we have

$$\left[A^{\gamma+1}(\varkappa)\right]^{\Delta} \le (\gamma+1)\,A^{\gamma}(\sigma(\varkappa))\;a^{-1/\gamma}(\varkappa). \tag{37}$$

Substituting (37) into (36) we get

$$\begin{aligned} A^{\gamma+1}(\zeta)\,w(\zeta) &\le A^{\gamma+1}(\zeta_2)\,w^{\Delta}(\zeta_2) \\ &\quad + (\gamma+1)\int_{\zeta_2}^{\zeta}\left[\frac{A(\sigma(\varkappa))}{A(\varkappa)}\right]^{\gamma} a^{-1/\gamma}(\varkappa)A^{\gamma}(\varkappa)\,w(\sigma(\varkappa))\,\Delta\varkappa \\ &\quad - k\int_{\zeta_2}^{\zeta} A^{\gamma+1}(\varkappa)\,\phi(\varkappa)q(\varkappa)\Delta\varkappa - \vartheta\left[A(\zeta)-A(\zeta_2)\right] \\ &\le A^{\gamma+1}(\zeta_2)\,w^{\Delta}(\zeta_2) + \frac{\gamma+1}{(l-\varepsilon)^{\gamma}}(a_*+\varepsilon)\left[A(\zeta)-A(\zeta_2)\right] \\ &\quad - k\int_{\zeta_2}^{\zeta} A^{\gamma+1}(\varkappa)\,\phi(\varkappa)q(\varkappa)\Delta\varkappa - \vartheta\left[A(\zeta)-A(\zeta_2)\right]. \end{aligned}$$

Dividing both sides by $A(\zeta)$, we have

$$\begin{aligned} A^{\gamma}(\zeta)\, w(\sigma(\zeta)) \;\le\; & A^{\gamma}(\zeta)\, w(\zeta) \le \frac{A^{\gamma+1}(\zeta_2)\, w^{\Delta}(\zeta_2)}{A(\zeta)} \\ & + \frac{(\gamma+1)}{(l-\varepsilon)^{\gamma}}(a_* + \varepsilon)\left[1 - \frac{A(\zeta_2)}{A(\zeta)}\right] \\ & - \frac{k}{A(\zeta)} \int_{\zeta_2}^{\zeta} A^{\gamma+1}(\varkappa)\, \phi(\varkappa) q(\varkappa) \Delta\varkappa - \vartheta\left[1 - \frac{A(\zeta_2)}{A(\zeta)}\right] \end{aligned}$$

Taking the lim sup of both sides as $\zeta \to \infty$ and by (2), we obtain

$$a_* \le \frac{(\gamma+1)}{(l-\varepsilon)^{\gamma}}(a_* + \varepsilon) - \liminf_{\zeta\to\infty} \frac{1}{A(\zeta)} \int_{\zeta_2}^{\zeta} a^{\gamma+1}(\varkappa)\, \phi(\varkappa) q(\varkappa) \Delta\varkappa - \vartheta.$$

Since $k,\ \varepsilon > 0$ are arbitrary, we get the required inequality

$$\liminf_{\zeta\to\infty} \frac{1}{A(\zeta)} \int_{\zeta_2}^{\zeta} A^{\gamma+1}(\varkappa)\, \phi(\varkappa) q(\varkappa) \Delta\varkappa \le a_* \left[\frac{\gamma+1}{l^{\gamma}} - 1\right] - \gamma a_*^{1+\frac{1}{\gamma}}. \tag{38}$$

From (30) and (38), we obtain

$$\liminf_{\zeta\to\infty} \frac{1}{A(\zeta)} \int_{\zeta_2}^{\zeta} A^{\gamma+1}(\varkappa) \phi(\varkappa) q(\varkappa) \Delta\varkappa \le \frac{1}{l^{\gamma(\gamma+1)}}\left(1 - \frac{l^{\gamma}}{\gamma+1}\right),$$

which is in contrast to the condition (35). The proof is accomplished. □

Example 1. *Consider the nonlinear dynamic equation of second order*

$$\left[\zeta^{\gamma-1}\varphi_{\gamma}\left(z^{\Delta}(\zeta)\right)\right]^{\Delta} + \frac{\delta\zeta^{\frac{1-\gamma}{\gamma}}}{\phi(\zeta)A^{\gamma+1}(\zeta)}\varphi_{\beta}\left(z(\eta(\zeta))\right) = 0, \tag{39}$$

where γ, β, *and* δ *are positive constants with* $\beta \ge \gamma$. *Here* $a(\zeta) = \zeta^{\gamma-1}$, *and* $q(\zeta) = \frac{\delta\zeta^{-(\gamma+1)}}{\phi(\zeta)A^{\gamma+1}(\zeta)}$, *then the condition* (2) *holds since*

$$\int^{\infty} \frac{\Delta\varkappa}{a^{\frac{1}{\gamma}}(\varkappa)} = \int^{\infty} \frac{\Delta\varkappa}{\varkappa^{1-\frac{1}{\gamma}}} = \infty$$

by Example 5.60 in [5]. In addition, a straightforward computation yields that

$$\liminf_{\zeta\to\infty} \frac{1}{A(\zeta)} \int_{\zeta}^{\zeta} A^{\gamma+1}(\varkappa)\phi(\varkappa)q(\varkappa)\Delta\varkappa = \delta \liminf_{\zeta\to\infty} \frac{1}{A(\zeta)} \int_{\zeta}^{\zeta} \frac{\Delta\varkappa}{\varkappa^{\gamma+1}} = \delta.$$

By Theorem 2, every solution of (39) *is oscillatory if*

$$\delta > \frac{1}{l^{\gamma(\gamma+1)}}\left(1 - \frac{l^{\gamma}}{\gamma+1}\right).$$

We present a Fite–Wintner type oscillation criterion for (1). The proof is similar to that in [7], and hence is omitted.

Theorem 3. *Let* (2) *holds, and*

$$\int_{\zeta_0}^{\infty} q(\varkappa)\Delta\varkappa = \infty. \tag{40}$$

Then every solution of Equation (1) *is oscillatory.*

From Theorem 3, we assume without loss of generality that

$$\int_{\zeta_0}^{\infty} \phi(\varkappa) q(\varkappa) \Delta\varkappa < \infty.$$

Otherwise, we have that (40) holds due to $\phi(\zeta) \leq 1$, which implies that Equation (1) is oscillatory by Theorem 3. The next theorem is generalized Hille type to the second order nonlinear dynamic Equation (1).

Theorem 4. *Let* (2) *holds, and*

$$\liminf_{\zeta \to \infty} A^{\gamma}(\zeta) \int_{\sigma(\zeta)}^{\infty} \phi(\varkappa) q(\varkappa) \Delta\varkappa > \frac{\gamma^{\gamma}}{l\gamma^2 (\gamma+1)^{\gamma+1}}. \tag{41}$$

Then every solutions of Equation (1) *is oscillatory.*

Proof. Assume $z(t)$ be a nonoscillatory solution of Equation (1) on $[\zeta_0, \infty)_{\mathbb{T}}$. Thus, without loss of generality, let $z(\zeta) > 0$ and $z(\eta(\zeta)) > 0$ on $[\zeta_0, \infty)_{\mathbb{T}}$. As depicted in the proof of Theorem 1, we obtain (24) for $\zeta \geq \zeta_1$, for some $\zeta_1 \in (\zeta_0, \infty)_{\mathbb{T}}$ such that $\eta(\zeta) \in (\zeta_0, \infty)_{\mathbb{T}}$ for $\zeta \geq \zeta_1$. Also for $\varepsilon > 0$, then we can pick $\zeta_2 \in [\zeta_1, \infty)_{\mathbb{T}}$, sufficiently large, so that (26) and (27) for $\zeta \in [\zeta_2, \infty)_{\mathbb{T}}$. Replacing ζ by $\varkappa$ in the inequality (24) and then integrating it from $\sigma(\zeta) \geq \zeta_2$ to $v \in [\zeta, \infty)_{\mathbb{T}}$ and using the fact $w > 0$, we have

$$\begin{aligned} -w(\sigma(\zeta)) &\leq w(v) - w(\sigma(\zeta)) \\ &\leq -k \int_{\sigma(\zeta)}^{v} \phi(\varkappa) q(\varkappa) \Delta\varkappa - \gamma \int_{\sigma(\zeta)}^{v} a^{-\frac{1}{\gamma}}(\varkappa) w^{1+\frac{1}{\gamma}}(\sigma(\varkappa)) \Delta\varkappa. \end{aligned}$$

Taking $v \to \infty$ we obtain

$$-w(\sigma(\zeta)) \leq -k \int_{\sigma(\zeta)}^{\infty} \phi(\varkappa) q(\varkappa) \Delta\varkappa - \gamma \int_{\sigma(\zeta)}^{\infty} a^{-1/\gamma}(\varkappa) w^{1+1/\gamma}(\sigma(\varkappa)) \Delta\varkappa. \tag{42}$$

Multiplying both sides of (42) by $A^{\gamma}(\zeta)$, we obtain

$$\begin{aligned} -A^{\gamma}(\zeta) w(\sigma(\zeta)) &\leq -k A^{\gamma}(\zeta) \int_{\sigma(\zeta)}^{\infty} \phi(\varkappa) q(\varkappa) \Delta\varkappa \\ &\quad -\gamma A^{\gamma}(\zeta) \int_{\sigma(\zeta)}^{\infty} a^{-1/\gamma}(\varkappa) w^{1+\frac{1}{\gamma}}(\sigma(\varkappa)) \Delta\varkappa \\ &= -k A^{\gamma}(\zeta) \int_{\sigma(\zeta)}^{\infty} \phi(\varkappa) q(\varkappa) \Delta\varkappa \\ &\quad -\gamma A^{\gamma}(\zeta) \int_{\sigma(\zeta)}^{\infty} \frac{A^{\Delta}(\varkappa)}{A^{\gamma+1}(\varkappa)} \left[A^{\gamma}(\varkappa) w(\sigma(\varkappa))\right]^{1+\frac{1}{\gamma}} \Delta\varkappa. \end{aligned}$$

It follows from (26) that

$$\begin{aligned} -A^{\gamma}(\zeta) w(\sigma(\zeta)) &\leq -k A^{\gamma}(\zeta) \int_{\sigma(\zeta)}^{\infty} \phi(\varkappa) q(\varkappa) \Delta\varkappa \\ &\quad - (a_* - \varepsilon)^{1+\frac{1}{\gamma}} A^{\gamma}(\zeta) \int_{\sigma(\zeta)}^{\infty} \gamma \frac{A^{\Delta}(\varkappa)}{A^{\gamma+1}(\varkappa)} \Delta\varkappa. \end{aligned} \tag{43}$$

By Pötzsche chain rule, we reach

$$\left(\frac{-1}{A^{\gamma}}\right)^{\Delta} = \gamma \int_0^1 \frac{1}{[A + h\mu(\varkappa) A^{\Delta}]^{\gamma+1}} dh \, A^{\Delta} \leq \gamma \frac{A^{\Delta}}{A^{\gamma+1}}. \tag{44}$$

Then from (43) and (44), we have

$$\begin{aligned} -A^{\gamma}(\zeta)\,w(\sigma(\zeta)) &\leq -k\,A^{\gamma}(\zeta)\int_{\sigma(\zeta)}^{\infty}\phi(\varkappa)q(\varkappa)\Delta\varkappa-(a_*-\varepsilon)^{1+\frac{1}{\gamma}}\left[\frac{A(\zeta)}{A(\sigma(\zeta))}\right]^{\gamma} \\ &\leq -k\,A^{\gamma}(\zeta)\int_{\sigma(\zeta)}^{\infty}\phi(\varkappa)q(\varkappa)\Delta\varkappa-(l-\varepsilon)^{\gamma}(a_*-\varepsilon)^{1+\frac{1}{\gamma}}, \end{aligned}$$

which yields

$$k\,A^{\gamma}(\zeta)\int_{\sigma(\zeta)}^{\infty}\phi(\varkappa)q(\varkappa)\Delta\varkappa\leq A^{\gamma}(\zeta)\,w(\sigma(\zeta))-(l-\varepsilon)^{\gamma}(a_*-\varepsilon)^{1+\frac{1}{\gamma}}.$$

By taking the lim inf of both sides as $\zeta\to\infty$ we obtain that

$$\liminf_{\zeta\to\infty}k\,A^{\gamma}(\zeta)\int_{\sigma(\zeta)}^{\infty}\phi(\varkappa)q(\varkappa)\Delta\varkappa\leq a_*-(l-\varepsilon)^{\gamma}(a_*-\varepsilon)^{1+\frac{1}{\gamma}}.$$

Since k and $\varepsilon>0$ are arbitrary, we achieve the following inequality

$$\liminf_{\zeta\to\infty}A^{\gamma}(\zeta)\int_{\sigma(\zeta)}^{\infty}\phi(\varkappa)q(\varkappa)\Delta\varkappa\leq a_*-l^{\gamma}\,a_*^{1+\frac{1}{\gamma}}.$$

Using the inequality (29) with $z=l^{\gamma}$, $Y=1$ and $u=a_*$, we get the desired inequality

$$\liminf_{\zeta\to\infty}A^{\gamma}(\zeta)\int_{\sigma(\zeta)}^{\infty}\phi(\varkappa)q(\varkappa)\Delta\varkappa\leq\frac{\gamma^{\gamma}}{l^{\gamma^2}(\gamma+1)^{\gamma+1}},$$

which is in contrast to the condition (41). The proof is accomplished in Theorem 4. □

Example 2. *Consider the nonlinear second order dynamic equation*

$$\left[\varphi_{\gamma}\left(z^{\Delta}(\zeta)\right)\right]^{\Delta}+\frac{\kappa\gamma}{L\zeta^{\gamma+1}}\varphi_{\beta}\left(z(\eta(\zeta))\right)=0, \tag{45}$$

where γ,β,κ *are positive constants, and* $L=\liminf_{\zeta\to\infty}\left(\frac{\zeta}{\sigma(\zeta)}\right)^{\gamma}$ *with* $\beta\geq\gamma$. *Here* $a(\zeta)=1$, $\eta(\zeta)\geq\zeta$ *and* $q(\zeta)=\frac{\eta\gamma}{L\zeta^{\gamma+1}}$, *then the condition* (2) *holds,* $A(\zeta)=\zeta-\zeta_0$ *and* $\phi(\zeta)=1$. *In addition,*

$$\begin{aligned} \liminf_{\zeta\to\infty}A^{\gamma}(\zeta)\int_{\sigma(\zeta)}^{\infty}\phi(\varkappa)q(\varkappa)\Delta\varkappa &= \frac{\kappa}{L}\liminf_{\zeta\to\infty}A^{\gamma}(\zeta)\int_{\sigma(\zeta)}^{\infty}\frac{\gamma\Delta\varkappa}{\varkappa^{\gamma+1}} \\ &\geq \frac{\kappa}{L}\liminf_{\zeta\to\infty}A^{\gamma}(\zeta)\int_{\sigma(\zeta)}^{\infty}\left(\frac{-1}{\varkappa^{\gamma}}\right)^{\Delta}\Delta\varkappa \\ &= \frac{\kappa}{L}\liminf_{\zeta\to\infty}\left(\frac{\zeta}{\sigma(\zeta)}-\frac{\zeta_0}{\sigma(\zeta)}\right)^{\gamma}=\kappa \end{aligned}$$

if $\kappa>\frac{\gamma^{\gamma}}{l^{\gamma^2}(\gamma+1)^{\gamma+1}}$. *Then by Theorem 4, all solutions of* (45) *are oscillatory if* $\kappa>\frac{\gamma^{\gamma}}{l^{\gamma^2}(\gamma+1)^{\gamma+1}}$.

Remark 1. *We could refer to the recent results due to [13,14] and others do not apply to Equations* (39) *and* (45).

Theorem 5. *Let* (2) *hold, and*

$$\limsup_{\zeta\to\infty}A^{\gamma}(\zeta)\int_{\zeta}^{\infty}\phi(\varkappa)q(\varkappa)\Delta\varkappa>1. \tag{46}$$

Then all solutions of Equation (1) *oscillate.*

Proof. Assume $z(t)$ is a nonoscillatory solution of Equation (1) on $[\zeta_0, \infty)_{\mathbb{T}}$. Thus, without loss of generality, let $z(\zeta) > 0$ and $z(\eta(\zeta)) > 0$ on $[\zeta_0, \infty)_{\mathbb{T}}$. Integrating both sides of the dynamic Equation (1) from ζ to $v \in [\zeta_0, \infty)_{\mathbb{T}}$, we obtain

$$\int_{\zeta}^{v} q(\backsim) z^{\beta}(\eta(\backsim))\Delta\backsim = a(\zeta)(z^{\Delta}(\zeta))^{\gamma} - a(v)(z^{\Delta}(v))^{\gamma} \leq a(\zeta)(z^{\Delta}(\zeta))^{\gamma}. \tag{47}$$

As shown in the proof of Theorem 1, there exists $\zeta_1 \in (\zeta_0, \infty)_{\mathbb{T}}$ satisfying $\eta(\zeta) \in (\zeta_0, \infty)_{\mathbb{T}}$ for $\zeta \geq \zeta_1$ such that for $\zeta \geq \zeta_1$

$$z^{\beta}(\eta(\zeta)) \geq k\, \phi(\zeta) z^{\gamma}(\zeta) \tag{48}$$

and

$$z^{\gamma}(\zeta) \geq a(\zeta)\left(z^{\Delta}(\zeta)\right)^{\gamma} A^{\gamma}(\zeta). \tag{49}$$

From (47) and (48), we obtain

$$k \int_{\zeta}^{v} \phi(\backsim) q(\backsim) z^{\gamma}(\backsim)\Delta\backsim \leq a(\zeta)(z^{\Delta}(\zeta))^{\gamma}.$$

Since $z^{\Delta}(\zeta) > 0$, we get that

$$k\, z^{\gamma}(\zeta) \int_{\zeta}^{v} \phi(\backsim) q(\backsim)\Delta\backsim \leq a(\zeta)(z^{\Delta}(\zeta))^{\gamma}. \tag{50}$$

From (49) and (50), we get

$$k\, A^{\gamma}(\zeta) \int_{\zeta}^{v} \phi(\backsim) q(\backsim)\Delta\backsim \leq 1.$$

Taking $v \to \infty$, we have

$$k\, A^{\gamma}(\zeta) \int_{\zeta}^{\infty} \phi(\backsim) q(\backsim)\Delta\backsim \leq 1.$$

Since $k > 0$ is arbitrary, we have

$$A^{\gamma}(\zeta) \int_{\zeta}^{\infty} \phi(\backsim) q(\backsim)\Delta\backsim \leq 1,$$

which gives us the contradiction

$$\limsup_{\zeta \to \infty} A^{\gamma}(\zeta) \int_{\zeta}^{\infty} \phi(\backsim) q(\backsim)\Delta\backsim \leq 1.$$

The proof of Theorem 5 is accomplished. □

3. Oscillation Criteria of (1) when $\beta \leq \gamma$

Assume that

$$z(\zeta) > 0,\ z(\eta(\zeta)) > 0,\ z^{\Delta}(\zeta) > 0,\ \left[a(\zeta)\varphi_{\gamma}\left(z^{\Delta}(\zeta)\right)\right]^{\Delta} < 0$$

eventually. Integrating Equation (1) from ζ to $v \in [\zeta, \infty)_{\mathbb{T}}$ and then using (22) and the fact that $z^{\Delta} > 0$, we obtain

$$\begin{aligned}-a(v)\varphi_\gamma\left(z^\Delta(v)\right)+a(\zeta)\varphi_\gamma\left(z^\Delta(\zeta)\right) &= \int_\zeta^v q(\varkappa)\,\varphi_\beta\left(z\left(\eta\left(\varkappa\right)\right)\right)\Delta\varkappa \\ &\geq \int_\zeta^v \phi(\varkappa)q(\varkappa)\,\varphi_\beta\left(z\left(\varkappa\right)\right)\Delta\varkappa \\ &\geq \varphi_\beta\left(z\left(\zeta\right)\right)\int_\zeta^v \phi(\varkappa)q(\varkappa)\,\Delta\varkappa,\end{aligned}$$

and $a(v)\varphi_\gamma\left(z^\Delta(v)\right) > 0$ gives

$$a(\zeta)\varphi_\gamma\left(z^\Delta(\zeta)\right) \geq \varphi_\beta\left(z\left(\zeta\right)\right)\int_\zeta^v \phi(\varkappa)q(\varkappa)\,\Delta\varkappa.$$

Hence by taking limits as $v \to \infty$ we have

$$a(\zeta)\varphi_\gamma\left(z^\Delta(\zeta)\right) \geq \varphi_\beta\left(z\left(\zeta\right)\right)\int_\zeta^\infty \phi(\varkappa)q(\varkappa)\,\Delta\varkappa. \tag{51}$$

Since $\left[a(\zeta)\varphi_\gamma\left(z^\Delta(\zeta)\right)\right]^\Delta < 0$ eventually, then

$$a(\zeta)\varphi_\gamma\left(z^\Delta(\zeta)\right) \leq a(\zeta_2)\varphi_\gamma\left(z^\Delta(\zeta_2)\right) =: b \quad \text{for } \zeta \geq \zeta_2,$$

and hence from (51), we have

$$b \geq a(\zeta)\varphi_\gamma\left(z^\Delta(\zeta)\right) \geq \varphi_\beta\left(z\left(\zeta\right)\right)\int_\zeta^\infty \phi(\varkappa)q(\varkappa)\,\Delta\varkappa,$$

and so

$$z^{\beta-\gamma}(\zeta) = \left[\varphi_\beta\left(z\left(\zeta\right)\right)\right]^{\frac{\beta-\gamma}{\beta}} \geq c\left[\int_\zeta^\infty \phi(\varkappa)q(\varkappa)\,\Delta\varkappa\right]^{\frac{\gamma-\beta}{\beta}},$$

where $c := b^{\frac{\beta-\gamma}{\beta}} > 0$. Combining all these we see that for every arbitrary $c > 0$,

$$z^{\beta-\gamma}(\zeta) \geq c\left[\int_\zeta^\infty \phi(\varkappa)q(\varkappa)\,\Delta\varkappa\right]^{\frac{\gamma-\beta}{\beta}}, \tag{52}$$

eventually. Let

$$Q(\zeta) := q(\zeta)\left[\int_\zeta^\infty \phi(\varkappa)q(\varkappa)\,\Delta\varkappa\right]^{\frac{\gamma-\beta}{\beta}}.$$

Therefore, by (52) and the definition of $Q(\zeta)$, as direct consequence of Theorems 1, 2, 4 and 5, we get oscillation criteria for Equation (1) with $\beta \leq \gamma$.

Theorem 6. *Let* (2) *hold, and*

$$\begin{aligned}\liminf_{\zeta\to\infty}\frac{1}{A(\zeta)}\int_T^\zeta A^{\gamma+1}(\varkappa)\,\phi(\varkappa)Q(\varkappa)\Delta\varkappa &> \frac{1}{l^{\gamma(\gamma+1)}}\left(1-\frac{l^\gamma}{\gamma l^\gamma+1}\right), \quad 0<\gamma\leq 1, \\ \liminf_{\zeta\to\infty}\frac{1}{A(\zeta)}\int_T^\zeta A^{\gamma+1}(\varkappa)\,\phi(\varkappa)Q(\varkappa)\Delta\varkappa &> \frac{\gamma}{l^{\gamma(\gamma+1)}\left(\gamma+l^\gamma\right)}, \quad \gamma\geq 1,\end{aligned} \tag{53}$$

for enough large $T \in [\zeta_0,\infty)_{\mathbb{T}}$. Then all solutions of Equation (1) *oscillate.*

Theorem 7. *Let* (2) *holds, and*

$$\liminf_{\zeta\to\infty} \frac{1}{A(\zeta)} \int_T^{\zeta} A^{\gamma+1}(\varkappa)\phi(\varkappa)Q(\varkappa)\Delta\varkappa > \frac{1}{l^{\gamma(\gamma+1)}}\left(1-\frac{l^{\gamma}}{\gamma+1}\right),$$

for enough large $T \in [\zeta_0,\infty)_{\mathbb{T}}$*. Then all solutions of Equation* (1) *oscillate.*

Theorem 8. *Let* (2) *holds, and*

$$\liminf_{\zeta\to\infty} A^{\gamma}(\zeta) \int_{\sigma(\zeta)}^{\infty} \phi(\varkappa)Q(\varkappa)\Delta\varkappa > \frac{\gamma^{\gamma}}{l^{\gamma^2}(\gamma+1)^{\gamma+1}}.$$

Then all solutions of Equation (1) *oscillate.*

Theorem 9. *Let* (2) *holds, and*

$$\limsup_{\zeta\to\infty} A^{\gamma}(\zeta) \int_{\zeta}^{\infty} \phi(\varkappa)Q(\varkappa)\Delta\varkappa > 1.$$

Then all solutions of Equation (1) *oscillate.*

4. Conclusions

(1) In this paper, several Nehari, Hille and Ohriska type oscillation criterion have been given. The applicability of these criteria for (1) on an arbitrary time scale is achieved. The reported results have extended related findings to the differential and dynamics equations of second order as follows:

(i) Condition (41) reduces to (7) in the case if $\mathbb{T}=\mathbb{R}$, $\gamma=\beta=1$, $a(\zeta)=1$, and $\eta(\zeta)=\zeta$;
(ii) Condition (41) reduces to (10) in the case when $\mathbb{T}=\mathbb{R}$, $\gamma=\beta=1$, $a(\zeta)=1$, and $g(\zeta)\le\zeta$;
(iii) Condition (41) reduces to (19) under the assumptions that $\gamma=\beta$, $a^{\Delta}(\zeta)\ge 0$, and $g(\zeta)\le\zeta$;
(iv) Conditions (46) reduces to (11) supposing that $\mathbb{T}=\mathbb{R}$, $\gamma=\beta=1$, $a(\zeta)=1$, and $g(\zeta)\le\zeta$.

(2) Several oscillation criteria for (1) have been derived in the cases: $\eta(\zeta)\le\zeta$, $\eta(\zeta)\ge\zeta$, $\beta\ge\gamma$, and $\beta\le\gamma$. In contrast to [13,14], the restrictive condition (18) is not imposed in the oscillation results of the presented case-study. This leads to a great improvement in comparison with the proceeding results.

Author Contributions: Conceptualization, T.S.H.; Data curation, A.A.M.; Formal analysis, T.S.H. and Y.S.; Project administration, Y.S.; Writing—original draft, T.S.H.; Resources, A.A.M.; Supervision, T.S.H. and Y.S.; Investigation, A.A.M.; Validation, T.S.H., Y.S. and A.A.M.; Writing—review & editing, T.S.H., Y.S. and A.A.M. All authors have read and agreed to the published version of the manuscript.

References

1. Hilger, S. Analysis on measure chains — A unified approach to continuous and discrete calculus. *Results Math.* **1990**, *18*, 18–56. [CrossRef]
2. Bohner, M.; Peterson, A. *Dynamic Equations on Time Scales: An Introduction with Applications*; Birkhäuser: Boston, MA, USA, 2001.
3. Kac, V.; Chueng, P. *Quantum Calculus*; Universitext: Ann Arbor, MI, USA, 2002.
4. Agarwal, R.P.; Bohner, M.; O'Regan, D.; Peterson, A. Dynamic equations on time scales: A survey. *J. Comput. Appl. Math.* **2002**, *141*, 1–26. [CrossRef]
5. Bohner, M.; Peterson, A. *Advances in Dynamic Equations on Time Scales*; Birkhäuser: Boston, MA, USA, 2003.
6. Nehari, Z. Oscillation criteria for second-order linear differential equations. *Trans. Am. Math. Soc.* **1957**, *85*, 428–445. [CrossRef]

7. Fite, W.B. Concerning the zeros of the solutions of certain differential equations. *Trans. Amer. Math. Soc.* **1918**, *19*, 341–352. [CrossRef]
8. Hille, E. Non-oscillation theorems. *Trans. Am. Math. Soc.* **1948**, *64*, 234–252. [CrossRef]
9. Wong, J.S. Second order oscillation with retarded arguments. In *Ordinary Differential Equations*; Academic Press: New York, NY, USA; London, UK, 1972; pp. 581–596.
10. Erbe, L. Oscillation criteria for second order nonlinear delay equations. *Can. Math. Bull.* **1973**, *16*, 49–56. [CrossRef]
11. Ohriska, J. Oscillation of second order delay and ordinary differential equations. *Czech. Math. J.* **1984**, *34*, 107–112. [CrossRef]
12. Thandapani, E.; Ravi, K.; Graef, J. Oscillation and comparison theorems for half-linear second order difference equations. *Comput. Math. Appl.* **2001**, *42*, 953–960. [CrossRef]
13. Erbe, L.; Hassan, T.S.; Peterson, A.; Saker, S.H. Oscillation criteria for half-linear delay dynamic equations on time scales. *Nonlinear Dyn. Syst. Theory* **2009**, *9*, 51–68.
14. Erbe, L.; Hassan, T.S.; Peterson, A.; Saker, S.H. Oscillation criteria for sublinear half-linear delay dynamic equations on time scales. *Int. J. Differ. Eq.* **2008**, *3*, 227–245.
15. Agarwal, R.P.; Bohner, M.; Li, T.; Zhang, C. Hille and Nehari type criteria for third order delay dynamic equations. *J. Differ. Eq. Appl.* **2013**, *19*, 1563–1579. [CrossRef]
16. Baculikova, B. Oscillation of second-order nonlinear noncanonical differential equations with deviating argument. *Appl. Math. Lett.* **2019**, *91*, 68–75. [CrossRef]
17. Bazighifan, O.; El-Nabulsi, E.M. Different techniques for studying oscillatory behavior of solution of differential equations. *Rocky Mountain Journal of Mathematics Volume forthcoming, Number Forthcoming (2020)*. Available online: https://projecteuclid.org/euclid.rmjm/1596037174 (accessed on 19 October 2020).
18. Bohner, M.; Hassan, T.S. Oscillation and boundedness of solutions to first and second order forced functional dynamic equations with mixed nonlinearities. *Appl. Anal. Discret. Math.* **2009**, *3*, 242–252. [CrossRef]
19. Bohner, M.; Hassan, T.S.; Li, T. Fite-Hille-Wintner-type oscillation criteria for second-order half-linear dynamic equations with deviating arguments. *Indag. Math.* **2018**, *29*, 548–560. [CrossRef]
20. Džurina, J.; Jadlovská, I. A sharp oscillation result for second-order half-linear noncanonical delay differential equations. *Electron. J. Qual. Theory* **2020**, *46*, 1–14.
21. Elabbasy, E.M.; El-Nabulsi, R.A.; Moaaz, O.; Bazighifan, O. Oscillatory properties of solutions of even order differential equations. *Symmetry* **2020**, *12*, 212. [CrossRef]
22. Erbe, L.; Peterson, A. Oscillation criteria for second-order matrix dynamic equations on a time scale. *J. Comput. Appl. Math.* **2002**, *141*, 169–185. [CrossRef]
23. Erbe, L.; Peterson, A. Boundedness and oscillation for nonlinear dynamic equations on a time scale. *Proc. Am. Math. Soc.* **2003**, *132*, 735–744. [CrossRef]
24. Erbe, L.; Hassan, T.S. New oscillation criteria for second order sublinear dynamic equations. *Dyn. Syst. Appl.* **2013**, *22*, 49–63.
25. Erbe, L.; Peterson, A.; Saker, S.H. Hille and Nehari type criteria for third order dynamic equations. *J. Math. Anal. Appl.* **2007**, *329*, 112–131. [CrossRef]
26. Grace, S.R.; Bohner, M.; Agarwal, R.P. On the oscillation of second-order half-linear dynamic equations. *J. Differ. Eq. Appl.* **2009**, *15*, 451–460. [CrossRef]
27. Hassan, T.S.; Agarwal, R.P.; Mohammed,W. Oscillation criteria for third-order functional half-linear dynamic equations. *Adv. Differ. Eq.* **2017**, *2017*, 111. [CrossRef]
28. Erbe, L.; Hassan, T.S.; Peterson, A. Oscillation criteria for second order sublinear dynamic equations with damping term. *J. Differ. Eq. Appl.* **2011**, *17*, 505–523.
29. Leighton, W. The detection of the oscillation of solutions of asecond order linear differential equation. *Duke J. Math.* **1950**, *17*, 57–62. [CrossRef]
30. Karpuz, B. Hille–Nehari theorems for dynamic equations with a time scale independent critical constant. *Appl. Math. Comput.* **2019**, *346*, 336–351. [CrossRef]
31. Řehak, P. New results on critical oscillation constants depending on a graininess. *Dyn. Syst. Appl.* **2010**, *19*, 271–288.
32. Sun, S.; Han, Z.; Zhao, P.; Zhang, C. Oscillation for a class of second-order Emden-Fowler delay dynamic equations on time scales. *Adv. Differ. Eq.* **2010** *2010*, 642356. [CrossRef]

33. Wintner, A. On the nonexistence of conjugate points. *Am. J. Math.* **1951**, *73*, 368–380. [CrossRef]
34. Sun, Y.; Hassan, T.S. Oscillation criteria for functional dynamic equations with nonlinearities given by Riemann-Stieltjes integral. *Abstr. Appl. Anal.* **2014**, *2014*, 697526. [CrossRef]
35. Zhang, Q.; Gao, L.; Wang, L. Oscillation of second-order nonlinear delay dynamic equations on time scales. *Comput. Math. Appl.* **2011**, *61*, 2342–2348. [CrossRef]

6

Establishing New Criteria for Oscillation of Odd-Order Nonlinear Differential Equations

Osama Moaaz [1,*], Jan Awrejcewicz [2,*] and Ali Muhib [1,3]

1 Department of Mathematics, Faculty of Science, Mansoura University, Mansoura 35516, Egypt; muhib39@yahoo.com
2 Department of Automation, Biomechanics and Mechatronics, Lodz University of Technology, 1/15 Stefanowski St., 90-924 Lodz, Poland
3 Department of Mathematics, Faculty of Education (Al-Nadirah), Ibb University, Ibb P.O. Box 70270, Yemen
* Correspondence: o_moaaz@mans.edu.eg (O.M.); jan.awrejcewicz@p.lodz.pl (J.A.)

Abstract: By establishing new conditions for the non-existence of so-called Kneser solutions, we can generate sufficient conditions to ensure that all solutions of odd-order equations are oscillatory. Our results improve and expand the previous results in the literature.

Keywords: odd-order differential equations; Kneser solutions; oscillation criteria

---

1. Introduction

In the 20th century, the extremely fast development of science led to applications in the fields of biology, population, chemistry, medicine dynamics, social sciences, genetic engineering, economics, and others. Many of these phenomena are modeled by delay differential equations. All these disciplines were promoted to a higher level and discoveries were made with the help of this kind of mathematical modeling.

The neutral differential equations are the differential equations in which the delayed argument occurs in the highest derivative of the state variable. The neutral equations appear in the modeling of the networks containing lossless transmission lines (as in high-speed computers where the lossless transmission lines are used to interconnect switching circuits); see [1].

Recently, an increasing interest in establishing conditions for the oscillatory behavior of different order of differential equations has been observed; see [2–9].

It is known that determination of the signs of the derivatives of the solution is necessary and causes a significant effect before studying the oscillation of delay differential equations. The other essential thing is to establish relationships between derivatives of different orders, which may lead to additional restrictions during the study. In odd-order differential equations, in some cases, it is difficult to find relationships between derivatives of different orders, which in turn is central to the study of oscillatory behavior. Therefore, it can very easily be observed that differential equations of odd-order received less attention than differential equations with even-order. Additionally, most studies are concerned with finding sufficient conditions that guarantee that every non-oscillating solution tends to zero; see [4,10–20].

In this paper, in Section 2, we offer some auxiliary lemmas that define the different cases of signs of derivatives and the relationships between derivatives of different orders. In Section 3, we establish a set of new criteria that ensure that there are no non-oscillating solutions in each case of derivatives separately. In Section 4, we establish new criteria for the oscillation of all solutions of the studied equation. Finally, in conclusion, we discuss the results and compare them to the related works.

In detail, we investigate the oscillatory properties of solutions to the odd-order neutral equation

$$\left(r(t)\left(\left(x(t)+p(t)x(\tau(t))\right)^{(n-1)}\right)^{\alpha}\right)' + q(t) f(x(g(t))) = 0, \tag{1}$$

where n is an odd natural number. Moreover, we suppose that

Hypothesis 1 (H1). *α is the ratio of odd positive integers, $r \in C^1(I_0, \mathbb{R}^+)$, $p \in C(I_0, [0, p_0])$, where p_0 is a positive constant, $\tau, g \in C^1(I_0, \mathbb{R})$, $q \in C(I_0, [0, \infty))$, $r'(t) \geq 0$, $g(t) \leq t$, $\lim_{t\to\infty} g(t) = \infty$, $\lim_{t\to\infty} \tau(t) = \infty$, $\int_{t_0}^{\infty} r^{-1/\alpha}(\rho)\, d\rho = \infty$, q is not eventually zero on any half line I_* for $t_* \geq t_0$, and $I_s := [t_s, \infty)$.*

Hypothesis 2 (H2). *$f \in C(\mathbb{R}, \mathbb{R})$ and there exists a positive constant k such that $f(x) \geq kx^{\alpha}$.*

Next, we present the basic definitions.

Definition 1. *The function $z(t) := x(t) + p(t)x(\tau(t))$ is called the corresponding function of x, and*

$$\phi(s,t) = \int_s^t r^{-1/\alpha}(\varrho)\, d\varrho$$

is called the canonical operator.

Definition 2. *Let x be a real-valued function defined for all t in a real interval I_x, $t_x \geq t_0$, and having a nth derivative for all $t \in I_x$. The function x is called a **solution** of the differential equation (Equation (1)) on I if x is continuous; $r\left(z^{(n-1)}\right)^{\alpha}$ is continuously differentiable and x satisfies (1), for all t in I_x.*

Definition 3. *A nontrivial solution x of (1) is said to be **oscillatory** if it has arbitrary large zeros; that is, there exists a sequence of zeros $\{t_n\}_{n=0}^{\infty}$ (i.e., $x(t_n) = 0$) of x such that $\lim_{n\to\infty} t_n = \infty$. Otherwise, it is said to be **non-oscillatory**.*

Notation 1. *The set of all eventually positive solutions of (1) is denoted by X^+.*

We restrict our discussion to those solutions x of (1) which satisfy $\sup\{|x(t)| : t_1 \leq t_0\} > 0$ for every $t_1 \in I_x$. All functional inequalities and properties, such as increasing, decreasing, positive, and so on, are assumed to hold eventually; that is, they are satisfied for all t large enough.

2. Preliminary Results

During this part of the paper, we provide auxiliary lemmas. These lemmas will be the cornerstone of the main results.

Notation 2. *For the sake of convenience, we use the following notation:*

$$\eta(t) := \frac{\lambda}{(n-2)!} \frac{g^{n-2}(t)\, g'(t)}{r^{1/\alpha}(t)},$$

$$\Theta(t) := kq(t)\left(1 - p(g(t))\right)^{\alpha}, \quad \widetilde{\Theta}(t) := \int_t^{\infty} \Theta(\varrho)\, d\varrho,$$

$$Q_1(t) := \min\{q(t), q(\tau(t))\}, \quad Q_2(t) := \min\left\{q\left(g^{-1}(t)\right), q\left(g^{-1}(\tau(t))\right)\right\},$$

$$\psi_1(s,t) := \int_s^t \phi(\varrho,t)\,\mathrm{d}\varrho, \quad \psi_{k+1}(s,t) := \int_s^t \psi_k(\varrho,t)\,\mathrm{d}\varrho, \; k = 1,2,...,n-2,$$

and

$$\mu := \begin{cases} 1 & \text{for } 0 < \alpha \leq 1; \\ 2^{\alpha-1} & \text{for } \alpha > 1. \end{cases}$$

Lemma 1. *([21], Lemma 1, Lemma 2) Assume that $u,\ v \in [0,\infty)$. Then,*

$$(u+v)^\alpha \leq \mu\left(u^\alpha + v^\alpha\right).$$

Lemma 2. *[22] Let $F \in C^n([t_0,\infty),(0,\infty))$. Assume that $F^{(n)}(t)$ is of fixed sign and not identically zero on I_0 and that there exists a $t_1 \geq t_0$ such that $F^{(n-1)}(t)\,F^{(n)}(t) \leq 0$ for all $t \geq t_1$. If $\lim_{t\to\infty} F(t) \neq 0$; then for every $\lambda \in (0,1)$ there exists $t_\mu \geq t_1$ such that*

$$F(t) \geq \frac{\lambda}{(n-1)!}t^{n-1}\left|F^{(n-1)}(t)\right| \text{ for } t \geq t_\mu.$$

The following lemma is a well-known result; see ([20], Lemma 2.4, Lemma 2.5); also see ([22], Lemma 2.2.1).

Lemma 3. *Suppose that $x \in X^+$. Then, there exists a sufficiently large $t_1 \geq t_0$ such that, for all $t \geq t_1$,*

$$z(t) > 0,\ z''(t) > 0,\ z^{(n-1)}(t) > 0 \text{ and } z^{(n)}(t) \leq 0.$$

Furthermore, there are only two cases:

$$\mathbf{P}: z'(t) > 0,$$

or

$$\mathbf{N}:\ (-1)^k z^{(k)}(t) > 0, \text{ for } k = 1,2,...,n-2.$$

Lemma 4. *Suppose that $x \in X^+$ and z satisfies **N**. Then*

$$z(\rho) \geq r^{1/\alpha}(\sigma)\,z^{(n-1)}(\sigma)\,\psi_{n-2}(\rho,\sigma) \tag{2}$$

for $\rho \leq \sigma$.

Proof. It follows from the monotonicity of $r(z^{(n-1)})(t)$ that

$$\begin{aligned} -z^{(n-2)}(\rho) &\geq z^{(n-2)}(\sigma) - z^{(n-2)}(\rho) = \int_\rho^\sigma \frac{1}{r^{1/\alpha}(s)} r^{1/\alpha}(s)\,z^{(n-1)}(s)\,\mathrm{d}s \\ &\geq r^{1/\alpha}(\sigma)\,z^{(n-1)}(\sigma)\,\phi(\rho,\sigma). \end{aligned} \tag{3}$$

Integrating (3) from ρ to σ, we have

$$-z^{(n-2)}(\sigma) + z^{(n-2)}(\rho) \geq r^{1/\alpha}(\sigma)\,z^{(n-1)}(\sigma)\int_\rho^\sigma \phi(s,\sigma)\,\mathrm{d}s$$

and so

$$z^{(n-3)}(\rho) \geq r^{1/\alpha}(\sigma)\,z^{(n-1)}(\sigma)\,\psi_1(\rho,\sigma). \tag{4}$$

Integrating (4) $n-3$ times from ρ to σ, we get

$$z(\rho) \geq r^{1/\alpha}(\sigma) z^{(n-1)}(\sigma) \psi_{n-2}(\rho,\sigma).$$

The proof is complete. □

Lemma 5. *Suppose that $x \in X^+$ and z satisfies* **P**. *If $p_0 < 1$, g is non-decreasing and*

$$w(t) := \delta(t) r(t) \left(\frac{z^{(n-1)}(t)}{z(g(t))}\right)^{\alpha}, \tag{5}$$

then

$$w'(t) \leq \frac{\delta'(t)}{\delta(t)} w(t) - \delta(t)\Theta(t) - \alpha\delta(t)\eta(t) w^{1+1/\alpha}(t), \tag{6}$$

where $\delta \in C^1(I_0,(0,\infty))$.

Proof. Assume that $x \in X^+$and z satisfies **P**. Then, there exists a $t_1 \geq t_0$ such that $x(t) > 0$, $x(\tau(t)) > 0$ and $x(g(t)) > 0$ for $t \in I_1$. Since $z(t) > x(t)$ and $z'(t) > 0$, it follows from the Definition 1 that $x(t) > (1-p(t)) z(t)$. Thus, (1) becomes

$$\begin{aligned} (r(t)(z^{(n-1)}(t))^{\alpha})' &= -q(t) f(x(g(t))) \leq -kq(t) x^{\alpha}(g(t)) \\ &\leq -kq(t)(1-p(g(t)))^{\alpha} z^{\alpha}(g(t)). \end{aligned} \tag{7}$$

Using Lemma 2 with $F = z'$, we obtain for every $\lambda \in (0,1)$,

$$(n-2)! z'(t) \geq \lambda t^{n-2} z^{(n-1)}(t)$$

which with the fact that $z^{(n)} \leq 0$ gives

$$z'(g(t)) \geq \frac{\lambda}{(n-2)!} g^{n-2}(t) z^{(n-1)}(g(t)) \geq \frac{\lambda}{(n-2)!} g^{n-2}(t) z^{(n-1)}(t). \tag{8}$$

Hence, from (5), (7) and (8), we get

$$\begin{aligned} w(t) &= \frac{\delta'(t)}{\delta(t)} w(t) + \delta(t) \frac{\left(r\left(z^{(n-1)}\right)^{\alpha}\right)'(t)}{z^{\alpha}(g(t))} - \delta(t) \frac{\left(r\left(z^{(n-1)}\right)^{\alpha}\right)(t)}{z^{\alpha+1}(g(t))} \alpha z'(g(t)) g'(t) \\ &\leq \frac{\delta'(t)}{\delta(t)} w(t) - \delta(t)\Theta(t) - \frac{\alpha\lambda}{(n-2)!}\delta(t) r(t) g^{n-2}(t) g'(t) \left(\frac{z^{(n-1)}(t)}{z(g(t))}\right)^{\alpha+1} \\ &\leq \frac{\delta'(t)}{\delta(t)} w(t) - \delta(t)\Theta(t) - \alpha\delta(t)\eta(t) w^{1+1/\alpha}(t). \end{aligned}$$

The proof is complete. □

Lemma 6. *Suppose that $x \in X^+$. If*

$$\tau'(t) \geq \tau_0 > 0, \tag{9}$$

then

$$\left((r(z^{(n-1)})^{\alpha})(t) + \frac{p_0^{\alpha}}{\tau_0}\left(r\left(z^{(n-1)}\right)^{\alpha}\right)(\tau(t))\right)' + kQ(t) z^{\alpha}(g(t)) \leq 0. \tag{10}$$

Moreover, if (9) holds and

$$\left(g^{-1}(t)\right)' \geq g_0 > 0, \tag{11}$$

then

$$\left(\frac{1}{g_0}\left(r\left(z^{(n-1)}\right)^{\alpha}\right)\left(g^{-1}(t)\right)+\frac{p_0^{\alpha}}{g_0\tau_0}\left(r\left(z^{(n-1)}\right)^{\alpha}\right)\left(g^{-1}(\tau(t))\right)\right)'+\frac{k}{\mu}Q_2(t)\,z^{\alpha}(t)\leq 0. \tag{12}$$

Proof. Let $x \in X^{+}$. Then, there exists a $t_1 \geq t_0$ such that $x(t) > 0$, $x(\tau(t)) > 0$ and $x(g(t)) > 0$ for $t \in I_1$. From (1), we get

$$\frac{1}{\tau'(t)}\left(r\left(z^{(n-1)}\right)^{\alpha}\right)'(\tau(t))+kq(\tau(t))\,x^{\alpha}(g(\tau(t)))\leq 0, \tag{13}$$

Combining (1) and (13) and taking into account that $\tau'(t) \geq \tau_0$, we obtain

$$(r(z^{(n-1)})^{\alpha})'(t)+\frac{p_0^{\alpha}}{\tau_0}\left(r\left(z^{(n-1)}\right)^{\alpha}\right)'(\tau(t))+kq(t)\,x^{\alpha}(g(t))+kp_0^{\alpha}q(\tau(t))\,x^{\alpha}(g(\tau(t)))\leq 0. \tag{14}$$

This implies that

$$\left((r(z^{(n-1)})^{\alpha})(t)+\frac{p_0^{\alpha}}{\tau_0}\left(r\left(z^{(n-1)}\right)^{\alpha}\right)(\tau(t))\right)'+kQ_1(t)\left(x^{\alpha}(g(t))+p_0^{\alpha}x^{\alpha}(g(\tau(t)))\right)\leq 0.$$

Using Lemma 1, we obtain

$$\left((r(z^{(n-1)})^{\alpha})(t)+\frac{p_0^{\alpha}}{\tau_0}\left(r\left(z^{(n-1)}\right)^{\alpha}\right)(\tau(t))\right)'+\frac{k}{\mu}Q_1(t)\left((x(g(t))+p_0x(g(\tau(t))))\right)^{\alpha}\leq 0.$$

From the definition of z, it is easy to conclude that

$$\left((r(z^{(n-1)})^{\alpha})(t)+\frac{p_0^{\alpha}}{\tau_0}\left(r\left(z^{(n-1)}\right)^{\alpha}\right)(\tau(t))\right)'+\frac{k}{\mu}Q_1(t)\,z^{\alpha}(g(t))\leq 0.$$

Next, from (1), we get

$$\frac{1}{(g^{-1}(t))'}\left(r\left(z^{(n-1)}\right)^{\alpha}\right)'\left(g^{-1}(t)\right)+kq\left(g^{-1}(t)\right)x^{\alpha}(t)\leq 0 \tag{15}$$

and

$$\frac{1}{(g^{-1}(\tau(t)))'}\left(r\left(z^{(n-1)}\right)^{\alpha}\right)'\left(g^{-1}(\tau(t))\right)+kq\left(g^{-1}(\tau(t))\right)x^{\alpha}(\tau(t))\leq 0. \tag{16}$$

Using (15) and (16) and taking into account (9) and (11), we obtain

$$\begin{aligned}\frac{1}{g_0}\left(r\left(z^{(n-1)}\right)^{\alpha}\right)'\left(g^{-1}(t)\right)+\frac{p_0^{\alpha}}{g_0\tau_0}\left(r\left(z^{(n-1)}\right)^{\alpha}\right)'\left(g^{-1}(\tau(t))\right)+kq\left(g^{-1}(t)\right)x^{\alpha}(t)&\\ +kq\left(g^{-1}(\tau(t))\right)x^{\alpha}(\tau(t))\leq 0.&\end{aligned} \tag{17}$$

By replacing (14) with (17), this part of proof is similar to that of the previous case and so we omit it. □

3. Nonexistence Criteria of Non-Oscillatory Solutions

At the beginning of this section, we define the following classes:

Notation 3. *The set of all positive solutions of (1) whose corresponding function z satisfies* **P** *or* **N** *is denoted by* X_P^+ *or* X_N^+*, respectively.*

Now, we create various criteria that ensure that there are no positive solutions of (1) whose corresponding function satisfies **P**.

Theorem 1. *If*

$$\frac{1}{\widetilde{\Theta}(t)}\int_t^\infty \eta(\varrho)\,\widetilde{\Theta}^{1+1/\alpha}(\varrho)\,d\varrho > \frac{1}{(1+\alpha)^{1+1/\alpha}}, \tag{18}$$

then X_P^+ *is an empty class.*

Proof. Assume the contrary that $x \in X_P^+$. Then, there exists a $t_1 \geq t_0$ such that $x(t) > 0$, $x(\tau(t)) > 0$ and $x(g(t)) > 0$ for $t \in I_1$. Using Lemma 5 with $\delta(t) := 1$, we arrive at

$$w'(t) \leq -\Theta(t) - \alpha\eta(t)\,w^{1+1/\alpha}(t) < 0.$$

By integrating the last inequality from t to ∞, we find

$$w(t) \geq \widetilde{\Theta}(t) + \alpha\int_t^\infty \eta(\varrho)\,w^{1+1/\alpha}(\varrho)\,d\varrho. \tag{19}$$

This implies that

$$\frac{w(t)}{\widetilde{\Theta}(t)} \geq 1 + \frac{\alpha}{\widetilde{\Theta}(t)}\int_t^\infty \eta(\varrho)\,\widetilde{\Theta}^{1+1/\alpha}(\varrho)\left(\frac{w(\varrho)}{\widetilde{\Theta}(\varrho)}\right)^{1+1/\alpha} d\varrho. \tag{20}$$

From (19), we note that $w(t) \geq \widetilde{\Theta}(t)$. Thus, we have

$$\beta := \inf\frac{w(t)}{\widetilde{\Theta}(t)} \geq 1. \tag{21}$$

Taking into account (18) and (21), (20) becomes

$$\beta \geq 1 + \alpha\left(\frac{\beta}{1+\alpha}\right)^{1+1/\alpha}$$

or

$$\frac{\beta}{\alpha+1} \geq \frac{1}{\alpha+1} + \frac{\alpha}{\alpha+1}\left(\frac{\beta}{\alpha+1}\right)^{1+1/\alpha},$$

which contradicts the expected value of $\beta > 1$ and $\alpha > 0$; therefore, the proof is complete. □

Now, let $\{S_m(t)\}_{m=0}^\infty$ be a sequence of continuous functions defined as follows: $S_0(t) = \widetilde{\Theta}(t)$ and

$$S_{m+1}(t) = S_0(t) + \alpha\int_t^\infty \eta(\varrho)\,S_m^{1+1/\alpha}(\varrho)\,d\varrho,\ m = 0, 1, \ldots. \tag{22}$$

By using the definition of $\{S_m(t)\}_{m=0}^\infty$, we can infer more new criteria as follows:

Theorem 2. *If*

$$\int_{t_0}^{\infty} \varphi(\varrho)\,\Theta(\varrho)\,d\varrho = \infty, \tag{23}$$

then X_P^+ *is an empty class, where*

$$\varphi(t) := \exp\left(\int_{t_1}^{t} \alpha\eta(\varrho)\,S_m^{1/\alpha}(\varrho)\,d\varrho\right).$$

Proof. Assume the contrary that $x \in X_P^+$. Then, there exists a $t_1 \geq t_0$ such that $x(t) > 0$, $x(\tau(t)) > 0$ and $x(g(t)) > 0$ for $t \in I_1$. From Theorem 1, we have that (19) holds. By induction, using (19), it is easy to see that the sequence $\{S_m(t)\}_{m=0}^{\infty}$ is non-decreasing and $w(t) \geq S_m(t)$. Thus the sequence $\{S_m(t)\}_{m=0}^{\infty}$ converges to $S(t)$. By the Lebesgue monotone convergence theorem and letting $m \to \infty$ in (22), we get

$$S(t) = S_0(t) + \alpha \int_{t}^{\infty} \eta(\varrho)\,S^{1+1/\alpha}(\varrho)\,d\varrho$$

which with $S(t) \geq S_m(t)$, gives

$$\begin{aligned} S'(t) &= -\Theta(t) - \alpha\eta(\varrho)\,S^{1+1/\alpha}(\varrho) \\ &\leq -\Theta(t) - \alpha\eta(\varrho)\,S(\varrho)\,S_m^{1/\alpha}(\varrho), \end{aligned}$$

and so

$$S'(t) + \left(\alpha\eta(\varrho)\,S_m^{1/\alpha}(\varrho)\right)S(\varrho) \leq -\Theta(t).$$

Thus, we get that

$$\varphi(t)\,S'(t) + \varphi(t)\left(\alpha\eta(\varrho)\,S_m^{1/\alpha}(\varrho)\right)S(\varrho) \leq -\varphi(t)\,\Theta(t)$$

or

$$(\varphi(t)\,S(t))' \leq -\varphi(t)\,\Theta(t). \tag{24}$$

Integrating (24) from $t_1$to t, we obtain

$$\varphi(t)\,S(t) \leq \varphi(t_1)\,S(t_1) - \int_{t_1}^{t} \varphi(\varrho)\,\Theta(\varrho)\,d\varrho.$$

However, letting $t \to \infty$ and using (23), the above inequality yields $\varphi(t)\,S(t) \to -\infty$, which contradicts the fact that $\varphi(t)\,S(t)$ is nonnegative. The proof is complete. □

Theorem 3. *If there exist some* $\lambda \in (0,1)$ *and* $S_m(t)$ *such that*

$$\limsup_{t\to\infty} \frac{1}{r(t)} g^{\alpha(n-1)}(t)\,S_m(t) > \left(\frac{(n-1)!}{\lambda}\right)^{\alpha}, \tag{25}$$

then X_P^+ *is an empty class.*

Proof. Assume the contrary that $x \in X_P^+$. Using Lemma 2 and taking into account the fact that $z^{(n-1)}(t)$ is non-increasing, we find

$$\begin{aligned} z(g(t)) &\geq \frac{\lambda}{(n-1)!} g^{n-1}(t) z^{(n-1)}(g(t)) \\ &\geq \frac{\lambda}{(n-1)!} g^{n-1}(t) z^{(n-1)}(t), \end{aligned}$$

for $\lambda \in (0,1)$. Then, from definition of $w(t)$ with $\delta(t) = 1$, we have

$$\frac{1}{w(t)} = \frac{1}{r(t)} \left(\frac{z(g(t))}{z^{(n-1)}(t)} \right)^{\alpha} \geq \frac{1}{r(t)} \left(\frac{\lambda g^{n-1}(t)}{(n-1)!} \right)^{\alpha},$$

and so

$$\left(\frac{(n-1)!}{\lambda} \right)^{\alpha} \geq \frac{1}{r(t)} g^{\alpha(n-1)}(t) w(t) \geq \frac{1}{r(t)} g^{\alpha(n-1)}(t) S_m(t),$$

which contradicts (25). The proof is complete. □

Corollary 1. *If there exist some $\lambda \in (0,1)$ such that*

$$\limsup_{t\to\infty} \frac{1}{r(t)} \left(g^{(n-1)}(t) \right)^{\alpha} \int_t^{\infty} \Theta(\varrho)\, d\varrho > \left(\frac{(n-1)!}{\lambda} \right)^{\alpha}, \tag{26}$$

then X_P^+ is an empty class.

Proof. Letting $m = 0$ in Theorem 3, we get (26). □

Next, by using comparison principles, we will create various criteria that ensure that there are no positive solutions of (1) whose corresponding function satisfies **N**.

Theorem 4. *If the first-order advanced inequality*

$$G'(t) + \frac{k\tau_0}{\tau_0 + p_0^{\alpha}} Q_1(t) \psi_{n-2}^{\alpha}(g(t), t) G\left(\tau^{-1}(t)\right) \leq 0, \tag{27}$$

then X_N^+ is an empty class.

Proof. Assume the contrary that $x \in X^+$ and z satisfy **N**. Then, there exists a $t_1 \geq t_0$ such that $x(t) > 0$, $x(\tau(t)) > 0$ and $x(g(t)) > 0$ for $t \in I_1$. From Lemmas 4 and 6, we arrive at (2) and (10), respectively. Now from (2), we get

$$z(g(t)) \geq r^{1/\alpha}(t) z^{(n-1)}(t) \psi_{n-2}(g(t), t) \tag{28}$$

which, by virtue of (10) yields that

$$0 \geq \left((r(z^{(n-1)})^{\alpha})(t) + \frac{p_0^{\alpha}}{\tau_0} \left(r \left(z^{(n-1)} \right)^{\alpha} \right) (\tau(t)) \right)' + kQ_1(t) r(t) \left(z^{(n-1)}(t) \psi_{n-2}(g(t), t) \right)^{\alpha}. \tag{29}$$

Now, set

$$G(t) := (r(z^{(n-1)})^{\alpha})(t) + \frac{p_0^{\alpha}}{\tau_0} \left(r \left(z^{(n-1)} \right)^{\alpha} \right) (\tau(t)) > 0. \tag{30}$$

Using the fact that $r(t)(z^{(n-1)}(t))$ is non-increasing, we obtain

$$G(t) \leq r(\tau(t)) \left(z^{(n-1)}(\tau(t))\right)^{\alpha} \left(1 + \frac{p_0^{\alpha}}{\tau_0}\right)$$

or equivalently,

$$r(t)(z^{(n-1)}(t))^{\alpha} \geq \frac{\tau_0}{\tau_0 + p_0^{\alpha}} G\left(\tau^{-1}(t)\right). \tag{31}$$

Using (31) in (29), we see that G is a positive solution of the inequality

$$G'(t) + \frac{k\tau_0}{\tau_0 + p_0^{\alpha}} Q_1(t) \psi_{n-2}^{\alpha}(g(t), t) G\left(\tau^{-1}(t)\right) \leq 0.$$

This a contradiction, and thus the proof is complete. □

Theorem 5. *If there exists a function $\vartheta(t) \in C(I_0, (0, \infty))$ satisfying*

$$g(t) \leq \vartheta(t), \ \tau^{-1}(\vartheta(t)) < t \tag{32}$$

and the first-order delay equation

$$G'(t) + \frac{k\tau_0}{\tau_0 + p_0^{\alpha}} Q_1(t) \psi_{n-2}^{\alpha}(g(t), \vartheta(t)) G\left(\tau^{-1}(\vartheta(t))\right) = 0 \tag{33}$$

is oscillatory, then X_N^+ is an empty class.

Proof. Assume the contrary that $x \in X^+$ and z satisfy **N**. Then, there exists a $t_1 \geq t_0$ such that $x(t) > 0$, $x(\tau(t)) > 0$ and $x(g(t)) > 0$ for $t \in I_1$. From Lemma 4 and Lemma 6, we arrive at (2) and (10), respectively. Now from (2), we get

$$z(g(t)) \geq r^{1/\alpha}(\vartheta(t)) z^{(n-1)}(\vartheta(t)) \psi_{n-2}(g(t), \vartheta(t)). \tag{34}$$

By replacing (28) with (34) and proceeding as in proof of Theorem 4, we arrive at G (defined as in (30)) which is a positive solution of the inequality

$$G'(t) + \frac{k\tau_0}{\tau_0 + p_0^{\alpha}} Q_1(t) \psi_{n-2}^{\alpha}(g(t), \vartheta(t)) G\left(\tau^{-1}(\vartheta(t))\right) \leq 0.$$

In view of ([23], Theorem 1), we have that (33) also has a positive solution, a contradiction. Thus, the proof is complete. □

Corollary 2. *If there exists a function $\vartheta(t) \in C(I_0, (0, \infty))$ satisfying (32) and*

$$\liminf_{t \to \infty} \int_{\tau^{-1}(\vartheta(t))}^{t} Q_1(\varrho) \psi_{n-2}^{\alpha}(g(t), \vartheta(t)) \, d\varrho > \frac{\tau_0 + p_0^{\alpha}}{ek\tau_0}, \tag{35}$$

then X_N^+ is an empty class.

Proof. By using Theorem 2 in [15], conditions (35) imply that (33) is oscillatory. □

Theorem 6. *Assume that $f(x(g(t))) := x^{\alpha}(t)$ and $p(t) < \widetilde{R}(t)$. If there exists a function $\theta \in C^1(I_0,(0,\infty))$ satisfying*

$$\theta'(t) \geq 0,\ \ \theta(t) > t,\ \ \tau\left(\theta^{n-1}(t)\right) < t \tag{36}$$

and the first-order delay equation

$$\omega'(t) + B_{n-2}(t)\,\omega\left(\tau\left(\theta^{n-1}(t)\right)\right) = 0 \tag{37}$$

is oscillatory, then X_N^+ is an empty class, where $\theta^{m-1}(t) := \theta\left(\theta^{m-2}(t)\right), \theta^0(t) := \theta(t)$,

$$R_0(t) := \left(\frac{1}{r(t)}\int_t^{\infty} kq(\varrho)\,d\varrho\right)^{1/\alpha},\quad R_m(t) := \int_t^{\infty} R_{m-1}(\varrho)\,d\varrho,$$

$$\widetilde{R}(t) := \exp\left(-\int_{\tau(t)}^{t} R_{n-2}(\varrho)\,d\varrho\right),$$

$$B_0(t) := \left(\frac{1}{r(t)}\int_t^{\theta(t)} q(t)\left(\widetilde{R}(\varrho) - p(\varrho)\right)^{\alpha} d\varrho\right)^{1/\alpha} \text{ and } B_m(t) := \int_t^{\theta(t)} B_{m-1}(\varrho)\,d\varrho,$$

for $m = 1, 2, ..., n-2$.

Proof. Assume the contrary that $x \in X^+$ and z satisfy **N**. Then, there exists a $t_1 \geq t_0$ such that $x(t) > 0$, $x(\tau(t)) > 0$ and $x(g(t)) > 0$ for $t \in I_1$. It is easy to notice that $\lim_{t\to\infty} z^{(j)} = 0$ for $j = 1, 2, ..., n-2$ and $\lim_{t\to\infty} r(t)\left(z^{(n-1)}(t)\right)^{\alpha} = 0$. Hence, by integrating (1) from t to ∞, we obtain

$$\begin{aligned} r(t)\left(z^{(n-1)}(t)\right)^{\alpha} &= \int_t^{\infty} q(\varrho)\,x^{\alpha}(\varrho)\,d\varrho \leq \int_t^{\infty} kq(\varrho)\,z^{\alpha}(\varrho)\,d\varrho \\ &\leq z^{\alpha}(t)\int_t^{\infty} kq(\varrho)\,d\varrho, \end{aligned}$$

and hence

$$z^{(n-1)}(t) \leq z(t)\left(\frac{1}{r(t)}\int_t^{\infty} kq(\varrho)\,d\varrho\right)^{1/\alpha} = z(t)\,R_0(t).$$

Integrating the last inequality $n-2$ times from t to ∞, we obtain

$$-z'(t) \leq z(t)\,R_{n-2}(t).$$

Thus, we get

$$z(v) \geq z(u)\exp\left(-\int_u^{v} R_{n-2}(\varrho)\,d\varrho\right),$$

for $u \leq v$. From the definition of z, we have

$$x(t) \geq z(t) - p(t)\,z(\tau(t)) \geq \left(\widetilde{R}(t) - p(t)\right) z(\tau(t))$$

which with (1) yields

$$\left(r(t)\left(z^{(n-1)}(t)\right)^{\alpha}\right)' = -q(t)\,x^{\alpha}(t) \leq -q(t)\left(\widetilde{R}(t) - p(t)\right)^{\alpha} z^{\alpha}(\tau(t)).$$

Integrating the last inequality from t to $\theta(t)$, we arrive at

$$\begin{aligned} z^{(n-1)}(t) &\geq \left(\frac{1}{r(t)}\int_t^{\theta(t)} q(t)\left(\widetilde{R}(\varrho)-p(\varrho)\right)^{\alpha} z^{\alpha}(\tau(\varrho))\,\mathrm{d}\varrho\right)^{1/\alpha} \\ &\geq z(\tau(\theta(t)))\,B_0(t). \end{aligned}$$

Integrating the last inequality $n-2$ times from t to $\theta(t)$, we get

$$z'(t)+z\left(\tau\left(\theta^{n-1}(t)\right)\right)B_{n-2}(t)\leq 0.$$

If we set

$$\omega(t):=\int_t^{\infty} z\left(\tau\left(\theta^{n-1}(t)\right)\right)B_{n-2}(t)>0,$$

then ω is a positive solution of the inequality $\omega'(t)+B_{n-2}(t)\,\omega\left(\tau\left(\theta^{n-1}(t)\right)\right)\leq 0$. In view of ([23], Theorem 1), we have that (37) also has a positive solution, a contradiction. The proof is complete. □

Corollary 3. *Assume that $f(x(g(t))):=x^{\alpha}(t)$ and $p(t)<\widetilde{R}(t)$. If there exists a function $\theta\in C^1(I_0,(0,\infty))$ satisfying (36) and*

$$\liminf_{t\to\infty}\int_{\tau(\theta^{n-1}(t))}^{t} B_{n-2}(\varrho)\,\mathrm{d}\varrho>\frac{1}{\mathrm{e}}, \tag{38}$$

then X_N^{+} is an empty class, where the functions $\widetilde{R}$, θ^{n-1} and B_{n-2} are defined as in Theorem 6.

Proof. By using Theorem 2 in [15], condition (38) implies that (37) is oscillatory. □

4. Asymptotic and Oscillatory Properties

Theorem 7. *Each non-oscillatory solution of (1) tends to zero if*

$$\lim_{\varrho\to\infty}\int_t^{\varrho}\left(\frac{1}{r(t)}\int_t^{\infty} q(\varrho)\,\mathrm{d}\varrho\right)^{1/\alpha}=\infty \tag{39}$$

and one of the conditions (18) or (26) is fulfilled.

Proof. Let x be a non-oscillatory solution of (1). Without loss of generality, we assume that $x\in X_+$. From Lemma 3, we have only two cases for z. Each of the conditions (18) or (26) contradicts that z fulfills **P**. Now, we suppose that z satisfies **N**. Since $z(t)>0$ and $z'(t)<0$, we get that $z\to c$ as $t\to\infty$, where $c\geq 0$. Suppose that $c>0$. Then, for every $\epsilon>0$, there exists a $T\geq t_0$ such that $c<z(t)<c+\epsilon$ for all $t>T$. By set $\varepsilon<(1-p)(c/p)$, we get that

$$\begin{aligned} x(t) &= z(t)-p(t)x(\tau(t))>c-pz(\tau(t)) \\ &> M(c+\epsilon)>Mz(t), \end{aligned}$$

where $M=(c-p(c+\epsilon))\,/\,(c+\epsilon)>0$. Thus, integrating from t to ∞, we have

$$\begin{aligned} r(t)\left(z^{(n-1)}(t)\right)^{\alpha} &\geq k\int_t^{\infty} q(\varrho)\,x^{\alpha}(g(\varrho))\,\mathrm{d}\varrho\geq kM^{\alpha}\int_t^{\infty} q(\varrho)\,z^{\alpha}(g(\varrho))\,\mathrm{d}\varrho \\ &\geq kM^{\alpha}z^{\alpha}(t)\int_t^{\infty} q(\varrho)\,\mathrm{d}\varrho>kM^{\alpha}c^{\alpha}\int_t^{\infty} q(\varrho)\,\mathrm{d}\varrho \end{aligned}$$

or

$$z^{(n-1)}(t) > k^{1/\alpha} Mc \left(\frac{1}{r(t)} \int_t^{\infty} q(\varrho)\, d\varrho \right)^{1/\alpha}.$$

By integrating from t to ϱ, we find

$$z^{(n-2)}(t) < z^{(n-2)}(\varrho) - k^{1/\alpha} Mc \int_t^{\varrho} \left(\frac{1}{r(t)} \int_t^{\infty} q(\varrho)\, d\varrho \right)^{1/\alpha}.$$

Taking the limit of both sides as $t \to \infty$ and using (39), we get that $z^{(n-2)}(t) \to -\infty$ as $t \to \infty$. But, z^{n-2} is a negative increasing function, this a contradiction. Therefore, $\lim_{t\to\infty} z(t) = 0$, which implies that $\lim_{t\to\infty} x(t) = 0$. The proof is complete. □

In the following, based on the fact that there are only two cases for the corresponding function z, we infer new criteria for oscillation of all solutions of the Equation (1). In each of the following theorems, we refer to two conditions through which it is possible to exclude the existence of solutions in X_P^+ or X_N^+. Thus, we rule out the existence of non-oscillatory solutions.

Theorem 8. *Assume that (18) or (26) holds. If there exists a function $\vartheta(t) \in C(I_0, (0,\infty))$ satisfying (32) and the first-order delay Equation (33) is oscillatory, then every solution of (1) is oscillatory.*

Theorem 9. *Assume that $f(x(g(t))) := x^{\alpha}(t)$, $p(t) < \widetilde{R}(t)$ and (18) hold. If there exists a function $\theta \in C^1(I_0, (0,\infty))$ satisfying (36) and the first-order delay Equation (37) is oscillatory, then every solution of (1) is oscillatory, where the functions $\widetilde{R}$, θ^{n-1} and B_{n-2} are defined as in Theorem 6.*

Corollary 4. *Assume that (18) or (26) holds. If there exists a function $\vartheta(t) \in C(I_0, (0,\infty))$ satisfying (32) and (35), then every solution of (1) is oscillatory.*

Corollary 5. *Assume that $f(x(g(t))) := x^{\alpha}(t)$, $p(t) < \widetilde{R}(t)$ and (18) (or (26)) hold. If there exists a function $\theta \in C^1(I_0, (0,\infty))$ satisfying (36) and (38), then every solution of (1) is oscillatory, where the functions $\widetilde{R}$, θ^{n-1} and B_{n-2} are defined as in Theorem 6.*

Example 1. *Consider the third-order neutral differential equation*

$$\left(\left((x(t) + p_0 x(\tau_0 t))'' \right)^{\alpha} \right)' + \frac{q_0}{t^{2\alpha+1}} x^{\alpha}(g_0 t) = 0, \tag{40}$$

where $p_0, \tau_0, g_0 \in (0,1)$ and $q_0 > 0$. From (40), we note that $n = 3$, $p(t) := p_0$, $\tau(t) := \tau_0 t$, $q(t) := q_0/t^{2\alpha+1}$, $g(t) := g_0 t$ and $r(t) = 1$. It is easy to verify that

$$\eta(t) = \lambda g_0^2 t, \quad \Theta(t) = q_0 (1-p_0)^{\alpha} \frac{1}{t^{2\alpha+1}}, \quad Q_1(t) := q_0/t^{2\alpha+1},$$

$$\phi(s,t) = (t-s), \quad \psi_1(s,t) = \frac{1}{2}(s-t)^2$$

and

$$\widetilde{\Theta}(t) = \frac{1}{2\alpha} q_0 (1-p_0)^{\alpha} \frac{1}{t^{2\alpha}}.$$

Thus, the condition (18) becomes:

$$q_0 (1 - p_0)^\alpha > \frac{1}{g_0^{2\alpha}} \left(\frac{2\alpha}{1+\alpha} \right)^{\alpha+1}.$$

The condition (26) simplifies to

$$q_0 (1 - p_0)^\alpha > \frac{\alpha 2^{\alpha+1}}{\lambda^\alpha g_0^{2\alpha}}.$$

By choosing $\vartheta(t) := (g_0 + \tau_0)(t/2)$*, where* $g_0 < 1$*, the condition (35) extends to*

$$q_0 (\tau_0 - g_0)^{2\alpha} \ln \frac{2\tau_0}{g_0 + \tau_0} > 2^{2\alpha+1} \frac{\tau_0 + p_0^\alpha}{e\tau_0}.$$

Using Corollary 4, Equation (40) is oscillatory if

$$q_0 > \max \left\{ \frac{1}{g_0^{2\alpha} (1 - p_0)^\alpha} \left(\frac{2\alpha}{1+\alpha} \right)^{\alpha+1}, \frac{2^{2\alpha+1} (\tau_0 + p_0^\alpha)}{e\tau_0 (\tau_0 - g_0)^{2\alpha}} \left(\ln \frac{2\tau_0}{g_0 + \tau_0} \right)^{-1} \right\} \tag{41}$$

or

$$q_0 > \max \left\{ \frac{\alpha 2^{\alpha+1}}{g_0^{2\alpha} (1 - p_0)^\alpha}, \frac{2^{2\alpha+1} (\tau_0 + p_0^\alpha)}{e\tau_0 (\tau_0 - g_0)^{2\alpha}} \left(\ln \frac{2\tau_0}{g_0 + \tau_0} \right)^{-1} \right\}.$$

Next, if we set $g(t) := t$*,* $\theta(t) := \gamma t$*,* $\gamma > 1$ *and* $p_0 < \tau_0^A$*, then the condition (38) becomes*

$$q_0^{1/\alpha} \left(\tau_0^A - p_0 \right) \frac{(\gamma - 1)^{-3}}{(2\alpha)^{1/\alpha}} \ln \left(\frac{1}{\gamma^2 \tau_0} \right) > \frac{1}{e},$$

where $A = (q_0/2\alpha)^{1/\alpha}$*. When* $g_0 = 1$*, by using Corollary 5, Equation (40) is oscillatory if*

$$q_0 > \max \left\{ \frac{1}{(1 - p_0)^\alpha} \left(\frac{2\alpha}{1+\alpha} \right)^{\alpha+1}, \frac{2\alpha (\gamma - 1)^{3\alpha}}{e (\tau_0^A - p_0)^\alpha (\ln 1/\gamma^2 \tau_0)^\alpha} \right\} \tag{42}$$

or

$$q_0 > \max \left\{ \frac{\alpha 2^{\alpha+1}}{(1 - p_0)^\alpha}, \frac{2\alpha (\gamma - 1)^{3\alpha}}{e (\tau_0^A - p_0)^\alpha (\ln 1/\gamma^2 \tau_0)^\alpha} \right\}.$$

5. Conclusions

When studying the oscillatory behavior of solutions of differential equations with odd-order, it is customary to find conditions that ensure solutions are either oscillatory or tend to zero. Dzurina et al. [5] and Vidhyaa et al. [24] established criteria for the oscillation of all solutions of a third-order linear and half-linear neutral differential equation, respectively. As an extension and also an improvement of these results, we obtained new oscillation criteria for the odd-order non-linear neutral Equation (1).

If we consider the third order differential equation

$$\left(x(t) + \frac{1}{10} x \left(\frac{1}{2} t \right) \right)''' + \frac{q_0}{t^3} x^\alpha \left(\frac{1}{10} t \right) = 0. \tag{43}$$

From Example 1 in [5], Equation (43) is oscillatory if $q_0 > 120$. However, by using our criterion (41), we get that (43) is oscillatory if $q_0 > 111.11$. Moreover, we consider the equation

$$\left(x(t) + \frac{1}{3}x\left(\frac{1}{2}t\right)\right)''' + \frac{q_0}{t^3}x^{\alpha}(t) = 0. \tag{44}$$

From Example 3 in [24], by choosing $\beta = 4/3$ Equation (44) is oscillatory if $q_0 > 4$. However, if we choose $\gamma = 4/3$, then our criterion (42) becomes $q_0 > 2$, and hence (44) is oscillatory. Thus, our results improve the results in [5,24]. In the future, we can try to study the advanced odd-order differential equations by the same approach.

Author Contributions: The authors claim to have contributed equally and significantly in this paper. All authors have read and agreed to the published version of the manuscript.

References

1. Hale, J.K. *Theory of Functional Differential Equations*; Springer: New York, NY, USA, 1977.
2. Bazighifan, O.; Ruggieri, M.; Scapellato, A. An Improved Criterion for the Oscillation of Fourth-Order Differential Equations. *Mathematics* **2020**, *8*, 610. [CrossRef]
3. Bohner, M.; Grace, S.R.; Jadlovska, I. Oscillation criteria for second-order neutral delay differential equations. *Electron. J. Qual. Theory Differ. Equ.* **2017**, 1–12. [CrossRef]
4. Chatzarakis, G.E.; Grace, S.R.; Jadlovska, I. Oscillation criteria for third-order delay differential equations. *Adv. Differ. Equ.* **2017**, *330*. [CrossRef]
5. Dzurina, J.; Grace, S.R.; Jadlovska, I. On nonexistence of Kneser solutions of third-order neutral delay differential equations. *Appl. Math. Lett.* **2019**, *88*, 193–200. [CrossRef]
6. Moaaz, O.; Baleanu, D.; Muhib, A. New Aspects for Non-Existence of Kneser Solutions of Neutral Differential Equations with Odd-Order. *Mathematics* **2020**, *8*, 494. [CrossRef]
7. Moaaz, O.; Chalishajar, D.; Bazighifan, O. Asymptotic behavior of solutions of the third order nonlinear mixed type neutral differential equations. *Mathematics* **2020**, *8*, 485. [CrossRef]
8. Moaaz, O.; Elabbasy, E.M.; Shaaban, E. Oscillation criteria for a class of third order damped differential equations. *Arab J. Math. Sci.* **2018**, *24*, 16–30. [CrossRef]
9. Moaaz, O.; Muhib, A. New oscillation criteria for nonlinear delay differential equations of fourth-order. *Appl. Math. Comput.* **2020**, *377*, 125192. [CrossRef]
10. Baculikova, B.; Dzurina, J. Oscillation of third-order neutral differential equations. *Math. Comput. Model.* **2010**, *52*, 215–226. [CrossRef]
11. Baculikova, B.; Dzurina, J. Oscillation of third-order nonlinear differential equations. *Appl. Math. Lett.* **2011**, *24*, 466–470. [CrossRef]
12. Dzurina, J.; Thandapani, E.; Tamilvanan, S. Oscillation of solutions to third-order half-linear neutral differential equations. *Electron. J. Differ. Equ.* **2012**, *2012*, 1–9.
13. Graef, J.R.; Tunc, E.; Grace, S.R. Oscillatory and asymptotic behavior of a third-order nonlinear neutral differential equation. *Opusc. Math.* **2017**, *37*, 839–852. [CrossRef]
14. Jiang, Y.; Li, T. Asymptotic behavior of a third-order nonlinear neutral delay differential equation. *J. Inequal. Appl.* **2014**, *2014*, 512. [CrossRef]
15. Kitamura, Y.; Kusano, T. Oscillation of first-order nonlinear differential equations with deviating arguments. *Proc. Am. Math. Soc.* **1980**, *78*, 64–68. [CrossRef]
16. Li, T.; Zhang, C.; Xing, G. Oscillation of third-order neutral delay differential equations. In *Abstract and Applied Analysis*; Hindawi: London, UK, 2012; Volume 2012.
17. Thandapani, E.; Tamilvanan, S.; Jambulingam, E.; Tech, V.T.M. Oscillation of third order half linear neutral delay differential equations. *Int. J. Pure Appl. Math.* **2012**, *77*, 359–368.
18. Tunc, E. Oscillatory and asymptotic behavior of third-order neutral differential equations with distributed deviating arguments. *Electron. J. Differ. Equ.* **2017**. [CrossRef]
19. Thandapani, E.; Li, T. On the oscillation of third-order quasi-linear neutral functional differential equations. *Arch. Math.* **2011** *47*, 181–199.

20. Xing, G.; Li, T.; Zhang, C. Oscillation of higher-order quasi-linear neutral differential equations. *Adv. Differ. Equ.* **2011**, *2011*, 45. [CrossRef]
21. Baculikova, B.; Dzurina, J. Oscillation theorems for second-order nonlinear neutral differential equations. *Comput. Math. Appl.* **2011**, *62*, 4472–4478. [CrossRef]
22. Agarwal, R.P.; Grace, S.R.; O'Regan, D. *Oscillation Theory for Difference and Functional Differential Equations*; Marcel Dekker, Kluwer Academic: Dordrecht, The Netherlands, 2000.
23. Philos, C. On the existence of nonoscillatory solutions tending to zero at ∞ for differential equations with positive delays. *Arch. Math.* **1981**, *36*, 168–178. [CrossRef]
24. Vidhyaa, K.S.; Graef , J.R.; Thandapani, E. New oscillation results for third-order half-linear neutral differential equations. *Mathematics* **2020**, *8*, 325. [CrossRef]

7

New Aspects for Non-Existence of Kneser Solutions of Neutral Differential Equations with Odd-Order

Osama Moaaz [1,†], **Dumitru Baleanu** [2,3,*,†] and **Ali Muhib** [1,4,†]

1 Department of Mathematics, Faculty of Science, Mansoura University, Mansoura 35516, Egypt; o_moaaz@mans.edu.eg (O.M.); muhib39@students.mans.edu.eg (A.M.)
2 Department of Mathematics and Computer Science, Faculty of Arts and Sciences, Çankaya University Ankara, 06790 Etimesgut, Turkey
3 Department of Medical Research, China Medical University Hospital, China Medical University, Taichung 40402, Taiwan
4 Department of Mathematics, Faculty of Education, Al-Nadirah, Ibb University, PO Box 70270 Ibb, Yemen
* Correspondence: dumitru@cankaya.edu.tr or Baleanu@mail.cmuh.org.tw
† These authors contributed equally to this work.

Abstract: Some new oscillatory and asymptotic properties of solutions of neutral differential equations with odd-order are established. Through the new results, we give sufficient conditions for the oscillation of all solutions of the studied equations, and this is an improvement of the relevant results. The efficiency of the obtained criteria is illustrated via example.

Keywords: odd-order differential equations; Kneser solutions; oscillatory solutions

1. Introduction

During this paper, we investigate the asymptotic properties of solutions to the odd-order neutral equation

$$(r(l)(z^{(n-1)}(l))^{\alpha})' + f(l,u(\eta(l))) = 0,\ l \geq l_0 > 0, \tag{1}$$

where $l \geq l_0$, $z(l) = u(l) + p(l)\,u(\theta(l))$, $0 \leq p(l) \leq p_0 < \infty$ and n is an odd natural number. Through the paper, we assume that

(I) α is a ratio of odd positive integers, $r, \eta, \theta \in C^1(I_0, \mathbb{R}^+)$, $r'(l) \geq 0$, $\eta(l) < l$, $\eta' > 0$, $(\eta^{-1}(l))' \geq \eta_0 > 0$, $\theta'(l) \geq \theta_0 > 0$, $\lim_{l\to\infty} \eta(l) = \infty$, $\lim_{l\to\infty} \theta(l) = \infty$, $I_\rho := [l_\rho, \infty)$, the function $f \in C(I_0 \times \mathbb{R}, \mathbb{R})$, and there exists a nonnegative function q such that $|f(l,u)| \geq q(l)\,|u|^{\alpha}$. Moreover, we study asymptotic behavior and oscillation of solutions of (1) in a canonical case, that is,

$$\int_{l_0}^{\infty} \frac{1}{r^{1/\alpha}(\varrho)} \mathrm{d}\varrho = \infty. \tag{2}$$

(II) $\theta(l) < l$ and $\theta \circ \eta = \eta \circ \theta$.

If there exists $l_u \geq l_0$ such that the real valued function u is continuous, $r\left(z^{(n-1)}\right)^{\alpha}$ is continuously differentiable and satisfies (1), for all $l \in I_u$; then, u is said to be a solution of (1). We restrict our discussion to those solutions u of (1) which satisfy $\sup\{|u(l)| : l_1 \leq l_0\} > 0$ for every $l_1 \in I_u$.

Definition 1. *A solution u of Equation (1) is called an* **N-Kneser solution** *if there exists a $l_* \in I_0$ such that $z(l)\,z'(l) < 0$ for all $l \in I_*$. The set of all eventually positive N-Kneser solutions of Equation (1) is denoted by $\Re$.*

Definition 2. *A solution u of (1) is said to be non-oscillatory if it is positive or negative, ultimately; otherwise, it is said to be oscillatory. The equation itself is termed oscillatory if all its solutions oscillate.*

There are many authors who studied the problem of oscillation of differential equations of a different order and presented many techniques in order to obtain criteria for oscillation of the studied equations, for example, [1–12].

For applications of odd-order equations in extrema, biology, and physics, we refer to the following examples. In 1701, James Bernoulli published the solution to the Isoperimetric Problem—a problem in which it is required to make one integral a maximum or minimum, while keeping constant the integral of a second given function, thus resulting in a differential equation of third-order (see [13]). In the early 1950s, Alan Lloyd Hodgkin and Andrew Huxley developed a mathematical model for the propagation of electrical pulses in the nerve of a squid. The Hodgkin–Huxley Model is a set of nonlinear ordinary differential equations. The model has played a seminal role in biophysics and neuronal modeling.

Recently, researchers have paid attention to neutral differential equations, as well as studying the oscillation behavior of their solutions. There is a practical side to study the problem of the oscillatory properties of solutions of neutral equations besides the theoretical side. For example, the neutral equations arise in applications to electric networks containing lossless transmission lines. Such networks appear in high-speed computers where lossless transmission lines are used to interconnect switching circuits. For more applications in science and technology, see [14–16].

Karpuz et al. [17] studied the higher-order neutral differential equations of the following type:

$$(u(l)+p(l)u(\theta(l)))^{(n)}+q(l)u(\eta(l))=0, \text{ for } l\in[l_0,\infty) \tag{3}$$

where oscillatory and asymptotic behaviors of all solutions of higher-order neutral differential equations are compared with first-order delay differential equations, depending on two different ranges of the coefficient associated with the neutral part

Xing et al. [18] established some oscillation criteria for certain higher-order quasi-linear neutral differential equation

$$(r(l)\left((u(l)+p(l)u(\theta(l)))^{(n-1)}\right)^{\alpha})'+q(l)u^{\alpha}(\eta(l))=0,\ n\geq 2 \tag{4}$$

where $\alpha\leq 1$ is the quotient of odd positive integers.

Li and Rogovchenko [19] concerned with the asymptotic behavior of solutions to a class of third-order nonlinear neutral differential equations

$$\left(r(t)\left((x(t)+p_0x(t-\varpi_0))''\right)^{\alpha}\right)'+q(t)x^{\alpha}(\tau(t))=0,$$

where $p_0\geq 0$, $p_0\neq 1$ and ϖ_0 are constants, $\varpi_0\geq 0$ (delayed argument) or $\varpi_0\leq 0$ (advanced argument).

Some results that are closely related to our work are presented as follows:

Theorem 1 ([17], Corollary 2, see [20], Theorem 3.1.1 and [21]). *Assume that p satisfies the condition*

$$p\in C\left([l_0,\infty),R^+\right) \text{ satisfies } \bar{l}_p:=\limsup_{l\to\infty}p(l)<1.$$

If

$$\limsup_{l\to\infty}\int_{\eta(l)}^{l}\frac{1}{(n-1)!}(\eta(\rho))^{n-1}q(\rho)\,\mathrm{d}\rho>1$$

or

$$\liminf_{l\to\infty}\int_{\eta(l)}^{l}\frac{1}{(n-1)!}(\eta(\rho))^{n-1}q(\rho)\,\mathrm{d}\rho>\frac{1}{\mathrm{e}}$$

holds, then (3) is almost oscillatory.

Theorem 2 ([18], Corollary 2.8). *Let n be odd, $\alpha \leq 1$, $\left(\eta^{-1}(l)\right)' \geq \eta_0 > 0$, $0 \leq p(l) \leq p_0 < \infty, \theta(l) \leq l$, and $\theta'(l) \geq \theta_0 > 0$, suppose that (2) holds. If $\theta^{-1}(\eta(l)) < l$ and*

$$\liminf_{l\to\infty} \int_{\theta^{-1}(\eta(l))}^{l} \frac{\Theta(\rho)\left(\rho^{n-1}\right)^{\alpha}}{r(\rho)} \mathrm{d}\rho > \left(\frac{1}{\eta_0} + \frac{p_0^{\alpha}}{\eta_0\theta_0}\right) \frac{((n-1)!)^{\alpha}}{\mathrm{e}},$$

where $\Theta(l) = \min\{q(\eta^{-1}(l)), q(\eta^{-1}(\theta(l)))$; then, every solution of (4) is oscillatory or tends to zero as $l \to \infty$.

Lemma 1 ([18,22]). *Assume that u_1, $u_2 \in [0, \infty)$. Then,*

$$(u_1 + u_2)^{\alpha} \leq \mu\left(u_1^{\alpha} + u_2^{\alpha}\right),$$

and

$$\mu = \begin{cases} 1 & \text{for } 0 < \alpha \leq 1; \\ 2^{\alpha-1} & \text{for } \alpha > 1. \end{cases}$$

Lemma 2 ([23]). *Let $u \in C^n([l_0, \infty), (0, \infty))$. Assume that $u^{(n)}(l)$ is of fixed sign and not identically zero on $[l_0, \infty)$ and that there exists a $l_1 \geq l_0$ such that $u^{(n-1)}(l)\, u^{(n)}(l) \leq 0$ for all $l \geq l_1$. If $\lim_{l\to\infty} u(l) \neq 0$, then, for every $\lambda \in (0, 1)$, there exists $l_\mu \geq l_1$ such that*

$$u(l) \geq \frac{\lambda}{(n-1)!} l^{n-1} \left|u^{(n-1)}(l)\right| \text{ for } l \geq l_\mu.$$

2. Main Results

For the sake of convenience, we use the following notation:

$$R_0(\varsigma, \varrho) = \int_{\varrho}^{\varsigma} r^{-1/\alpha}(\rho)\,\mathrm{d}\rho,\ R_k(\varsigma, \varrho) = \int_{\varrho}^{\varsigma} R_{k-1}(\varsigma, \rho)\,\mathrm{d}\rho,\ k = 1, 2, ..., n-2$$

and

$$Q(l) = \min\{q(l), q(\theta(l))\},\ Q_1(l) = \min\{q\left(\eta^{-1}(l)\right), q\left(\eta^{-1}(\theta(l))\right)\}.$$

The following lemma is a direct conclusion from Lemmas 2.1 and 2.4 in [18], so its proof was neglected.

Lemma 3. *Assume that u is an eventually positive solution of (1). Then, there exists a sufficiently large $l_1 \geq l_0$ such that, for all $l \geq l_1$, either*

$$\text{Case (1)}: z(l) > 0,\ z'(l) > 0,\ z^{(n-1)}(l) > 0,\ (r(l)(z^{(n-1)}(l))^{\alpha})' < 0$$

or

$$\text{Case (2)}: (-1)^k z^{(k)}(l) > 0, \text{ for } k = 0, 1, 2, ..., n.$$

Now, in the following theorem, we will provide a new criterion for non-existence of N-Kneser solutions of (1) by using the comparison theorem.

Theorem 3. *Assume (I) and (II) holds. If there exists a function $\zeta(l) \in C([l_0, \infty), (0, \infty))$ satisfying $\eta(l) < \zeta(l)$ and $\theta^{-1}(\zeta(l)) < l$, such that the differential equation*

$$G'(l) + \frac{1}{\mu} \frac{\theta_0}{\theta_0 + p_0^{\alpha}} R_{n-2}^{\alpha}(\zeta(l), \eta(l))\, Q(l)\, G\left(\theta^{-1}(\zeta(l))\right) = 0 \tag{5}$$

is oscillatory, then $\Re$ is an empty set.

Proof. Let u be a N-Kneser solution of (1), say $u(l) > 0$ and $u(\eta(l)) > 0$ for $l \geq l_1 \geq l_0$. This implies that

$$(-1)^k z^{(k)}(l) > 0, \text{ for } k = 0, 1, 2, ..., n. \tag{6}$$

From (1), we see that

$$\begin{aligned} 0 &\geq \frac{p_0^\alpha}{\theta'(l)} \left(r(\theta(l)) \left(z^{(n-1)}(\theta(l)) \right)^\alpha \right)' + p_0^\alpha q(\theta(l)) u^\alpha(\eta(\theta(l))) \\ &\geq \frac{p_0^\alpha}{\theta_0} \left(r(\theta(l)) \left(z^{(n-1)}(\theta(l)) \right)^\alpha \right)' + p_0^\alpha q(\theta(l)) u^\alpha(\eta(\theta(l))) \\ &= \frac{p_0^\alpha}{\theta_0} \left(r(\theta(l)) \left(z^{(n-1)}(\theta(l)) \right)^\alpha \right)' + p_0^\alpha q(\theta(l)) u^\alpha(\theta(\eta(l))). \end{aligned} \tag{7}$$

Combining (1) and (7), we obtain

$$\begin{aligned} 0 &\geq (r(l)(z^{(n-1)}(l))^\alpha)' + \frac{p_0^\alpha}{\theta_0} \left(r(\theta(l)) \left(z^{(n-1)}(\theta(l)) \right)^\alpha \right)' + q(l) u^\alpha(\eta(l)) \\ &+ p_0^\alpha q(\theta(l)) u^\alpha(\theta(\eta(l))) \\ &\geq (r(l)(z^{(n-1)}(l))^\alpha)' + \frac{p_0^\alpha}{\theta_0} \left(r(\theta(l)) \left(z^{(n-1)}(\theta(l)) \right)^\alpha \right)' \\ &+ Q(l) \left(u^\alpha(\eta(l)) + p_0^\alpha u^\alpha(\theta(\eta(l))) \right). \end{aligned} \tag{8}$$

From definition of z and using (I), we have

$$z(\eta(l)) = u(\eta(l)) + p(\eta(l)) u(\theta(\eta(l))) \leq u(\eta(l)) + p_0 u(\theta(\eta(l))).$$

By using the latter inequality in (8), we get

$$\begin{aligned} 0 &\geq (r(l)(z^{(n-1)}(l))^\alpha)' + \frac{p_0^\alpha}{\theta_0} \left(r(\theta(l)) \left(z^{(n-1)}(\theta(l)) \right)^\alpha \right)' \\ &+ Q(l) \left(u(\eta(l)) + p_0 u(\theta(\eta(l))) \right)^\alpha \\ &\geq (r(l)(z^{(n-1)}(l))^\alpha)' + \frac{p_0^\alpha}{\theta_0} \left(r(\theta(l)) \left(z^{(n-1)}(\theta(l)) \right)^\alpha \right)' + \frac{1}{\mu} Q(l) z^\alpha(\eta(l)), \end{aligned}$$

that is,

$$0 \geq \left(r(l)(z^{(n-1)}(l))^\alpha + \frac{p_0^\alpha}{\theta_0} r(\theta(l)) \left(z^{(n-1)}(\theta(l)) \right)^\alpha \right)' + \frac{1}{\mu} Q(l) z^\alpha(\eta(l)). \tag{9}$$

On the other hand, it follows from the monotonicity of $r(l)(z^{(n-1)}(l))$ that

$$\begin{aligned} -z^{(n-2)}(\varrho) &\geq z^{(n-2)}(\varsigma) - z^{(n-2)}(\varrho) = \int_\varrho^\varsigma \frac{r^{1/\alpha}(\rho) z^{(n-1)}(\rho)}{r^{1/\alpha}(\rho)} d\rho \\ &\geq r^{1/\alpha}(\varsigma) z^{(n-1)}(\varsigma) R_0(\varsigma, \varrho). \end{aligned} \tag{10}$$

Integrating (10) from ϱ to ς, we have

$$-z^{(n-3)}(\varrho) \leq z^{(n-3)}(\varsigma) - z^{(n-3)}(\varrho) = r^{1/\alpha}(\varsigma) z^{(n-1)}(\varsigma) R_1(\varsigma, \varrho). \tag{11}$$

Integrating (11) $n-3$ times from ϱ to ς and using (6), we get

$$z(\varrho) \geq r^{1/\alpha}(\varsigma) z^{(n-1)}(\varsigma) R_{n-2}(\varsigma, \varrho). \tag{12}$$

Thus, we have

$$z(\eta(l)) \geq r^{1/\alpha}(\zeta(l))\, z^{(n-1)}(\zeta(l))\, R_{n-2}(\zeta(l), \eta(l)),$$

which, by virtue of (9), yields that

$$\begin{aligned} 0 \;\geq\; & (r(l)(z^{(n-1)}(l))^{\alpha} + \frac{p_0^{\alpha}}{\theta_0} r(\theta(l)) \left(z^{(n-1)}(\theta(l))\right)^{\alpha})' \\ & + \frac{1}{\mu} Q(l)\, r(\zeta(l)) \left(z^{(n-1)}(\zeta(l))\, R_{n-2}(\zeta(l), \eta(l))\right)^{\alpha}. \end{aligned} \tag{13}$$

Now, set

$$G(l) = r(l)(z^{(n-1)}(l))^{\alpha} + \frac{p_0^{\alpha}}{\theta_0} r(\theta(l)) \left(z^{(n-1)}(\theta(l))\right)^{\alpha} > 0.$$

From (I) and the fact that $r(l)(z^{(n-1)}(l))$ is non-increasing, we have

$$G(l) \leq r(\theta(l)) \left(z^{(n-1)}(\theta(l))\right)^{\alpha} \left(1 + \frac{p_0^{\alpha}}{\theta_0}\right)$$

or equivalently,

$$r(\zeta(l))(z^{(n-1)}(\zeta(l)))^{\alpha} \geq \frac{\theta_0}{\theta_0 + p_0^{\alpha}} G\left(\theta^{-1}(\zeta(l))\right). \tag{14}$$

Using (14) in (13), we see that G is a positive solution of the differential inequality

$$G'(l) + \frac{1}{\mu} \frac{\theta_0}{\theta_0 + p_0^{\alpha}} R_{n-2}^{\alpha}(\zeta(l), \eta(l))\, Q(l)\, G\left(\theta^{-1}(\zeta(l))\right) \leq 0.$$

In view of [24], Theorem 1, we have that (5) also has a positive solution, a contradiction. Thus, the proof is complete. □

In the following theorem, we establish a hille and nehari type condition that confirms the non-existence of N-Kneser solutions of (1).

Theorem 4. *Assume (I) and (II) hold. If there exists a function $\delta(l) \in C([l_0, \infty), (0, \infty))$ satisfying $\delta(l) < l$ and $\eta(l) < \theta(\delta(l))$ such that*

$$\limsup_{l \to \infty} \frac{1}{\mu} \frac{R_{n-2}^{\alpha}(\theta(\delta(l)), \eta(l))}{r(\theta(\delta(l)))} \int_{\delta(l)}^{l} Q(\rho)\, d\rho > \frac{\theta_0 + p_0^{\alpha}}{\theta_0}, \tag{15}$$

then $\Re$ is an empty set.

Proof. By using the same method in proof of Theorem 3, we obtain (9). Integrating (9) from $\delta(l)$ to l and using the fact that z is decreasing, we get

$$\begin{aligned} & r(\delta(l))(z^{(n-1)}(\delta(l)))^{\alpha} + \frac{p_0^{\alpha}}{\theta_0} r(\theta(\delta(l))) \left(z^{(n-1)}(\theta(\delta(l)))\right)^{\alpha} \\ \geq\; & r(l)(z^{(n-1)}(l))^{\alpha} + \frac{p_0^{\alpha}}{\theta_0} r(\theta(l)) \left(z^{(n-1)}(\theta(l))\right)^{\alpha} + \frac{1}{\mu} z^{\alpha}(\eta(l)) \int_{\delta(l)}^{l} Q(\rho)\, d\rho \\ \geq\; & \frac{1}{\mu} z^{\alpha}(\eta(l)) \int_{\delta(l)}^{l} Q(\rho)\, d\rho. \end{aligned}$$

Since $\theta(\delta(l)) < \theta(l)$ and $r(l)(z^{(n-1)}(l))$ is non-increasing, we have

$$r(\theta(\delta(l)))(z^{(n-1)}(\theta(\delta(l))))^{\alpha}\left(1+\frac{p_0^{\alpha}}{\theta_0}\right) \geq \frac{1}{\mu}z^{\alpha}(\eta(l))\int_{\delta(l)}^{l} Q(\rho)\,\mathrm{d}\rho. \tag{16}$$

By using (12) with $\varsigma = \theta(\delta(l))$ and $\varrho = \eta(l)$ in (16), we obtain

$$\begin{aligned} & r(\theta(\delta(l)))(z^{(n-1)}(\theta(\delta(l))))^{\alpha}\left(1+\frac{p_0^{\alpha}}{\theta_0}\right) \\ \geq\ & \frac{1}{\mu}\left(z^{(n-1)}(\theta(\delta(l)))\right)^{\alpha} R_{n-2}^{\alpha}(\theta(\delta(l)),\eta(l))\int_{\delta(l)}^{l} Q(\rho)\,\mathrm{d}\rho, \end{aligned}$$

that is,

$$\frac{\theta_0 + p_0^{\alpha}}{\theta_0} \geq \frac{1}{\mu}\frac{R_{n-2}^{\alpha}(\theta(\delta(l)),\eta(l))}{r(\theta(\delta(l)))}\int_{\delta(l)}^{l} Q(\rho)\,\mathrm{d}\rho.$$

Now, we take the lim sup of both sides of the previous inequality, and we obtain a contradiction to (15). The proof is complete. □

In the following theorem, we will provide another criterion for the non-existence of N-Kneser solutions of (1) using the comparison theorem.

Theorem 5. *Assume (I), (II), and $\eta(\theta(l)) < l$ hold. If the differential equation*

$$\Psi'(l) + Q_1(l) R_{n-2}^{\alpha}(\theta(l), l)\left(\frac{\eta_0\theta_0}{\theta_0 + p_0^{\alpha}}\right)\Psi(\eta(l)) = 0 \tag{17}$$

is oscillatory, then $\Re$ is an empty set.

Proof. Let u be a N-Kneser solution of (1), say $u(l) > 0$, $u(\theta(l)) > 0$ and $u(\eta(l)) > 0$ for $l \geq l_1 \geq l_0$. This implies that

$$(-1)^k z^{(k)}(l) > 0, \text{ for } k = 0, 1, 2, ..., n.$$

By using (1) and **(I)**, we see that

$$\begin{aligned} 0 \ &\geq\ \frac{1}{(\eta^{-1}(l))'}\left(r\left(\eta^{-1}(l)\right)\left(z^{(n-1)}\left(\eta^{-1}(l)\right)\right)^{\alpha}\right)' + q\left(\eta^{-1}(l)\right)u^{\alpha}(l) \\ &\geq\ \frac{1}{\eta_0}\left(r\left(\eta^{-1}(l)\right)\left(z^{(n-1)}\left(\eta^{-1}(l)\right)\right)^{\alpha}\right)' + q\left(\eta^{-1}(l)\right)u^{\alpha}(l), \end{aligned}$$

and, similarly,

$$\begin{aligned} 0 \ &\geq\ \frac{p_0^{\alpha}}{(\eta^{-1}(\theta(l)))'}\left(r\left(\eta^{-1}(\theta(l))\right)\left(z^{(n-1)}\left(\eta^{-1}(\theta(l))\right)\right)^{\alpha}\right)' \\ &\quad + p_0^{\alpha}q\left(\eta^{-1}(\theta(l))\right)u^{\alpha}(\theta(l)) \\ &\geq\ \frac{p_0^{\alpha}}{\eta_0\theta_0}\left(r\left(\eta^{-1}(\theta(l))\right)\left(z^{(n-1)}\left(\eta^{-1}(\theta(l))\right)\right)^{\alpha}\right)' \\ &\quad + p_0^{\alpha}q\left(\eta^{-1}(\theta(l))\right)u^{\alpha}(\theta(l)). \end{aligned}$$

Combining the above inequalities yields that

$$\begin{aligned} 0 \geq & \frac{1}{\eta_0}\left(r\left(\eta^{-1}(l)\right)\left(z^{(n-1)}\left(\eta^{-1}(l)\right)\right)^{\alpha}\right)' \\ & +\frac{p_0^{\alpha}}{\eta_0\theta_0}\left(r\left(\eta^{-1}(\theta(l))\right)\left(z^{(n-1)}\left(\eta^{-1}(\theta(l))\right)\right)^{\alpha}\right)' \\ & +q\left(\eta^{-1}(l)\right)u^{\alpha}(l)+p_0^{\alpha}q\left(\eta^{-1}(\theta(l))\right)u^{\alpha}(\theta(l)), \end{aligned}$$

that is,

$$\begin{aligned} 0 \geq & \left(\frac{1}{\eta_0}r\left(\eta^{-1}(l)\right)\left(z^{(n-1)}\left(\eta^{-1}(l)\right)\right)^{\alpha}+\frac{p_0^{\alpha}}{\eta_0\theta_0}r\left(\eta^{-1}(\theta(l))\right)\left(z^{(n-1)}\left(\eta^{-1}(\theta(l))\right)\right)^{\alpha}\right)' \\ & +Q_1(l)\,z^{\alpha}(l). \end{aligned} \tag{18}$$

Now, we set

$$\Psi(l)=\frac{1}{\eta_0}r\left(\eta^{-1}(l)\right)\left(z^{(n-1)}\left(\eta^{-1}(l)\right)\right)^{\alpha}+\frac{p_0^{\alpha}}{\eta_0\theta_0}r\left(\eta^{-1}(\theta(l))\right)\left(z^{(n-1)}\left(\eta^{-1}(\theta(l))\right)\right)^{\alpha}. \tag{19}$$

From (II) and the fact that $r(l)\left(z^{(n-1)}(l)\right)$ is non-increasing, it is easy to see that

$$\Psi(l)\leq\frac{r\left(\left(\eta^{-1}(\theta(l))\right)\right)\left(z^{(n-1)}\left(\eta^{-1}(\theta(l))\right)\right)^{\alpha}}{\eta_0}\left(1+\frac{p_0^{\alpha}}{\theta_0}\right). \tag{20}$$

By using (12) with $\varsigma=\theta(l)$ and $\varrho=l$ and (20), we have

$$z^{\alpha}(l)\geq r(\theta(l))\left(z^{(n-1)}(\theta(l))\right)^{\alpha}R_{n-2}^{\alpha}(\theta(l),l)\geq\Psi(\eta(l))\,R_{n-2}^{\alpha}(\theta(l),l)\left(\frac{\eta_0\theta_0}{\theta_0+p_0^{\alpha}}\right).$$

From definition Ψ and using the above inequality in (18), we get

$$0\geq\Psi'(l)+Q_1(l)\,R_{n-2}^{\alpha}(\theta(l),l)\left(\frac{\eta_0\theta_0}{\theta_0+p_0^{\alpha}}\right)\Psi(\eta(l)).$$

In view of [24], Theorem 1, we have that (17) also has a positive solution, a contradiction. Thus, the proof is complete. □

3. New Oscillation Criteria

In the following lemma, we present criteria that ensure that non-existence of solutions satisfies case (1).

Lemma 4. *Assume that u be an eventually positive solution of (1) and the differential equation*

$$\Phi'(l)+\frac{Q(l)}{\left(1+\frac{p_0^{\alpha}}{\theta_0}\right)}\left(\frac{\lambda_0}{(n-1)!r^{1/\alpha}(\eta(l))}(\eta(l))^{n-1}\right)^{\alpha}\Phi\left(\eta\left(\theta^{-1}(l)\right)\right)=0 \tag{21}$$

or

$$\phi'(l)+\frac{Q_1(l)}{\left(\frac{1}{\eta_0}+\frac{p_0^{\alpha}}{\eta_0\theta_0}\right)}\left(\frac{\lambda_1}{(n-1)!r^{1/\alpha}(l)}l^{n-1}\right)^{\alpha}\phi\left(\theta^{-1}(\eta(l))\right)=0 \tag{22}$$

is oscillatory, then z does not satisfy the following case:

$$z(l)>0,\ z'(l)>0,\ z^{(n-1)}(l)>0 \text{ and } z^{(n)}(l)\leq 0. \tag{23}$$

Proof. Assume on the contrary that u is an eventually positive solution of (1) and z satisfies (23). Proceeding as in the proof of Theorem 3, we obtain (9). By using Lemma 2, we get

$$z(l) \geq \frac{\lambda}{(n-1)!r^{1/\alpha}(l)} l^{n-1} r^{1/\alpha}(l) z^{(n-1)}(l). \tag{24}$$

Therefore, by setting $w(l) = r(l)(z^{(n-1)}(l))^{\alpha}$ in (9) and utilizing (24), we see that w is a positive solution of the equation

$$\left(w(l) + \frac{p_0^{\alpha}}{\theta_0} w(\theta(l))\right)' + Q(l)\left(\frac{\lambda}{(n-1)!r^{1/\alpha}(\eta(l))}(\eta(l))^{n-1}\right)^{\alpha} w(\eta(l)) = 0. \tag{25}$$

Since $w(l) = r(l)(z^{(n-1)}(l))^{\alpha}$ is non-increasing and it satisfies (25), let us denote

$$\Phi(l) = w(l) + \frac{p_0^{\alpha}}{\theta_0} w(\theta(l)).$$

It follows from $\theta(l) < l$

$$\Phi(l) \leq w(\theta(l))\left(1 + \frac{p_0^{\alpha}}{\theta_0}\right).$$

Substituting these terms into (25), we get that Φ is a positive solution of

$$\Phi'(l) + \frac{Q(l)}{\left(1 + \frac{p_0^{\alpha}}{\theta_0}\right)}\left(\frac{\lambda}{(n-1)!r^{1/\alpha}(\eta(l))}(\eta(l))^{n-1}\right)^{\alpha} \Phi\left(\eta\left(\theta^{-1}(l)\right)\right) \leq 0.$$

In view of [24], Theorem 1, we have that (21) also has a positive solution, which is a contradiction (21).

Now, proceeding as in the proof of Theorem 5, we obtain (18). In the same style as the first part, we have

$$\begin{aligned} 0 \geq & \left(\frac{1}{\eta_0} r\left(\eta^{-1}(l)\right)\left(z^{(n-1)}\left(\eta^{-1}(l)\right)\right)^{\alpha} + \frac{p_0^{\alpha}}{\eta_0\theta_0} r\left(\eta^{-1}(\theta(l))\right)\left(z^{(n-1)}\left(\eta^{-1}(\theta(l))\right)\right)^{\alpha}\right)' \\ & + Q_1(l) z^{\alpha}(l). \end{aligned}$$

By using Lemma 2, we get

$$z(l) \geq \frac{\lambda}{(n-1)!r^{1/\alpha}(l)} l^{n-1} r^{1/\alpha}(l) z^{(n-1)}(l).$$

Therefore, by setting $U(l) = r(l)(z^{(n-1)}(l))^{\alpha}$ in (18) and utilizing (24), we see that U is a positive solution of the equation

$$\left(\frac{1}{\eta_0} U\left(\eta^{-1}(l)\right) + \frac{p_0^{\alpha}}{\eta_0\theta_0} U\left(\eta^{-1}(\theta(l))\right)\right)' + Q_1(l)\left(\frac{\lambda}{(n-1)!r^{1/\alpha}(l)} l^{n-1}\right)^{\alpha} U(l) = 0. \tag{26}$$

Since $U(l) = r(l)(z^{(n-1)}(l))^{\alpha}$ is non-increasing and it satisfies (26), let us denote

$$\phi(l) = \frac{1}{\eta_0} U\left(\eta^{-1}(l)\right) + \frac{p_0^{\alpha}}{\eta_0\theta_0} U\left(\eta^{-1}(\theta(l))\right).$$

It follows from $\theta(l) < l$

$$\phi(l) \leq U\left(\eta^{-1}(\theta(l))\right)\left(\frac{1}{\eta_0} + \frac{p_0^{\alpha}}{\eta_0\theta_0}\right).$$

Substituting these terms into (26), we get that ϕ is a positive solution of

$$\phi'(l) + \frac{Q_1(l)}{\left(\frac{1}{\eta_0} + \frac{p_0^\alpha}{\eta_0\theta_0}\right)} \left(\frac{\lambda}{(n-1)! r^{1/\alpha}(l)} l^{n-1}\right)^\alpha \phi\left(\theta^{-1}(\eta(l))\right) \leq 0.$$

In view of [24], Theorem 1, we have that (22) also has a positive solution, which is a contradiction (22). Thus, the proof is complete. □

The following theorems give the criteria for oscillation for all solutions of Equation (1).

Theorem 6. *If (5) and (21) are oscillatory, then (1) is oscillatory.*

Proof. Assume on the contrary that u is an eventually positive solution of (1). Then, from Lemma 3, we conclude that there are two possible cases for the behavior of z and its derivatives. By using Theorem 3 and Lemma 4, conditions (5) and (21) ensure that there are no solutions for Equation (1) satisfy case (1) and case (2) respectively. Thus, the proof is complete. □

Theorem 7. *If (17) and (21) are oscillatory, then (1) is oscillatory.*

Proof. Assume on the contrary that u is an eventually positive solution of (1). Then, from Lemma 3, we conclude that there are two possible cases for the behavior of z and its derivatives. By using Theorem 5 and Lemma 4, conditions (17) and (21) ensure that there are no solutions for Equation (1) satisfying case (1) and case (2), respectively. Thus, the proof is complete. □

The following corollaries provided criteria for the oscillation of the first-order equations that were used in the comparison.

Corollary 1. *If there exists a function $\zeta(l) \in C([l_0,\infty), (0,\infty))$ satisfying $\eta(l) < \zeta(l)$ and $\theta^{-1}(\zeta(l)) < l$, such that*

$$\liminf_{l\to\infty} \int_{\theta^{-1}(\zeta(l))}^{l} R_{n-2}^\alpha(\zeta(\rho), \eta(\rho)) \frac{Q(\rho)}{\mu} \mathrm{d}\rho \geq \frac{\theta_0 + p_0^\alpha}{\theta_0 \mathrm{e}} \tag{27}$$

and

$$\liminf_{l\to\infty} \int_{\eta(\theta^{-1}(l))}^{l} Q(l) \left(\frac{\lambda_0}{(n-1)! r^{1/\alpha}(\eta(l))} (\eta(l))^{n-1}\right)^\alpha \mathrm{d}\rho \geq \frac{\theta_0 + p_0^\alpha}{\theta_0 \mathrm{e}} \tag{28}$$

hold, then (1) is oscillatory.

Corollary 2. *Let $\delta(l) = \theta(l)$ in Theorem 4. If $\eta(l) < \theta(\theta(l))$, such that (28) and*

$$\limsup_{l\to\infty} \frac{1}{\mu} \frac{R_{n-2}^\alpha(\theta(\theta(l)), \eta(l))}{r(\theta(\theta(l)))} \int_{\theta(l)}^{l} Q(\rho)\, \mathrm{d}\rho > \frac{\theta_0 + p_0^\alpha}{\theta_0} \tag{29}$$

hold, then (1) is oscillatory.

Corollary 3. *If $\eta(\theta(l)) < l$, such that (28) and*

$$\liminf_{l\to\infty} \int_{\eta(\theta(l))}^{l} Q_1(\rho) R_{n-2}^\alpha(\theta(\rho), \rho) > \frac{\theta_0 + p_0^\alpha}{\eta_0 \theta_0 \mathrm{e}} \tag{30}$$

hold, then (1) is oscillatory.

Example 1. *Consider the differential equation*

$$\left(\left(\left(u\left(l\right)+pu\left(\delta l\right)\right)^{(n-1)}\right)^{\alpha}\right)'+\frac{q_0}{l^{\alpha(n-1)+1}}u^{\alpha}\left(\lambda l\right)=0,\ l\geq 1 \tag{31}$$

From (31), we have $r\left(l\right)=1$, $p\left(l\right)=p$, $\theta\left(l\right)=\delta l$, $\eta\left(l\right)=\lambda l$ *and* $q\left(l\right)=q_0/l^{\alpha(n-1)+1}$. *Using some mathematical operations. By using Corollary 1, we find that (31) is oscillatory if*

$$q_0\ln\left(\frac{2\delta}{\delta+\lambda}\right)>\frac{\mu\left(\theta_0+p_0^{\alpha}\right)}{\theta_0\mathrm{e}}\left((n-1)!\left(\frac{2}{\delta-\lambda}\right)^{n-1}\right)^{\alpha}$$

and

$$q_0\ln\left(\frac{\lambda}{\delta}\right)>\frac{\left(\theta_0+p_0^{\alpha}\right)}{\theta_0\mathrm{e}}\frac{\left((n-1)!\right)^{\alpha}}{\lambda_0^{\alpha}\lambda^{\alpha(n-1)}}.$$

By using Corollary 3, we find that (31) is oscillatory if

$$q_0\ln\left(\frac{\lambda}{\delta}\right)>\frac{\left(\theta_0+p_0^{\alpha}\right)}{\theta_0\mathrm{e}}\frac{\left((n-1)!\right)^{\alpha}}{\lambda_0^{\alpha}\lambda^{\alpha(n-1)}}$$

and

$$q_0\ln\left(\frac{1}{\delta\lambda}\right)>\frac{\left(\theta_0+p_0^{\alpha}\right)}{\eta_0\theta_0\mathrm{e}}\frac{\left((n-1)!\right)^{\alpha}}{\lambda^{\alpha(n-1)+1}\left(\delta-1\right)^{\alpha(n-1)}}.$$

4. Conclusions

This article is concerned with oscillatory properties of solutions for the odd-order neutral equation. Many works have studied the oscillatory properties of solutions of an odd-order equation; see [17,18]. However, in these works, we find sufficient conditions to ensure that every non-oscillatory solution tends to zero, that is, conditions that guarantee that all solutions are oscillatory or tend to zero. Unusually, in this paper, we presented new criteria ensuring that all solutions of (1) are oscillatory, which in turn is an improvement and extension of the results in [17,18]. For this purpose, we used the comparison technique with first-order equations. For ease of application in the examples, Corollaries 1–3 provided criteria for the oscillation of the first-order equations that were used in the comparison.

Author Contributions: Formal analysis, D.B. and A.M.; Investigation, O.M.; Supervision, O.M.; Writing—original draft, A.M.; Writing—review and editing, O.M., D.B. and A.M. The authors claim to have contributed equally and significantly in this paper. All authors have read and agreed to the published version of the manuscript.

Acknowledgments: The authors thank the reviewers for for their useful comments, which led to the improvement of the content of the paper.

References

1. Grace, S.R. Oscillation theorems for nth-order differential equations with deviating arguments. *J. Math. Appl. Anal.* **1984**, *101*, 268–296. [CrossRef]
2. Bazighifan, O.; Cesarano, C. A Philos-Type Oscillation Criteria for Fourth-Order Neutral Differential Equations. *Symmetry* **2020**, *12*, 379. [CrossRef]
3. Bazighifan, O.; Elabbasy, E.M.; Moaaz, O. Oscillation of higher-order differential equations with distributed delay. *J. Inequal. Appl.* **2019**, *55*, 55. [CrossRef]
4. Bazighifan, O.; Cesarano, C. Some New Oscillation Criteria for Second-Order Neutral Differential Equations with Delayed Arguments. *Mathematics* **2019**, *7*, 619. [CrossRef]
5. Moaaz, O. New criteria for oscillation of nonlinear neutral differential equations. *Adv. Differ. Equ.* **2019**, *2019*, 484. [CrossRef]
6. Moaaz, O.; Elabbasy, E.M.; Bazighifan, O. On the asymptotic behavior of fourth-order functional differential equations. *Adv. Differ. Equ.* **2017** *2017*, 261. [CrossRef]

7. Moaaz, O.; Elabbasy, E.M.; Muhib, A. Oscillation criteria for even-order neutral differential equations with distributed deviating arguments. *Adv. Differ. Equ.* **2019**, *2019*, 297. [CrossRef]
8. Li, T.; Rogovchenko, Y.V. *Asymptotic Behavior of Higher-order Quasilinear Neutral Differential Equations*; Hindawi Publishing Corporation: London, UK, 2014; 11p.
9. Moaaz, O.; Muhib, A. New oscillation criteria for nonlinear delay differential equations of fourth-order. *Appl. Math. Comput.* **2020**, *377*, 125192. [CrossRef]
10. Moaaz, O.; Park, C.; Muhib, A.; Bazighifan, O. Oscillation criteria for a class of even-order neutral delay differential equations. *J. Appl. Math. Comput.* **2020**. [CrossRef]
11. Moaaz, O.; Furuichi, S.; Muhib, A. New Comparison Theorems for the Nth Order Neutral Differential Equations with Delay Inequalities. *Mathematics* **2020**, *8*, 454. [CrossRef]
12. Agarwal, R.; Grace, S.; O'Regan, D. Oscillation criteria for certain nth order differential equations with deviating arguments. *J. Math. Appl. Anal.* **2001**, *262*, 601–622. [CrossRef]
13. Leibniz, G. *Acta Eruditorm, A Source Book in Mathematics*, 1200–1800 ed.; Struik, D.J., Ed.; Prenceton Unversity Press: Princeton, NJ, USA, 1986.
14. Hale, J.K. *Theory of Functional Differential Equations*; Springer: New York, NY, USA, 1977.
15. Treanta, S.; Varsan, C. Weak small controls and approximations associated with controllable affine control systems. *J. Differ. Equ.* **2013**, *255*, 1867–1882. [CrossRef]
16. Doroftei, M.-M.; Treanta, S. Higher order hyperbolic equations involving a finite set of derivations. *Balk. J. Geom. Its Appl.* **2012**, *17*, 22–33.
17. Karpuz, B.; Ocalan, O.; Ozturk, S. Comparison theorems on the oscillation and asymptotic behaviour of higher-order neutral differential equations. *Glasgow Math. J.* **2010**, *52*, 107–114. [CrossRef]
18. Xing, G.; Li, T.; Zhang, C. Oscillation of higher-order quasi-linear neutral differential equations. *Adv. Differ. Equ.* **2011**, *2011*, 45. [CrossRef]
19. Li, T.; Rogovchenko, Y.V. On the asymptotic behavior of solutions to a class of third-order nonlinear neutral differential equations. *Appl. Math. Lett.* **2020**, *105*, 106293. [CrossRef]
20. Gyori; Ladas, G. *Oscillation Theory of Delay Differential Equations: With Applications*; Oxford University Press: New York, NY, USA, 1991.
21. Ladas, G.; Laskhmikantham, V.; Papadakis, J.S. *Oscillations of Higher-order Retarded Differential Equations G by the Retarded Argument, in Delay and Functional Differential Equations and Their Applications*; Schmitt, K., Ed.; Academic: New York, NY, USA, 1972; pp. 219–231.
22. Hilderbrandt, T.H. *Introduction to the Theory of Integration*; Academic Press: New York, NY, USA, 1963.
23. Agarwal, R.P.; Grace, S.R.; O'Regan, D. *Oscillation Theory for Difference and Functional Differential Equations*; Kluwer Academic, Dordrecht, The Netherlands, 2000.
24. Philos, C. On the existence of nonoscillatory solutions tending to zero at ∞ for differential equations with positive delays. *Arch. Math.* **1981**, *36*, 168–178. [CrossRef]

8

Oscillation Theorems for Nonlinear Differential Equations of Fourth-Order

Osama Moaaz [1,†], Ioannis Dassios [2,*,†], Omar Bazighifan [3,4,†] and Ali Muhib [5,†]

1 Department of Mathematics, Faculty of Science, Mansoura University, Mansoura 35516, Egypt; o_moaaz@mans.edu.eg
2 AMPSAS, University College Dublin, D4 Dublin, Ireland
3 Department of Mathematics, Faculty of Science, Hadhramout University, Hadhramout 50512, Yemen; o.bazighifan@gmail.com
4 Department of Mathematics, Faculty of Education, Seiyun University, Hadhramout 50512, Yemen
5 Department of Mathematics, Faculty of Education, Ibb University, Ibb, Yemen; muhib39@yahoo.com
* Correspondence: ioannis.dassios@ucd.ie
† These authors contributed equally to this work.

Abstract: We study the oscillatory behavior of a class of fourth-order differential equations and establish sufficient conditions for oscillation of a fourth-order differential equation with middle term. Our theorems extend and complement a number of related results reported in the literature. One example is provided to illustrate the main results.

Keywords: deviating argument; fourth order; differential equation; oscillation

---

1. Introduction

In this paper, we are concerned with the oscillation and the asymptotic behavior of solutions of the following two fourth-order differential equations. The nonlinear differential equation:

$$\left(r(t)\left(x'''(t)\right)^{\alpha}\right)' + q(t)\, x^{\beta}(\sigma(t)) = 0, \tag{1}$$

and the differential equation with the middle term of the form:

$$\left(r(t)\left(x'''(t)\right)^{\alpha}\right)' + p(t)\left(x'''(t)\right)^{\alpha} + q(t)\, x^{\beta}(\sigma(t)) = 0, \tag{2}$$

where α and β are quotient of odd positive integers, $r,\ q \in C([t_0,\infty),[0,\infty))$, $r(t) > 0, q(t) > 0, \sigma(t) \in C([t_0,\infty),\mathbb{R})$, $\sigma(t) \leq t$, $\lim_{t\to\infty}\sigma(t) = \infty$. Moreover, we study Equation (1) under the condition

$$\int_{t_0}^{\infty} \frac{1}{r^{1/\alpha}(s)} \mathrm{d}s = \infty \tag{3}$$

and Equation (2) under the conditions $p \in C([t_0,\infty),[0,\infty))$, $r'(t) + p(t) \geq 0$ and

$$\int_{t_0}^{\infty} \left[\frac{1}{r(s)} \exp\left(-\int_{t_0}^{s} \frac{p(u)}{r(u)} du\right)\right]^{1/\alpha} ds = \infty. \tag{4}$$

We aim for a solution of Equation (1) or Equation (2) as a function $x(t) : [t_x,\infty) \to \mathbb{R}, t_x \geq t_0$ such that $x(t)$ and $r(t)\left(x'''(t)\right)^{\alpha}$ are continuously differentiable for all $t \in [t_x,\infty)$ and $\sup\{|x(t)| : t \geq T\} > 0$ for any $T \geq t_x$. We assume that Equation (1) or Equation (2) possesses such a solution. A solution of

Equation (1) or Equation (2) is called oscillatory if it has arbitrarily large zeros on $[t_x, \infty)$. Otherwise, it is called non-oscillatory. Equation (1) or Equation (2) is said to be oscillatory if all its solutions are oscillatory. The equation itself is called oscillatory if all of its solutions are oscillatory.

In mechanical and engineering problems, questions related to the existence of oscillatory and non-oscillatory solutions play an important role. As a result, there has been much activity concerning oscillatory and asymptotic behavior of various classes of differential and difference equations (see, e.g., [1–34], and the references cited therein).

Zhang et al. [30] considered Equation (1) where $\alpha = \beta$ and obtained some oscillation criteria. Baculikova et al. [5] proved that the equation

$$\left[r(t)\left(x^{(n-1)}(t)\right)^{\alpha}\right]' + q(t) f(x(\tau(t))) = 0$$

is oscillatory if the delay differential equations

$$y'(t) + q(t) f\left(\frac{\delta \tau^{n-1}(t)}{(n-1)! r^{\frac{1}{\alpha}}(\tau(t))}\right) f\left(y^{\frac{1}{\alpha}}(\tau(t))\right) = 0$$

is oscillatory and under the assumption that Equation (3) holds, and obtained some comparison theorems.

In [15], El-Nabulsi et al. studied the asymptotic properties of the solutions of equation

$$\left(r(t)\left(x'''(t)\right)^{\alpha}\right)' + q(t) x^{\alpha}(\sigma(t)) = 0, \tag{5}$$

where α is ratios of odd positive integers and under the condition (3).

Elabbasy et al. [14] proved that Equation (2) where $\alpha = \beta = 1$ is oscillatory if

$$\int_{t_0}^{\infty}\left(\rho(s) q(s) \frac{\mu}{2} \tau^2(s) - \frac{1}{4\rho(s) r(s)}\left[\frac{\rho'_+(s)}{\rho(s)} - \frac{p(s)}{r(s)}\right]^2\right) ds = \infty,$$

for some $\mu \in (0,1)$, and

$$\int_{t_0}^{\infty}\left[\vartheta(s) \int_s^{\infty}\left[\frac{1}{r(v)} \int_v^{\infty} q(v)\left(\frac{\tau^2(v)}{v^2}\right) dv\right] dv - \frac{(\vartheta'(s))^2}{4\vartheta(s)}\right] ds = \infty$$

where positive functions $\rho, \vartheta \in C^1([v_0, \infty), \mathbb{R})$ and under the condition in Equation (4).

The motivation in studying this paper improves results in [15]. An example is presented in the last section to illustrate our main results.

We firstly provide the following lemma, which is used as a tool in the proofs our theorems.

Lemma 1 ([10]). *Let $h \in C^n([t_0, \infty), (0, \infty))$. Suppose that $h^{(n)}(t)$ is of a fixed sign, on $[t_0, \infty)$, $h^{(n)}(t)$ not identically zero and that there exists a $t_1 \geq t_0$ such that, for all $t \geq t_1$,*

$$h^{(n-1)}(t) h^{(n)}(t) \leq 0.$$

If we have $\lim_{t\to\infty} h(t) \neq 0$, then there exists $t_\lambda \geq t_0$ such that

$$h(t) \geq \frac{\lambda}{(n-1)!} t^{n-1} \left|h^{(n-1)}(t)\right|,$$

for every $\lambda \in (0,1)$ and $t \geq t_\lambda$.

Lemma 2 ([26]). *If the function x satisfies $x^{(i)}(t) > 0, i = 0, 1, ..., n$, and $x^{(n+1)}(t) < 0$, then*

$$\frac{x(t)}{t^n/n!} \geq \frac{x'(t)}{t^{n-1}/(n-1)!}.$$

Lemma 3 ([27] Lemma 1.2). *Assume that α is a quotient of odd positive integers, $V > 0$ and U are constants. Then,*

$$Uy - Vy^{(\alpha+1)/\alpha} \leq \frac{\alpha^\alpha}{(\alpha+1)^{\alpha+1}} U^{\alpha+1} V^{-\alpha}. \tag{6}$$

2. Oscillation Results

Firstly we establish oscillation results for Equation (1). For convenience, we denote

$$G(t) := \frac{\lambda^\beta}{6^\beta} \frac{q(t)\sigma^{3\beta}(t)}{r^{\beta/\alpha}(\sigma(t))},$$

$$R(t) := \int_t^\infty \left(\frac{1}{r(u)} \int_u^\infty q(s)\,\mathrm{d}s\right)^{1/\alpha} \mathrm{d}u$$

and

$$\widetilde{R}(t) := \mu^{\beta/\alpha} \int_t^\infty \left(\frac{1}{r(u)} \int_u^\infty q(s) \left(\frac{\sigma(s)}{s}\right)^\beta \mathrm{d}s\right)^{1/\alpha} \mathrm{d}u,$$

where $\lambda, \mu \in (0, 1)$.

Lemma 4. *Assume that Equation (3) holds. If x is an eventually positive solution of Equation (1); then, $x' > 0$ and $x''' > 0$.*

Proof. Assume that x is an eventually positive solution of Equation (1); then, $x(t) > 0$ and $x(\sigma(t)) > 0$ for $t \geq t_1$. From Equation (1), we get

$$\left(r(t)\left(x'''(t)\right)^\alpha\right)' = -q(t)\,x^\beta(\sigma(t)) < 0.$$

Hence, $r(t)(x'''(t))^\alpha$ is decreasing of one sign. Thus, we see that

$$x'''(t) > 0.$$

From Equation (1), we obtain

$$\left(r(t)\left(x'''(t)\right)^\alpha\right)' = r'(t) + \alpha r(t)\left(x'''(t)\right)^{\alpha-1} x^{(4)}(t) \leq 0,$$

from which it follows that $x^{(4)}(t) \leq 0$, hence $x'(t) > 0,\ t \geq t_1$. The proof is complete. □

Theorem 1. *Assume that Equation (3) holds. If the differential equation*

$$u'(t) + G(t)\,u^{\beta/\alpha}(\sigma(t)) = 0 \tag{7}$$

is oscillatory for some $\lambda \in (0, 1)$, then Equation (1) is oscillatory.

Proof. Assume to the contrary that Equation (1) has a nonoscillatory solution in $[t_0, \infty)$. Without loss of generality, we only need to be concerned with positive solutions of Equation (1). Then, there exists a $t_1 \geq t_0$ such that $x(t) > 0$ and $x(\sigma(t)) > 0$ for $t \geq t_1$. Let

$$u(t) := r(t)\left(x'''(t)\right)^{\alpha} > 0 \text{ [from Lemma 4]},$$

which with Equation (1) gives

$$u'(t) + q(t)\, x^{\beta}(\sigma(t)) = 0. \tag{8}$$

Since x is positive and increasing, we have $\lim_{t\to\infty} x(t) \neq 0$. Thus, from Lemma 1, we get

$$x^{\beta}(\sigma(t)) \geq \frac{\lambda^{\beta}}{6^{\beta}} \sigma^{3\beta}(t)\left(x'''(\sigma(t))\right)^{\beta}, \tag{9}$$

for all $\lambda \in (0,1)$. By Equations (8) and (9), we see that

$$u'(t) + \frac{\lambda^{\beta}}{6^{\beta}} q(t)\, \sigma^{3\beta}(t)\left(x'''(\sigma(t))\right)^{\beta} \leq 0.$$

Thus, we note that u is positive solution of the differential inequality

$$u'(t) + G(t)\, u^{\beta/\alpha}(\sigma(t)) \leq 0.$$

In view of [25] (Theorem 1), the associated Equation (7) also has a positive solution, which is a contradiction. The theorem is proved. □

Corollary 1. *Assume that $\alpha = \beta$ and Equation (3) holds. If*

$$\liminf_{t\to\infty} \int_{\sigma(t)}^{t} G(s)\, \mathrm{d}s > \frac{1}{\mathrm{e}}, \tag{10}$$

for some $\lambda \in (0,1)$, then Equation (1) is oscillatory.

Proof. It is well-known (see [28] (Theorem 2.1.1)) that Equation (10) implies the oscillation of Equation (11). □

Lemma 5. *Assume that Equation (3) holds and x is an eventually positive solution of Equation (1). If*

$$\int_{t_0}^{\infty} \left(M^{\beta-\alpha} \rho(t)\, q(t) \frac{\sigma^{3\alpha}(t)}{t^{3\alpha}} - \frac{2^{\alpha}}{(\alpha+1)^{\alpha+1}} \frac{r(t)\left(\rho'(t)\right)^{\alpha+1}}{\mu^{\alpha} t^{2\alpha} \rho^{\alpha}(t)} \right) \mathrm{d}s = \infty, \tag{11}$$

for some $\mu \in (0,1)$, then $x'' < 0$.

Proof. Assume to the contrary that $x''(t) > 0$. Using Lemmas 2 and 1, we obtain

$$\frac{x(\sigma(t))}{x(t)} \geq \frac{\sigma^{3}(t)}{t^{3}} \tag{12}$$

and

$$x'(t) \geq \frac{\mu}{2} t^{2} x'''(t), \tag{13}$$

for all $\mu \in (0,1)$ and every sufficiently large t. Now, we define a function ψ by

$$\psi(t) := \rho(t) \frac{r(t)\left(x'''(t)\right)^{\alpha}}{x^{\alpha}(t)} > 0.$$

By differentiating and using Equations (12) and (13), we obtain

$$\psi'(t) \leq \frac{\rho'(t)}{\rho(t)} \omega(t) - \rho(t)\, q(t) \frac{\sigma^{3\alpha}(t)}{t^{3\alpha}} x^{\beta-\alpha}(\sigma(t)) - \frac{\alpha\mu}{2} \frac{t^{2}}{\rho^{1/\alpha}(t)\, r^{1/\alpha}(t)} \psi^{1+1/\alpha}(t). \tag{14}$$

Since $x'(t) > 0$, there exist a $t_2 \geq t_1$ and a constant $M > 0$ such that $x(t) > M$, for all $t \geq t_2$. Using the inequality in Equation (6) with $U = \rho'/\rho$, $V = \alpha\mu t^2/\left(2r^{1/\alpha}(t)\,\rho^{1/\alpha}(t)\right)$ and $y = \psi$, we get

$$\psi'(t) \leq -M^{\beta-\alpha}\rho(t)\,q(t)\,\frac{\sigma^{3\alpha}(t)}{t^{3\alpha}} + \frac{2^{\alpha}}{(\alpha+1)^{\alpha+1}}\frac{r(t)\,(\rho'(t))^{\alpha+1}}{\mu^{\alpha}t^{2\alpha}\rho^{\alpha}(t)}.$$

This implies that

$$\int_{t_1}^{t}\left(M^{\beta-\alpha}\rho(t)\,q(t)\,\frac{\sigma^{3\alpha}(t)}{t^{3\alpha}} - \frac{2^{\alpha}}{(\alpha+1)^{\alpha+1}}\frac{r(t)\,(\rho'(t))^{\alpha+1}}{\mu^{\alpha}t^{2\alpha}\rho^{\alpha}(t)}\right)\mathrm{d}s \leq \psi(t_1),$$

which contradicts Equation (11). The proof is complete. □

Theorem 2. *Assume that $\beta \geq \alpha$ and Equations (3) and (11) hold, for some $\mu \in (0,1)$. If*

$$y''(t) + M^{\beta-\alpha}\widetilde{R}(t)\,y(t) = 0 \tag{15}$$

is oscillatory, then Equation (1) is oscillatory.

Proof. Assume to the contrary that Equation (1) has a nonoscillatory solution in $[t_0,\infty)$. Without loss of generality, we only need to be concerned with positive solutions of Equation (1). Then, there exists a $t_1 \geq t_0$ such that $x(t) > 0$ and $x(\sigma(t)) > 0$ for $t \geq t_1$. From Lemmas 4 and 1, we have that

$$x'(t) > 0,\ x''(t) < 0 \text{ and } x'''(t) > 0, \tag{16}$$

for $t \geq t_2$, where t_2 is sufficiently large. Now, integrating Equation (1) from t to l, we have

$$r(l)\,(x'''(l))^{\alpha} = r(t)\,(x'''(t))^{\alpha} - \int_t^l q(s)\,x^{\beta}(\sigma(s))\,\mathrm{d}s. \tag{17}$$

Using Lemma 3 from [29] with Equation (16), we get

$$\frac{x(\sigma(t))}{x(t)} \geq \lambda\frac{\sigma(t)}{t},$$

for all $\lambda \in (0,1)$, which with Equation (17) gives

$$r(l)\,(x'''(l))^{\alpha} - r(t)\,(x'''(t))^{\alpha} + \lambda^{\beta}\int_t^l q(s)\left(\frac{\sigma(s)}{s}\right)^{\beta}x^{\beta}(s)\,\mathrm{d}s \leq 0.$$

It follows by $x' > 0$ that

$$r(l)\,(x'''(l))^{\alpha} - r(t)\,(x'''(t))^{\alpha} + \lambda^{\beta}x^{\beta}(t)\int_t^l q(s)\left(\frac{\sigma(s)}{s}\right)^{\beta}\mathrm{d}s \leq 0. \tag{18}$$

Taking $l \to \infty$, we have

$$-r(t)\,(x'''(t))^{\alpha} + \lambda^{\beta}x^{\beta}(t)\int_t^{\infty} q(s)\left(\frac{\sigma(s)}{s}\right)^{\beta}\mathrm{d}s \leq 0,$$

that is

$$x'''(t) \geq \frac{\lambda^{\beta/\alpha}}{r^{1/\alpha}(t)}x^{\beta/\alpha}(t)\left(\int_t^{\infty} q(s)\left(\frac{\sigma(s)}{s}\right)^{\beta}\mathrm{d}s\right)^{1/\alpha}.$$

Integrating the above inequality from t to ∞, we obtain

$$-x''(t) \geq \lambda^{\beta/\alpha} x^{\beta/\alpha}(t) \int_t^\infty \left(\frac{1}{r(u)} \int_u^\infty q(s) \left(\frac{\sigma(s)}{s} \right)^\beta ds \right)^{1/\alpha} du,$$

hence

$$x''(t) \leq -\widetilde{R}(t)\, x^{\beta/\alpha}(t). \tag{19}$$

Now, if we define ω by

$$\omega(t) = \frac{x'(t)}{x(t)},$$

then $\omega(t) > 0$ for $t \geq t_1$, and

$$\omega'(t) = \frac{x''(t)}{x(t)} - \left(\frac{x'(t)}{x(t)} \right)^2.$$

By using Equation (19) and definition of $\omega(t)$, we see that

$$\omega'(t) \leq -\widetilde{R}(t) \frac{x^{\beta/\alpha}(t)}{x(t)} - \omega^2(t). \tag{20}$$

Since $x'(t) > 0$, there exists a constant $M > 0$ such that $x(t) \geq M$, for all $t \geq t_2$, where t_2 is sufficiently large. Then, Equation (20) becomes

$$\omega'(t) + \omega^2(t) + M^{\beta-\alpha}\widetilde{R}(t) \leq 0. \tag{21}$$

It is well known (see [3]) that the differential equation in Equation (15) is nonoscillatory if and only if there exists $t_3 > \max\{t_1, t_2\}$ such that Equation (21) holds, which is a contradiction. Theorem is proved. □

Theorem 3. *Assume that $\beta \geq \alpha$ and $\sigma'(t) > 1$ and Equations (3) and (11) hold, for some $\mu \in (0,1)$. If*

$$\left(\frac{1}{\sigma'(t)} y'(t) \right)' + M^{\beta/\alpha - 1} R(t)\, y(t) = 0 \tag{22}$$

is oscillatory, then Equation (1) is oscillatory.

Proof. Proceeding as in the proof of Theorem 2, we obtain Equation (17). Thus, it follows from $\sigma'(t) \geq 0$ and $x'(t) \geq 0$ that

$$r(l)\left(x'''(l)\right)^\alpha - r(t)\left(x'''(t)\right)^\alpha + x^\beta(\sigma(t)) \int_t^l q(s)\, ds \leq 0. \tag{23}$$

Thus, Equation (16) becomes

$$x''(t) \leq -R(t)\, x^{\beta/\alpha}(\sigma(t)). \tag{24}$$

Now, if we define w by

$$w(t) = \frac{x'(t)}{x(\sigma(t))},$$

then $w(t) > 0$ for $t \geq t_1$, and

$$\begin{aligned} w'(t) &= \frac{x''(t)}{x(\sigma(t))} - \frac{x'(t)}{x^2(\sigma(t))}x'(\sigma(t))\sigma'(t) \\ &\le \frac{x''(t)}{x(\sigma(t))} - \sigma'(t)\left(\frac{x'(t)}{x(\sigma(t))}\right)^2. \end{aligned}$$

By using Equation (24) and definition of $w(t)$, we see that

$$w'(t) + M^{\beta/\alpha-1}R(t) + \sigma'(t)w^2(t) \le 0. \tag{25}$$

It is well known (see [3]) that the differential equation in Equation (22) is nonoscillatory if and only if there exists $t_3 > \max\{t_1, t_2\}$ such that Equation (25) holds, which is a contradiction. Theorem is proved. □

There are many results concerning the oscillation of Equations (15) and (22), which include Hille–Nehari types, Philos type, etc. On the basis of [33,34], we have the following corollary, respectively.

Corollary 2. *Assume that $\beta = \alpha$ and Equations (3) and (11) hold, for some $\mu \in (0,1)$. If*

$$\lim_{t\to\infty} \frac{1}{H(t,t_0)} \int_{t_0}^{t} \left(H(t,s)\widetilde{R}(s) - \frac{1}{4}h^2(t,s)\right) ds = \infty$$

or

$$\liminf_{t\to\infty} \int_{t}^{\infty} \widetilde{R}(s)\, ds > \frac{1}{4}, \tag{26}$$

then Equation (1) is oscillatory.

Corollary 3. *Assume that $\beta = \alpha$ and Equations (3) and (11) hold, for some $\mu \in (0,1)$. If there exists a constant $\kappa \in (0,1/4]$ such that*

$$t^2\widetilde{R}(s) \ge \kappa$$

and

$$\limsup_{t\to\infty} \left(t^{\kappa-1}\int_{t_0}^{t} s^{2-\kappa}\widetilde{R}(s)\, ds + t^{1-\widetilde{\kappa}}\int_{t}^{\infty} s^{\widetilde{\kappa}}\widetilde{R}(s)\, ds\right) > 1,$$

where $\widetilde{\kappa} = \frac{1}{2}\left(1-\sqrt{1-4\kappa}\right)$, then Equation (1) is oscillatory.

We will now define the following notation:

$$\eta_{t_0}(t) := \exp\left(\int_{t_0}^{t} \frac{p(u)}{r(u)} du\right)$$

and

$$\widehat{R}(t) := \mu_1^{\beta/\alpha} \int_{t}^{\infty} \left(\frac{1}{r(u)\eta_{t_0}(t)} \int_{u}^{\infty} \eta_{t_0}(t) q(s) \left(\frac{\sigma(s)}{s}\right)^{\beta} ds\right)^{1/\alpha} du,$$

where $\mu_1 \in (0,1)$. We establish oscillation results for Equation (2) by converting into the form of Equation (1). It is not difficult to see that

$$\begin{aligned} \frac{1}{\eta_{t_0}(t)}\frac{d}{dt}\left(\mu(t)r(t)(x'''(t))^{\alpha}\right) &= \frac{1}{\eta_{t_0}(t)}\left[\eta_{t_0}(t)\left(r(t)(x'''(t))^{\alpha}\right)' + \eta'_{t_0}(t)r(t)(x'''(t))^{\alpha}\right] \\ &= \left(r(t)(x'''(t))^{\alpha}\right)' + \frac{\eta'_{t_0}(t)}{\eta_{t_0}(t)}r(t)(x'''(t))^{\alpha} \\ &= \left(r(t)(x'''(t))^{\alpha}\right)' + p(t)(x'''(t))^{\alpha}, \end{aligned}$$

which with Equation (2) gives

$$\left(\eta_{t_0}(t)\, r(t)\left(x'''(t)\right)^{\alpha}\right)' + \eta_{t_0}(t)\, q(t)\, x^{\beta}(\sigma(t)) = 0.$$

Corollary 4. *Assume that $\alpha = \beta$ and Equation (4) holds. If*

$$\liminf_{t\to\infty} \int_{\sigma(t)}^{t} \widehat{G}(s)\, \mathrm{d}s > \frac{1}{\mathrm{e}},$$

for some $\lambda \in (0,1)$, where

$$\widehat{G}(t) := \frac{\lambda^{\beta}}{6^{\beta}} \frac{\eta_{t_0}(t)\, q(t)\, \sigma^{3\beta}(t)}{\eta_{t_0}^{\beta/\alpha}(\sigma(t))\, r^{\beta/\alpha}(\sigma(t))},$$

then Equation (2) is oscillatory.

Corollary 5. *Assume that $\beta = \alpha$, Equation (4) and*

$$\int_{t_0}^{\infty} \left(M^{\beta-\alpha} \rho(t)\, \eta_{t_0}(t)\, q(t) \frac{\sigma^{3\alpha}(t)}{t^{3\alpha}} - \frac{2^{\alpha}}{(\alpha+1)^{\alpha+1}} \frac{r(t)\, \eta_{t_0}(t) \left(\rho'(t)\right)^{\alpha+1}}{\mu^{\alpha} t^{2\alpha} \rho^{\alpha}(t)} \right) \mathrm{d}s = \infty, \qquad (27)$$

hold, for some $\mu \in (0,1)$. If

$$\lim_{t\to\infty} \frac{1}{H(t,t_0)} \int_{t_0}^{t} \left(H(t,s)\, \widehat{R}(s) - \frac{1}{4} h^2(t,s) \right) \mathrm{d}s = \infty$$

or

$$\liminf_{t\to\infty} \int_{t}^{\infty} \widehat{R}(s)\, \mathrm{d}s > \frac{1}{4},$$

then Equation (2) is oscillatory.

Corollary 6. *Assume that $\beta = \alpha$ and Equations (4) and (27) hold, for some $\mu \in (0,1)$. If there exists a constant $\kappa \in (0, 1/4]$ such that*

$$t^2 \widehat{R}(s) \geq \kappa$$

and

$$\limsup_{t\to\infty} \left(t^{\kappa-1} \int_{t_0}^{t} s^{2-\kappa} \widehat{R}(s)\, \mathrm{d}s + t^{1-\widetilde{\kappa}} \int_{t}^{\infty} s^{\widetilde{\kappa}} \widehat{R}(s)\, \mathrm{d}s \right) > 1,$$

where $\widetilde{\kappa}$ is defined as Corollary 3, then Equation (2) is oscillatory.

3. Example

In this section, we give the following example to illustrate our main results.

Example 1. *For $t \geq 1$, consider a differential equation:*

$$\left(t^3 \left(x'''(t)\right)^3\right)' + \frac{q_0}{t^7} x^3(\gamma t) = 0, \qquad (28)$$

where $\gamma \in (0,1]$ and $q_0 > 0$. We note that $\alpha = \beta = 3$, $r(t) = t^3$, $\sigma(t) = \gamma t$ and $q(t) = q_0/t^7$. Thus, it is easy to verify that

$$G(t) = \frac{\lambda^3 \gamma^6}{6^3} \frac{q_0}{t} \text{ and } \widetilde{R}(t) = \lambda \left(\frac{q_0}{6}\right)^{1/3} \gamma \frac{1}{2t^2}.$$

By using Corollary 1, we see that Equation (28) is oscillatory if

$$q_0 > \frac{6^3}{e\left(\ln \frac{1}{\gamma}\right)\gamma^6}. \tag{29}$$

This result can be obtained from [5].

For using Corollary 2, we see that the conditions in Equations (11) and (26) become

$$q_0 > \left(\frac{3^4}{2}\right)\frac{1}{\gamma^9}$$

and

$$q_0 > 6\left(\frac{1}{4\gamma}\right)^3$$

respectively. Thus, Equation (28) is oscillatory if

$$q_0 > \max\left\{\left(\frac{3^4}{2}\right)\frac{1}{\gamma^9}, 6\left(\frac{1}{4\gamma}\right)^3\right\} = \left(\frac{3^4}{2}\right)\frac{1}{\gamma^9}. \tag{30}$$

Remark 1. *By applying equation Equation (30) on the work in [15] where $\gamma = 1/2$, we find*

$$q_0 > 20736.$$

Therefore, our result improves results [15].

4. Conclusions

In this article, we study the oscillatory behavior of a class of non-linear fourth-order differential equations and establish sufficient conditions for oscillation of a fourth-order differential equation with middle term. The outcome of this article extends a number of related results reported in the literature.

Author Contributions: O.M., O.B. and A.M.: Writing original draft, and writing review and editing. I.D.: Formal analysis, writing review and editing, funding and supervision. All authors have read and agreed to the published version of the manuscript.

Acknowledgments: The authors thank the reviewers for for their useful comments, which led to the improvement of the content of the paper.

References

1. Agarwal, R.; Grace, S.; O'Regan, D. *Oscillation Theory for Difference and Functional Differential Equations*; Kluwer Acad. Publ.: Dordrecht, The Netherlands, 2000.
2. Agarwal, R.P.; Brzdek, J.; Chudziak, J. Stability problem for the composite type functional equations. *Expo. Math.* **2018**, *36*, 178–196. [CrossRef]
3. Agarwal, R.; Shieh, S.L.; Yeh, C.C. Oscillation criteria for second order retard differential equations. *Math. Comput. Model.* **1997**, *26*, 1–11. [CrossRef]
4. Agarwal, R.P.; Zhang, C.; Li, T. Some remarks on oscillation of second order neutral differential equations. *Appl. Math. Compt.* **2016**, *274*, 178–181 [CrossRef]
5. Baculikova, B.; Dzurina, J.; Graef, J.R. On the oscillation of higher-order delay differential equations. *Math. Slovaca* **2012**, *187*, 387–400. [CrossRef]
6. Bahyrycz, A.; Brzdek, J. A note on d'Alembert's functional equation on a restricted domain. *Aequationes Math.* **2014**, *88*, 169–173. [CrossRef]
7. El-hady, E.; Brzdek, J.; Nassar, H. On the structure and solutions of functional equations arising from queueing models. *Aequationes Math.* **2017**, *91*, 445–477. [CrossRef]

8. Bazighifan, O.; Cesarano, C. Some New Oscillation Criteria for Second-Order Neutral Differential Equations with Delayed Arguments. *Mathematics* **2019**, *7*, 619. [CrossRef]
9. Bazighifan, O.; Elabbasy, E.M.; Moaaz, O. Oscillation of higher-order differential equations with distributed delay. *J. Inequal. Appl.* **2019**, *55*, 1–9. [CrossRef]
10. Bazighifan, O.; Cesarano, C. A Philos-Type Oscillation Criteria for Fourth-Order Neutral Differential Equations. *Symmetry* **2020**, *12*, 379. [CrossRef]
11. Cesarano, C.; Pinelas, S.; Al-Showaikh, F.; Bazighifan, O. Asymptotic Properties of Solutions of Fourth-Order Delay Differential Equations. *Symmetry* **2019**, *11*, 628. [CrossRef]
12. Cesarano, C.; Bazighifan, O. Oscillation of fourth-order functional differential equations with distributed delay. *Axioms* **2019**, *8*, 61. [CrossRef]
13. Cesarano, C.; Bazighifan, O. Qualitative behavior of solutions of second order differential equations. *Symmetry* **2019**, *11*, 777. [CrossRef]
14. Elabbasy, E.M.; Thandpani, E.; Moaaz, O.; Bazighifan, O. Oscillation of solutions to fourth-order delay differential equations with middle term. *Open J. Math. Sci.* **2019**, *3*, 191–197. [CrossRef]
15. El-Nabulsi, R.; Moaaz, O.; Bazighifan, O. New Results for Oscillatory Behavior of Fourth-Order Differential Equations. *Symmetry* **2020**, *12*, 136. [CrossRef]
16. Grace, S.; Lalli, B. Oscillation theorems for nth order nonlinear differential equations with deviating arguments. *Proc. Am. Math. Soc.* **1984**, *90*, 65–70. [CrossRef]
17. Grace, S.; Agarwal, R.; Graef, J. Oscillation theorems for fourth order functional differential equations. *J. Appl. Math. Comput.* **2009**, *30*, 75–88. [CrossRef]
18. Gyori, I.; Ladas, G. *Oscillation Theory of Delay Differential Equations with Applications*; Clarendon Press: Oxford, UK, 1991.
19. Liu, S.; Zhang, Q.; Yu, Y. Oscillation of even-order half-linear functional differential equations with damping. *Comput. Math. Appl.* **2011**, *61*, 2191–2196. [CrossRef]
20. Li, T.; Baculikova, B.; Dzurina, J.; Zhang, C. Oscillation of fourth order neutral differential equations with p-Laplacian like operators. *Bound. Value Probl.* **2014**, *56*, 41–58. [CrossRef]
21. Moaaz, O.; Furuichi, S.; Muhib, A. New Comparison Theorems for the Nth Order Neutral Differential Equations with Delay Inequalities. *Mathematics* **2020**, *8*, 454. [CrossRef]
22. Moaaz, O.; Elabbasy, E.M.; Bazighifan, O. On the asymptotic behavior of fourth-order functional differential equations. *Adv. Differ. Equ.* **2017**, *261*, 1–13. [CrossRef]
23. Moaaz, O.; Elabbasy, E.M.; Muhib, A. Oscillation criteria for even-order neutral differential equations with distributed deviating arguments. *Adv. Differ. Equ.* **2019**, *2019*, 297. [CrossRef]
24. Moaaz, O.; Dassios, I.; Bazighifan, O. Oscillation Criteria of Higher-order Neutral Differential Equations with Several Deviating Arguments. *Mathematics* **2020**, *8*, 412. [CrossRef]
25. Philos, C. On the existence of nonoscillatory solutions tending to zero at ∞ for differential equations with positive delay. *Arch. Math. (Basel)* **1981**, *36*, 168–178. [CrossRef]
26. Kiguradze, I.T.; Chanturiya, T.A. *Asymptotic Properties of Solutions of Nonautonomous Ordinary Differential Equations*; Kluwer Acad. Publ.: Dordrecht, The Netherlands, 1993.
27. Moaaz, O. New criteria for oscillation of nonlinear neutral differential equations. *Adv. Differ. Equ.* **2019**, *2019*, 484. [CrossRef]
28. Ladde, G.S.; Lakshmikantham, V.; Zhang, B.G. *Oscillation Theory of Differential Equations with Deviating Arguments*; Marcel Dekker: New York, NY, USA, 1987.
29. Baculikova, B.; Dzurina, J. Oscillation theorems for second-order nonlinear neutral differential equations. *Comput. Math. Appl.* **2011**, *62*, 4472–4478. [CrossRef]
30. Zhang, C.; Li, T.; Saker, S. Oscillation of fourth-order delay differential equations. *J. Math. Sci.* **2014**, *201*, 296–308. [CrossRef]
31. Zhang, C.; Agarwal, R.P.; Bohner, M.; Li, T. New results for oscillatory behavior of even-order half-linear delay differential equations. *Appl. Math. Lett.* **2013**, *26*, 179–183 [CrossRef]
32. Zhang, C.; Li, T.; Suna, B.; Thandapani, E. On the oscillation of higher-order half-linear delay differential equations. *Appl. Math. Lett.* **2011**, *24*, 1618–1621. [CrossRef]
33. Philos, C.G. Oscillation theorems for linear differential equation of second order. *Arch. Math.* **1989**, *53*, 483–492. [CrossRef]
34. Hille, E. Non-oscillation theorems. *Trans. Am. Math. Soc.* **1948**, *64*, 234–253. [CrossRef]

9

Multi-Wavelets Galerkin Method for Solving the System of Volterra Integral Equations

Hoang Viet Long [1,2], Haifa Bin Jebreen [3,*] and Stefania Tomasiello [4]

1 Division of Computational Mathematics and Engineering, Institute for Computational Science, Ton Duc Thang University, Ho Chi Minh City 70000, Vietnam; hoangvietlong@tdtu.edu.vn
2 Faculty of Mathematics and Statistics, Ton Duc Thang University, Ho Chi Minh City 70000, Vietnam
3 Department of Mathematics, College of Science, King Saud University, P.O. Box 2455, Riyadh 11451, Saudi Arabia
4 Institute of Computer Science, University of Tartu, 50090 Tartu, Estonia; stefania.tomasiello@ut.ee
* Correspondence: hjebreen@ksu.edu.sa

Abstract: In this work, an efficient algorithm is proposed for solving the system of Volterra integral equations based on wavelet Galerkin method. This problem is reduced to a set of algebraic equations using the operational matrix of integration and wavelet transform matrix. For linear type, the computational effort decreases by thresholding. The convergence analysis of the proposed scheme has been investigated and it is shown that its convergence is of order $O(2^{-Jr})$, where J is the refinement level and r is the multiplicity of multi-wavelets. Several numerical tests are provided to illustrate the ability and efficiency of the method.

Keywords: Volterra integral equations; operational matrix of integration; multi-wavelets

---

1. Introduction

In this paper, we study and construct a novel numerical algorithm for the system of Volterra integral equations of the second kind

$$\mathbf{u}(x) = \mathbf{f}(x) + \int_0^x \mathbf{g}(x,t,\mathbf{u}(t))dt, \quad x \in \Omega := [0,1], \tag{1}$$

where $\mathbf{f} : \Omega \to \mathbb{R}^n$ $(n \in \mathbb{N})$ is a given real-valued continuous function, $\mathbf{u} : \Omega \to \mathbb{R}^n$ is the unknown function that will be determined and the function $\mathbf{g} : S \to \mathbb{R}^n$ with $S = \{(x,t) : x,t \in \Omega\}$ is a given linear or nonlinear function of $\mathbf{u}$ which satisfies the following Lipschitz condition with respect to the third variable: for all $x,t \in [0,1]$ and for all $\mathbf{u}_1, \mathbf{u}_2 \in \mathbb{R}^n$,

$$|\mathbf{g}(x,t,\mathbf{u}_1(t)) - \mathbf{g}(x,t,\mathbf{u}_2(t))| \leq A|\mathbf{u}_1 - \mathbf{u}_2|. \tag{2}$$

Therefore, the functions $\mathbf{f}$ and $\mathbf{g}$ are considered so that the Equation (1) has a unique solution.

Equation (1) is the general form of second-order Volterra integral equation and appears in scientific applications in chemistry, engineering, mathematics, and physics [1–4]. Numerical and analytical solutions of linear and nonlinear Volterra integral equations have been investigated in many papers. A useful method to solve such equations is the Adomian decomposition method. This method was used to investigate the existence and uniqueness of solutions of this type of equation [5,6]. One of the best paper which utilizes the multi-wavelets for solving integro-differential equations was presented by Saray [7]. In [7], an efficient algorithm was proposed for solving the Volterra integro-differential equation. This method outperforming former approaches. Golbabai et al. [8] developed a general method based on radial basis function networks to solve the system of Volterra integral equations. The modified homotopy perturbation method for solving this type of equation has been proposed by Aminikhah et al. [5,9]. Kılıçman et al. [10] used Simpson's 3/8 rule to solve this equation. Aguilar and Brunner used collocation techniques based on spline polynomials [11]. The umbral calculus and the Laplace transform methods were used as solution approaches as well [12].

Wavelets and specially multi-wavelets Galerkin method represent an efficient way to solve a variety of equations, including ordinary differential equations (ODEs), partial differential equations (PDEs), and integral equations [7,13,14]. Due to the discrete-time characterization of wavelet coefficient decay, the sparse form of the coefficients matrices arises. This property is very useful to reduce the computational cost. In this work, we aim to solve the system of the Volterra integral equation using Alpert's multi-wavelets by exploiting the above-mentioned property. Some results are formally proved and supported by numerical experiments.

The paper is structured as follows. A brief introduction of the Alpert's multi-wavelets is provided in Section 2. In Section 3, the wavelet Galerkin method is used to approximate the solution of the problem, and the convergence analysis is investigated. Some numerical experiments are performed to illustrate the efficiency and accuracy of the proposed method.

2. Alpert's Multi-Wavelets and Multiresolution Analysis

Alpert et al. [13,15] introduced a class of multi-wavelets for L^2, which are indexed by a parameter $r \geq 0$ and built via Lagrange polynomials of degree less than r. These multi-wavelets are piecewise polynomials that are locally supported and orthonormal. The multiresolution analysis (MRA) framework, introduced and developed by Mallat [16] and Meyer [17], is useful to construct these bases.

According to MRA, a set of primal scaling functions $\{\phi^0_{0,0}, \dots, \phi^{r-1}_{0,0}\}$ is introduced for primal subspace $V^r_0 \in L^2[0,1]$. By translation and dilation of primal scaling functions $\{\phi^k\}, k = 0, \dots, r-1$, we determine a space V^r_j,

$$V^r_j = Span\{\phi^k_{j,b} := \mathcal{D}_{2^j}\mathcal{T}_b\phi^k, b \in \mathcal{B}_j, k = 0, 1, \dots, r-1\},$$

where $\mathcal{D}_a$ and $\mathcal{T}_b$ are the dilation and translation operators, respectively such that for a given function h, $\mathcal{D}_a h(x) = a^{\frac{1}{2}} h(ax)$ and $\mathcal{T}_b h(x) = h(x-b)$, also $\mathcal{B}_j := \{0, 1, \dots, 2^j - 1\}$ for $j \in \mathbb{Z}^+ \cup \{0\}$. Therefore, $\phi^k_{j,b}$ is a polynomial of degree less than k which is restricted to $I_{j,b} = [x_{j,b}, x_{j,b+1}]$ where $x_{j,b} := 2^{-j}b$ and $\Omega := [0,1] = \cup_{b \in \mathcal{B}_j} I_{j,b}$.

For a fixed integer $J \geq 0$, the orthogonal projection $\mathcal{P}^r_J h$ of $h \in L^2[0,1]$ onto V^r_J is determined by

$$h \approx \mathcal{P}^r_J(h) = \sum_{b \in \mathcal{B}_J} \sum_{k=0}^{r-1} \langle h, \phi^k_{J,b} \rangle \phi^k_{J,b}. \tag{3}$$

The coefficients $h^k_{J,b} = \langle f, \phi^k_{J,b} \rangle$ are determined by $h^k_{J,b} = \int_{I_{J,b}} h(x)\phi^k_{J,b}(x)dx$ [18–20]. To avoid computing integrals, the Gauss–Legendre Quadrature are applied as follows

$$h^k_{J,b} \approx 2^{-J/2}\sqrt{\frac{\omega_k}{2}} h\left(2^{-J}(\frac{\tau_k + 1}{2} + b)\right), \quad k = 0, ..., r-1,\ b \in \mathcal{B}_J, \tag{4}$$

where ω_k and τ_k are the Gauss-Legendre Quadrature weights and the roots of Legendre polynomial of order r, respectively which are introduced in [21–23]. The projection $\mathcal{P}^r_J h$ converges to h if the function $h \in C^r(\Omega)$ (r times continuously differentiable) [15]. $\mathcal{P}^r_J h$ approximates h with mean error bounded as follows

$$\|\mathcal{P}^r_J(h) - h\| \leq 2^{-Jr} \frac{2}{4^r r!} \sup_{x \in \Omega} |h^{(r)}(x)|. \tag{5}$$

Assume that $\Phi_J^r := [\Phi_{r,J,0}, \ldots, \Phi_{r,J,2^J-1}]^T$, where $\Phi_{r,J,b} := [\phi_{J,b}^0, \ldots, \phi_{J,b}^{r-1}]$. Φ_J^r refers to the function vector called a multi-scaling function. In fact, Φ_J^r is a function vector which includes the scaling function in the space V_J^r. By this definition, one can rewrite (3) viz, $\mathcal{P}_J^r(h) = H^T \Psi_J^r$ where H is a vector with entries $H_{br+k+1} := h_{J,b}^k$ and has dimension $N := r2^J$.

This construction of multi-wavelets for $L^2(\Omega)$ can be extended to the two–dimensional space including $L^2(\Omega)^2$. Let us introduce the space $V_J^{r,2} := V_J^r \times V_J^r$ which is spanned by orthonormal bases $\{\phi_{J,b}^k \phi_{J,b'}^{k'} : b, b' \in \mathcal{B}_J, \quad k, k' = 0, 1, \ldots, r-1\}$. Therefore, any function $h \in L^2(\Omega)^2$ can be projected onto the $V_J^{r,2}$ by the projection $\mathcal{P}_J^r$ via,

$$h \approx \mathcal{P}_J^r h = \sum_{b \in \mathcal{B}_j} \sum_{k'=0}^{r-1} \sum_{k=0}^{r-1} \sum_{b' \in \mathcal{B}_j} H_{rb+(k+1), rb'+(k'+1)} \phi_{J,b}^k(x) \phi_{J,b'}^{k'}(y) = \Phi_J^{r\,T}(x) H \Phi_J^r(y), \tag{6}$$

where H is an $(N \times N)$ matrix whose elements are obtained by

$$H_{rb+(k+1), rb'+(k'+1)} \approx 2^{-J} \sqrt{\frac{\omega_k}{2}} \sqrt{\frac{\omega_{k'}}{2}} h\left(2^{-J}(\hat{\tau}_k + b), 2^{-J}(\hat{\tau}_{k'} + b')\right), \tag{7}$$

where $\hat{\tau}_k = (\tau_k + 1)/2$. By the following theorem, it is possible to bound the error for such projection, if the function h is sufficiently smooth.

Theorem 1 ([15]). *Suppose that the function $h : [0,1]^2 \to \mathbb{R}^2$ has continuous partial derivatives of order r and mixed partial derivative of order $2r$. Then*

$$\|\mathcal{P}_J^r h - h\| \leq \mathcal{M}_{\max} \frac{2^{1-rJ}}{4^r r!} \left(2 + \frac{2^{1-Jr}}{4^r r!}\right), \tag{8}$$

where

$$\mathcal{M}_{\max} = \max\left\{ \sup_{\xi \in [0,1)} \left|\frac{\partial^r}{\partial x^r} h(\xi, y)\right|, \sup_{\eta \in [0,1)} \left|\frac{\partial^r}{\partial y^r} h(x, \eta)\right|, \sup_{\xi', \eta' \in [0,1)} \left|\frac{\partial^{2r}}{\partial x^r \partial y^r} h(\xi', \eta')\right| \right\}.$$

As the subspaces V_j^r are nested, there exist complementary orthogonal subspaces W_j^r such that

$$V_{j+1}^r = V_j^r \bigoplus W_j^r, \qquad j \in \mathbb{Z} \cup \{0\}, \tag{9}$$

here and in the following $\bigoplus$ denotes orthogonal sums. There is a family of other bases such that the dilations and translations of these bases span the complementary subspaces W_j^r, namely,

$$W_j^r = Span\{\psi_{j,b}^k := \mathcal{D}_{2^j} \mathcal{T}_b \psi^k, b \in \mathcal{B}_j, k = 0, 1, \ldots, r-1\}.$$

The functions $\psi_{j,b}^k$ are called multi-wavelets. Due to (9), the multi-scale decomposition can be inductively found, $V_J^r = V_0^r \oplus (\oplus_{j=0}^{J-1} W_j^r)$. This decomposition gives rise to the multi-scale projection operator $\mathcal{M}_J^r$ that maps $L^2(\Omega)$ onto V_J^r via

$$h \approx \mathcal{M}_J^r(h) = (\mathcal{P}_0^r + \sum_{j=0}^{J-1} \mathcal{Q}_j^r)(h), \tag{10}$$

where $\mathcal{Q}_j^r$ is the orthonormal projection operator that maps $L^2(\Omega)$ onto W_j^r. In fact, by using multi-scale projection operator, any function $h \in L^2(\Omega)$ can be approximated by multi-wavelets of higher levels $W_j^r, j = 0, 1, \dots, J-1$ and the multi-scaling functions of the coarse space V_0^r viz,

$$h \approx \mathcal{M}_J^r(h) = \sum_{k=0}^{r-1} h_{0,0}^k \phi_{0,0}^k + \sum_{j=0}^{J-1} \sum_{b \in \mathcal{B}_j} \sum_{k=0}^{r-1} \tilde{h}_{j,b}^k \psi_{j,b}^k, \tag{11}$$

where

$$h_{0,0}^k := \langle h, \phi_{0,0}^k \rangle, \quad \tilde{h}_{j,b}^k := \langle h, \psi_{j,b}^k \rangle. \tag{12}$$

Note that the single-scale coefficients $h_{0,0}^k$ can be determined by (4). However, for evaluating the multi-wavelets coefficients $\tilde{h}_{j,b}^k$ of higher levels, in many cases, they have to be calculated numerically. To avoid such numerical computations, the wavelet transform matrix T_J can be applied, as introduced in [14,24]. This matrix is useful to find the multi-wavelets by using the scaling functions

$$\Psi_J^r = T_J \Phi_J^r, \tag{13}$$

where $\Psi_J^r := [\Phi_{r,0,b}, \Psi_{r,0,b}, \Psi_{r,1,b}, \dots, \Psi_{r,J-1,b}]^T$ and $\Psi_{r,j,b} := [\psi_{j,b}^0, \dots, \psi_{j,b}^{r-1}]$, $b \in \mathcal{B}_j$. Using the vector function Ψ_J^r and (11), we can write

$$h \approx \mathcal{M}_J^r(h) = \tilde{H}_J^T \Psi_J^r, \tag{14}$$

where $\tilde{H}_J$ is an N-dimensional vector with entries $h_{0,0}^k$ and $\tilde{h}_{j,b}^k$ for $b \in \mathcal{B}_j$, $j = 0, \dots, J-1$ and $k = 0, \dots, r-1$. Besides, it is obvious that $\tilde{H}_J = {T_J^{-1}}^T H_J$.

Thresholding

Alpert's multi-wavelets provide vanishing moments of order $N_\psi^k = k + r - 1$ for $k = 0, 1, \dots, r-1$, i.e.,

$$\mathcal{N}_p^k = \int_{-\infty}^{\infty} x^p \psi_{0,0}^k(x) dx, \qquad 0 \le p < N_\psi^k, \text{ and } k = 0, 1, \dots, r-1. \tag{15}$$

Furthermore, Alpert's multi-wavelets are uniformly bounded concerning to L_∞ and L_1, i.e.,

$$\|\psi_{J,b}^k\|_{L_\infty(\Omega)} \lesssim 1, \quad \|\psi_{J,b}^k\|_{L_1(\Omega)} \lesssim 1. \tag{16}$$

The vanishing moments and normalization (16) imply that the detail coefficients $\tilde{h}_{J,b}^k$ become small when the underlying function is locally smooth. Therefore it is possible to obtain [25]

$$\tilde{h}_{J,b}^k = |\langle h, \psi_{J,b}^k \rangle| \le \inf_{P \in \Pi_{N_\psi^k}} |\langle h - P, \psi_{J,b}^k \rangle| \lesssim 2^{-JN_\psi^k} \|h\|_{W^{1,N_\psi^k}(\Omega)}. \tag{17}$$

So the detail coefficients decay at the rate of $2^{-JN_\psi^k}$ and in the regions where the function is smooth, most of the detail coefficients may be discarded when the refinement level J increases. This gives rise to thresholding. The thresholding operator $\mathcal{T}_{D_\varepsilon}$ is introduced by

$$\mathcal{T}_{D_\varepsilon}(\tilde{H}_J) = \mathcal{H}_J, \tag{18}$$

where $D_\varepsilon := \{(J, b, k) : |\tilde{h}_{J,b}^k| > \varepsilon\}$ and the elements of $\mathcal{H}_J$ are defined by

$$\bar{h}_{j,b}^k := \begin{cases} \tilde{h}_{j,b}^k, & (j, b, k) \in D_\varepsilon, \\ 0, & \text{else}, \end{cases} \qquad b \in \mathcal{B}_j,\ j = 0, \dots, J-1, k = 0, \dots, r-1. \tag{19}$$

Note that the thresholding affects only the detail coefficients while the coarse scale coefficients remain unchanged.

The approximation error due to the thresholding can be estimated similarly to the classical wavelets. Let $\mathcal{A}_{D_\varepsilon}$ be the approximation operator $\mathcal{A}_{D_\varepsilon} := \mathcal{M}_J^{r-1}\mathcal{T}_{D_\varepsilon}\mathcal{M}_J^r$. The approximation error due to the thresholding can be bounded as stated by the following proposition [25].

Proposition 1. *(Approximation error). Let Ω be bounded and $\varepsilon_j = \bar{a}^{j-J}\varepsilon$ with $\bar{a} > 1$. Then the approximation error concerning to the set of significant details D_ε is uniformly bounded concerning to $L^q(\Omega)$, $q \in [1,\infty]$, i.e.,*

$$\|\mathcal{P}_J^r h - \mathcal{P}_{J,D_\varepsilon}^r h\|_{L^q(\Omega)} \le C_{thr}\varepsilon, \tag{20}$$

for some constant $C_{thr} > 0$ independent of J,ε. Here $\mathcal{P}_J^r h$ and $\mathcal{P}_{J,D_\varepsilon}^r h$ are the projections according to (11) corresponding to the coefficients $\tilde{H}_J$ and $\mathcal{A}_{D_\varepsilon}\tilde{H}_J$.

3. Multi-Wavelets Galerkin Method

In this section, we use the wavelet Galerkin method to solve the system of the Volterra integral Equation (1). To this end, we will apply the interpolation property of scaling functions to reach an efficient algorithm. Assume that the solution $\mathbf{u}(x)$ of Equation (1) can be expanded using multi-scale projection operator $\mathcal{M}_J^r$ based on multi-wavelets as follows

$$\mathbf{u}(x) \approx \mathcal{M}_J^r(\mathbf{u})(x) = (\mathcal{P}_0^r + \sum_{j=0}^{J-1} \mathcal{Q}_j^r)(\mathbf{u})(x) = \tilde{\mathbf{U}}^T \otimes \Psi_J^r(x), \tag{21}$$

where $\otimes$ is the Kronecker product and $\tilde{\mathbf{U}} = (\tilde{U}_1^T, \ldots, \tilde{U}_n^T)^T$ is a $(1 \times nr2^J)$ vector whose elements are n unknown sub-vectors $\tilde{U}_i$ with a dimension of $(r2^J \times 1)$ such that

$$u_i(x) := \tilde{U}_i^T \Psi_J^r(x), \quad i = 1,2,\ldots,n.$$

One can imagine two types of equations, linear and nonlinear. For the linear type, the i-th component of the vector function $\mathbf{g}(x,t,\mathbf{u}(t))$ has the form

$$\mathbf{g}_j(x,t,\mathbf{u}(t)) := \sum_{i=1}^{n} a_{ji}(x,t)u_i(t), \quad j = 1,2,\ldots,n, \tag{22}$$

and it can be approximated by multi-scale operator, i.e.,

$$\begin{aligned}\sum_{i=1}^{n} a_{ji}(x,t)u_i(t) \approx \sum_{i=1}^{n} \mathcal{P}_J^r(a_{ji}u_i)(x,t) &= \sum_{i=1}^{n} \Phi_J^{r\,T}(x)A_{ji}^T\Phi_J^r(t)\\ &\sum_{i=1}^{n} \Psi_J^{r\,T}(x)T_J A_{ji}^T T_J^{-1}\Psi_J^r(t), \quad j = 1,2,\ldots,n,\end{aligned}$$

where A_{ji} $(j,i = 1:n)$ are $(r2^J \times r2^J)$ matrices. Integrating from 0 to x, we get

$$\begin{aligned}\int_0^x \mathbf{g}_j(x,t,\mathbf{u}(t))dt &= \sum_{i=1}^{n} \underbrace{\Psi_J^{r\,T}(x)T_J A_{ji}^T T_J^{-1} I_\psi \Psi_J^r(x)}_{p_j(x)}\\ &= \sum_{i=1}^{n} P_i^T T_J^{-1}\Psi_J^r(x) = \sum_{i=1}^{n} \tilde{U}_i^T A_{ji} T_J^{-1}\Psi_J^r(t),\end{aligned} \tag{23}$$

where $\mathcal{A}_{ji}$ $(j,i = 1:n)$ are $(r2^J \times r2^J)$ matrices and the rest are $(r2^J \times 1)$ vectors. But if the j-th component of vector function $\mathbf{g}(x,t,\mathbf{u}(t))$ is nonlinear, one can consider the following expansion

$$\begin{aligned}\int_0^x g_j(x,t,\mathbf{u}(t))dt &\approx \mathcal{P}_J^r\left(\int_0^x g_j\left(x,t,\mathcal{M}_J^r(\mathbf{u})(t)\right)dt\right)\\ G_j^T\Phi_J^r(x) &= G_j^T T_J^{-1}\Psi_J^r(x), \quad j=1,2,\ldots,n,\end{aligned} \tag{24}$$

where G_j is a $(r2^J \times 1)$ vector whose elements are nonlinear algebraic equations. In view of the Equations (23) and (24), and using operational matrix I_ψ of integration for multi-wavelets introduced in [7,14,15], one can write

$$\int_0^x \mathbf{g}(x,t,\mathbf{u}(t))dt \approx \begin{cases} \tilde{\mathbf{U}}^T\Gamma^T \otimes T_J^{-1}\Psi_J^r(x), & \text{linear,}\\ \mathbf{G}^T \otimes T_J^{-1}\Psi_J^r(x), & \text{nonlinear,}\end{cases} \tag{25}$$

with $\Gamma := (\mathcal{A})_{ji}$, $(j,i = 1:n)$ and $\mathbf{G} := (G_1^T, G_2^T, \ldots, G_n^T)^T$.

Such an approximation can be considered for the j-th element of $\mathbf{f}$ viz,

$$f_j(x) \approx \mathcal{P}_J^r(f_j)(x) = F_j^T\Phi_J^r(x) = F_j^T T_J^{-1}\Psi_J^r(x),$$

and thus by putting $\mathbf{F} := (F_1^T, F_2^T, \ldots, F_n^T)^T$ we have

$$\mathbf{f} \approx \mathbf{F}^T \otimes T_J^{-1}\Psi_J^r(x). \tag{26}$$

Now, we introduce the residual as

$$r_J^r(x) = \tilde{\mathbf{U}}^T \otimes \Psi_J^r(x) - \mathbf{F}^T \otimes T_J^{-1}\Psi_J^r(x) - \begin{cases} \tilde{\mathbf{U}}^T\Gamma^T \otimes T_J^{-1}\Psi_J^r(x), & \text{linear,}\\ \mathbf{G}^T \otimes T_J^{-1}\Psi_J^r(x), & \text{nonlinear.}\end{cases} \tag{27}$$

To apply the Galerkin method, it is necessary that $\langle r_J^r, \Psi_J^r\rangle = 0$. Thus we have

$$\tilde{\mathbf{U}}^T - \mathbf{F}^T \otimes T_J^{-1} - \begin{cases} \tilde{\mathbf{U}}^T\Gamma^T \otimes T_J^{-1} \\ \mathbf{G}^T \otimes T_J^{-1}\end{cases} = 0, \quad \begin{array}{l}\text{linear,}\\ \text{nonlinear.}\end{array} \tag{28}$$

By solving this system of linear and nonlinear equations using restarted generalized minimal residual method (GMRES) and Newton methods, respectively, we obtain the approximate solution of the Equation (1). Note that because we use the Galerkin method with orthogonal bases, such a system will have a unique solution [26].

Convergence Analysis

Theorem 2. *Suppose that $\mathbf{e}_J = \mathbf{u} - \mathcal{M}_J^r(\mathbf{u})$ where $\mathbf{u}$ and $\mathcal{M}_J^r(\mathbf{u})$ are the exact and approximate solutions of nonlinear system (1), respectively. Let X be an open set in $\mathbb{R}$ and let $g : \Omega \times X \to \mathbb{R}$ be a function such that $g(x,t,\mathbf{u}(x)) \in C^r(\Omega)$ for any $\mathbf{u} \in X$ and the condition (2) is satisfied. Furthermore, presume that $\mathbf{f} \in C^r(\Omega)$. Furthermore, assume that the residual $\mathbf{e} := \tilde{r} - (r_J^r)$ where the residual $\tilde{r}$ is specified as*

$$\tilde{r}(x) = u(x) - f(x) - \int_0^x g(x,t,u(t))dt, \tag{29}$$

and r_J^r is introduced in (27).

If $\mathbf{u} \in C^r(\Omega)$, and the method used to solve system (28) is convergent then one has

$$\|\mathbf{e}\|_2^2 \le \frac{2^{1-Jr}}{4^r r!}\left(|1 + \sqrt{nr2^J}\kappa| \sup_{x\in\Omega}|\mathbf{u}(x)| + \sup_{x\in\Omega}|\mathbf{f}(x)|\right),$$

where κ is a positive constant. Consequently, $\mathbf{e} \to 0$ *when* $J \to \infty$.

Proof. Using (27), (29) and the hypotheses of the theorem, we can write

$$\mathbf{e}(x) = \mathbf{e}_J(x) - \Big(\mathbf{f}(x) - \mathcal{M}_J^r(\mathbf{f})(x)\Big) - \Big(\int_0^x \mathbf{g}(x,t,\mathbf{u}(t))dt - \mathcal{M}_J^r\big(\int_0^x \mathbf{g}(x,t,\mathcal{M}_J^r(\mathbf{u}))(t)dt\big)\Big), \tag{30}$$

where $\mathbf{e} := \tilde{r} - (r_J^r)$. Since the function $\mathbf{g}$ satisfies the Lipschitz condition (2), Equation (30) can be reduced to

$$\mathbf{e}(x) = \mathbf{e}_J(x) - \Big(\mathbf{f}(x) - \mathcal{M}_J^r(\mathbf{f})(x)\Big) - A\int_0^x \mathbf{e}_J(t)dt.$$

Now, suppose that

$$\mathbf{e} \approx \mathcal{E} \otimes \Psi_J^r, \quad \mathbf{e}_J \approx \mathcal{E}_J \otimes \Psi_J^r,$$

where $\mathcal{E}$ and $\mathcal{E}_J$ are the $(1 \times nr2^J)$ vectors and thus, one can write

$$\mathcal{E} \otimes \Psi_J^r = \mathcal{E}_J \otimes \Psi_J^r - \Big(\mathbf{f}(x) - \mathcal{M}_J^r(\mathbf{f})(x)\Big) - A\mathcal{E}_J \otimes I_\psi \Psi_J^r,$$

Taking L_2-norm from both sides and using the triangle inequality yields

$$\begin{aligned}\|\mathcal{E} \otimes \Psi_J^r\|_2^2 &\le \|\mathcal{E}_J \otimes \Psi_J^r\|_2^2 + \|\mathbf{f}(x) - \mathcal{M}_J^r(\mathbf{f})(x)\|_2^2 + A\|\mathcal{E}_J \otimes I_\psi \Psi_J^r\|_2^2 \\ &= \|\mathcal{E}_J\|_2^2\|\Psi_J^r\|_2^2 + A\|\mathcal{E}_J\|_2^2\|I_\psi \Psi_J^r\|_2^2 + \|\mathbf{f}(x) - \mathcal{M}_J^r(\mathbf{f})(x)\|_2^2 \\ &\le \|\mathcal{E}_J\|_2^2\|\Psi_J^r\|_2^2 + A\|\mathcal{E}_J\|_2^2\|I_\psi\|_2^2\|\Psi_J^r\|_2^2 + \|\mathbf{f}(x) - \mathcal{M}_J^r(\mathbf{f})(x)\|_2^2\end{aligned}$$

where the second row comes from theorem 8 in [27]. Since Alpert multi-wavelets are orthonormal, one can write

$$\begin{aligned}\|\mathcal{E}\|_2^2 &\le \|\mathcal{E}_J\|_2^2 + A\|\mathcal{E}_J\|_2^2\|I_\psi\|_2^2 + \|\mathbf{f}(x) - \mathcal{M}_J^r(\mathbf{f})(x)\|_2^2 \\ &= \|\mathcal{E}_J\|_2^2\left(\|I_{nr2^J}\|_2^2 + A\|I_\psi\|_2^2\right) + \|\mathbf{f}(x) - \mathcal{M}_J^r(\mathbf{f})(x)\|_2^2.\end{aligned}$$

According to the previous section, when Ψ_J^r has high vanishing moments and the function h is smooth, $\langle h, \Psi_J^r\rangle$ decays fast in $J \to \infty$. By means of vanishing moments of Alpert's multi-wavelets and the matrix norms inequalities, we get

$$\|\mathcal{E}\|_2^2 \le \|\mathcal{E}_J\|_2^2|1 + \sqrt{nr2^J}\kappa| + \|\mathbf{f}(x) - \mathcal{M}_J^r(\mathbf{f})(x)\|_2^2,$$

where $\kappa = A\|I_\psi\|_\infty^2$. Now Equation (5) leads to the desired result

$$\|\mathcal{E}\|_2^2 \le \frac{2^{1-Jr}}{4^r r!}\left(|1 + \sqrt{nr2^J}\kappa|\sup_{x\in\Omega}|\mathbf{u}(x)| + \sup_{x\in\Omega}|\mathbf{f}(x)|\right).$$

□

Theorem 3. *Let the condition of Theorem 2 be valid. Suppose that* $\mathbf{u}_J$ *is the approximate solution obtained from solving (28) using restarted GMRES or Newton methods. If these methods solve (28) with proper accuracy, the error can be estimated from*

$$\|\mathbf{u} - \mathbf{u}_J\|_2^2 \le (1 - A\|I_\psi\|_2^2)^{-1}\frac{2^{1-Jr}}{4^r r!}\sup_{x\in\Omega}|\mathbf{f}(x)| + \eta, \tag{31}$$

where η *is a small positive number that desire to zero.*

Proof. Taking $\mathcal{M}_J^r(\mathbf{u})$ as the approximate solution of (1) and $\mathbf{u}_J$ as the approximate solution obtained from solving (28) using restarted GMRES or Newton methods, the convergence can be concluded from

$$\|\mathbf{u}-\mathbf{u}_J\| \leq \|\mathbf{u}-\mathcal{M}_J^r(\mathbf{u})\| + \|\mathcal{M}_J^r(\mathbf{u})-\mathbf{u}_J\|. \tag{32}$$

The approximate solution of (1) satisfies

$$\mathcal{M}_J^r(\mathbf{u})(x) = \mathcal{M}_J^r(\mathbf{f})(x) + \mathcal{M}_J^r\left(\int_0^x \mathbf{g}(x,t,\mathcal{M}_J^r(\mathbf{u})(t)dt\right). \tag{33}$$

Subtracting (33) from (1), and using the Lipschits condition (2), one can write

$$\mathbf{e}_J \leq \mathbf{f} - \mathcal{M}_J^r(\mathbf{f}) + A\int_0^x \mathbf{e}_J dt. \tag{34}$$

Let us consider $\mathbf{e}_J \approx \mathcal{E}_J \otimes \Psi_J^r$ where $\mathcal{E}_J$ is the $(1 \times nr2^J)$ vector and thus, we have

$$\mathcal{E}_J \otimes \Psi_J^r = \Big(\mathbf{f}(x) - \mathcal{M}_J^r(\mathbf{f})(x)\Big) + A\mathcal{E}_J \otimes I_\psi \Psi_J^r. \tag{35}$$

Taking L_2-norm from both sides of (35) and using Theorem 8 in [27] yields

$$\begin{aligned}\|\mathcal{E}_J \otimes \Psi_J^r\|_2^2 &\leq \|\mathbf{f}(x) - \mathcal{M}_J^r(\mathbf{f})(x)\|_2^2 + A\|\mathcal{E}_J \otimes I_\psi \Psi_J^r\|_2^2 \\ &\leq \|\mathbf{f}(x) - \mathcal{M}_J^r(\mathbf{f})(x)\|_2^2 + A\|\mathcal{E}_J\|_2^2\|I_\psi \Psi_J^r\|_2^2\end{aligned}$$

Since Alpert multi-wavelets are orthonormal, one can write

$$\|\mathcal{E}_J\|_2^2 \leq \|\mathbf{f}(x) - \mathcal{M}_J^r(\mathbf{f})(x)\|_2^2 + A\|\mathcal{E}_J\|_2^2\|I_\psi\|_2^2. \tag{36}$$

Therefore one can bound the error of $\|\mathbf{u} - \mathcal{M}_J^r(\mathbf{u})\|$ via

$$\|\mathcal{E}_J\|_2^2 \leq (1 - A\|I_\psi\|_2^2)^{-1}\|\mathbf{f}(x) - \mathcal{M}_J^r(\mathbf{f})(x)\|_2^2. \tag{37}$$

According to the theorem hypotheses, the methods used to solve the obtained system are convergent. So $\eta := \|\mathcal{M}_J^r(\mathbf{u}) - \mathbf{u}_J\|$ will be very small. Inequality (31) is obtained using (5) and (32), i.e.,

$$\|\mathbf{u}-\mathbf{u}_J\|_2^2 \leq (1 - A\|I_\psi\|_2^2)^{-1}\frac{2^{1-Jr}}{4^r r!}\sup_{x\in\Omega}|\mathbf{f}(x)| + \eta.$$

□

4. Numerical Examples

In this section, some numerical experiments are considered to illustrate the convergence and efficiency of the proposed method. To this end, we report the L^2 errors of the solution which is defined by

$$\xi_u := \|u - \mathcal{M}_J^r(u)\|_2 = \left(\int_\Omega |u(x) - \mathcal{M}_J^r(u)(x)|^2 dx\right)^{1/2},$$

where u and $\mathcal{M}_J^r(u)$ are the exact and approximate solution of systems (1), respectively. In order to get the sparse coefficients matrix in the linear type, all the entries of this matrix that are less than a small positive number ε are set to zero. Finally, one can find the rate of sparsity S_ε which is defined by [28]

$$S_\varepsilon = \frac{N_0 - N_\varepsilon}{N_0} \times 100\%.$$

where N_0 is the total number of elements and N_ε the number of elements remaining after thresholding.

Example 1. *Let us run the proposed method on the following linear Volterra integral equation [5,29]*

$$u_1(x) = -\int_0^x e^{-(s-x)} u_1(s)ds - \int_0^x \cos(s-x) u_2(s)ds + \cosh(x) + x\sin(x),$$
$$u_2(x) = -\int_0^x e^{s+x} u_1(s)ds - \int_0^x x\cos(s) u_2(s)ds + 2\sin(x) + x(\sin^2(x) + e^x).$$

The exact solution is $\mathbf{u}(x) = (e^{-x}, 2\sin(x))$. The effect of thresholding on L_2-errors and sparsity percentage is reported in Table 1 for different values of r, J ad ε. To illustrate the effect of the refinement level J and the multiplicity parameter r on L^2 error, Figure 1 is plotted. Figure 2 shows sparse matrix when $r = 5$, $J = 3$ and $\varepsilon = 10^{-4}, 10^{-2}$.

Table 1. Effects of parameters r, J and ε on sparsity and L^2-error for Example 1.

		Without Thresholding		$\boldsymbol{\varepsilon = 10^{-5}}$		$\boldsymbol{\varepsilon = 10^{-3}}$	
r	J	S_ε	L^2-**Error**	S_ε	L^2-**Error**	S_ε	L^2-**Error**
3	2	6.25	$\xi_{u_1} = 3.24 \times 10^{-5}$ $\xi_{u_2} = 8.41 \times 10^{-5}$	24.48	$\xi_{u_1} = 3.30 \times 10^{-5}$ $\xi_{u_2} = 8.41 \times 10^{-5}$	49.83	$\xi_{u_1} = 2.42 \times 10^{-4}$ $\xi_{u_2} = 3.04 \times 10^{-4}$
	3	17.19	$\xi_{u_1} = 4.05 \times 10^{-6}$ $\xi_{u_2} = 1.05 \times 10^{-5}$	53.30	$\xi_{u_1} = 6.02 \times 10^{-6}$ $\xi_{u_2} = 1.26 \times 10^{-5}$	74.13	$\xi_{u_1} = 2.41 \times 10^{-4}$ $\xi_{u_2} = 5.15 \times 10^{-4}$
5	2	6.25	$\xi_{u_1} = 6.38 \times 10^{-9}$ $\xi_{u_2} = 1.66 \times 10^{-8}$	45.69	$\xi_{u_1} = 4.62 \times 10^{-6}$ $\xi_{u_2} = 1.28 \times 10^{-6}$	70.69	$\xi_{u_1} = 1.53 \times 10^{-4}$ $\xi_{u_2} = 6.51 \times 10^{-4}$
	3	17.19	$\xi_{u_1} = 2.00 \times 10^{-10}$ $\xi_{u_2} = 5.19 \times 10^{-10}$	71.42	$\xi_{u_1} = 1.24 \times 10^{-5}$ $\xi_{u_2} = 8.54 \times 10^{-6}$	86.75	$\xi_{u_1} = 1.53 \times 10^{-4}$ $\xi_{u_2} = 6.52 \times 10^{-4}$

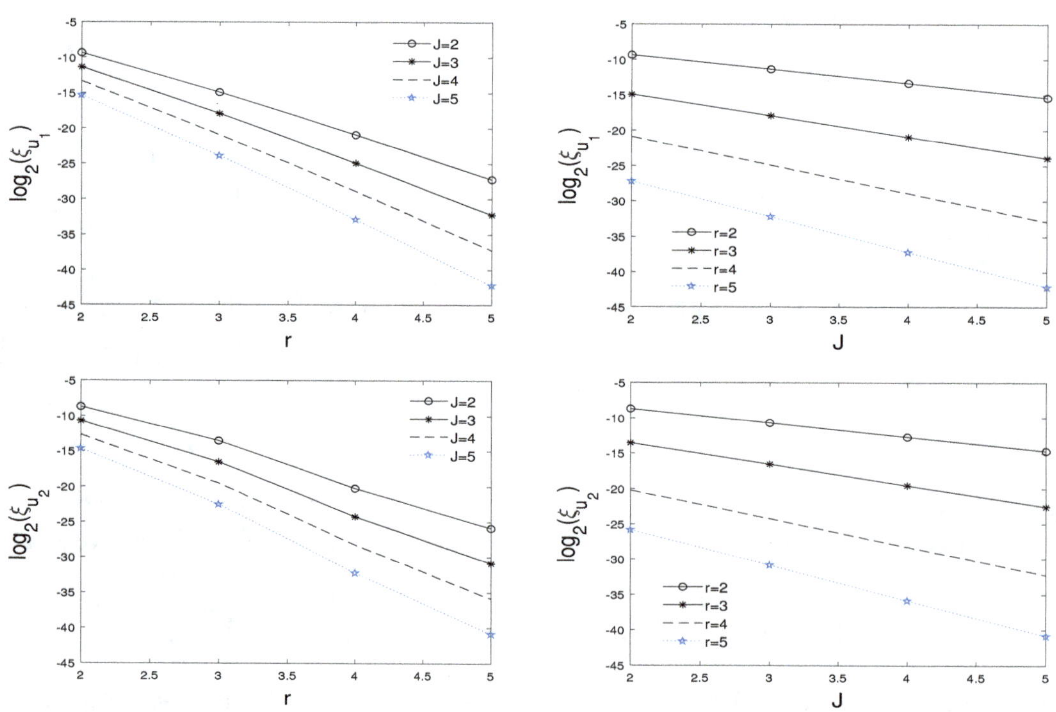

Figure 1. Effects of the refinement level J (**left**) and the multiplicity parameter r (**right**) on L^2 error when $r = 5$ and $J = 2$ for Example 1.

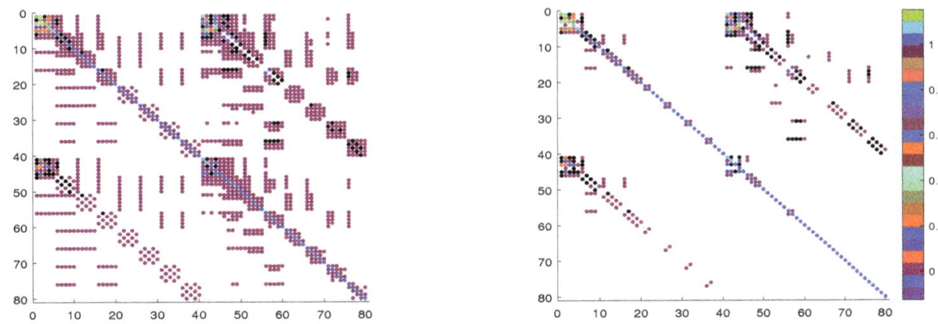

Figure 2. Plot of sparse matrix after thresholding with $\varepsilon = 10^{-4}$ (**left**) and $\varepsilon = 10^{-2}$ (**right**) for Example 1.

Example 2. *Let us consider the following system of Volterra integral equation as a further example*

$$u_1(x) = -\frac{x^5}{3} - \frac{x^4}{4} + \frac{x^3}{3} + x + \int_0^x (x^2 - s)(u_1(s) + u_2(s))ds$$
$$u_2(x) = \frac{x^3}{2} - \frac{x^4}{3} + x^2 - \int_0^x x(u_1(s) - u_2(s))ds.$$

The solution is reported in [5,30] and is $\mathbf{u} = (x, x^2)$. *To illustrate the effect of thresholding on the coefficients matrix obtained from proposed method, the graph in Figure 3 is provided. Figure 4 shows the effect of parameters r and J.*

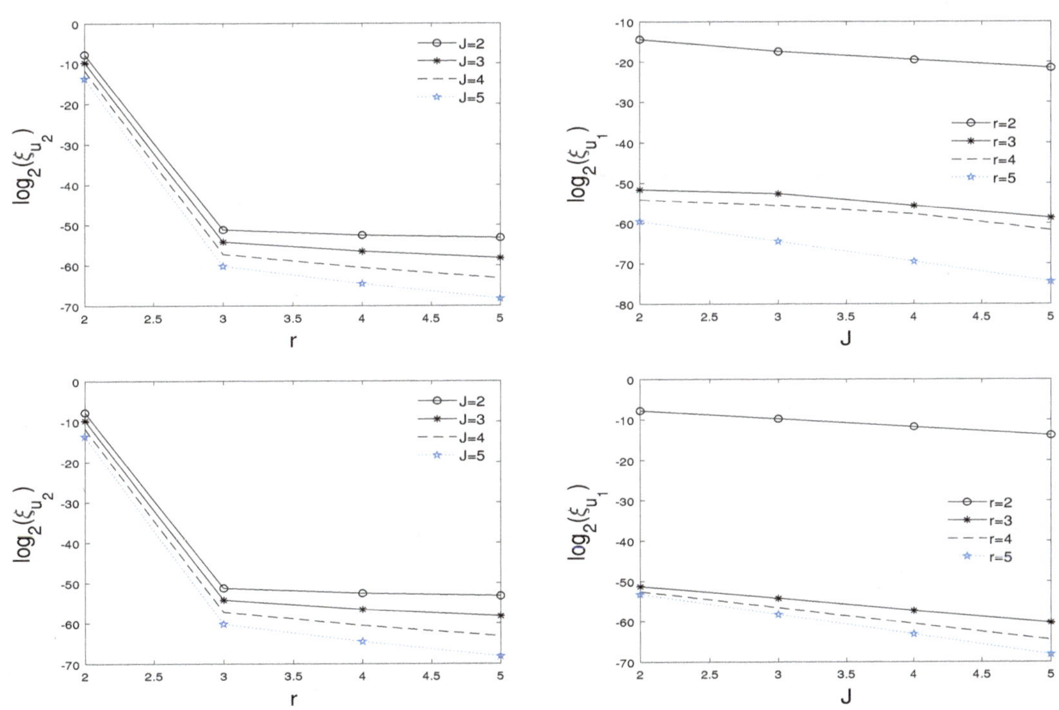

Figure 3. Effects of the refinement level J (**left**) and the multiplicity parameter r (**right**) on L^2 error when $r = 4$ and $J = 3$ for Example 2.

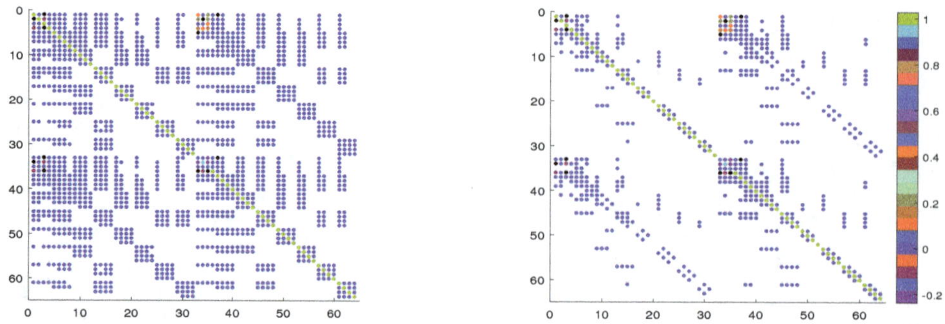

Figure 4. Plot of sparse matrix after thresholding with $\varepsilon = 10^{-5}$ **(left)** and $\varepsilon = 10^{-3}$ **(right)** for Example 2.

Example 3. *Let us consider the following nonlinear system*

$$u_1(x) = \sin(x) - x + \int_0^x (u_1^2(s) + u_2^2(s))ds$$
$$u_2(x) = \cos(x) - \frac{1}{2}\sin^2(x) + \int_0^x u_1(s)u_2(s)ds.$$

The exact solutions of this system are $u_1 = \sin(x)$ and $u_2 = \cos(x)$ [8].

Table 2 is reported to show the efficiency and accuracy of the proposed method. We observe when the refinement level J and multiplicity parameter r increase, the L_2-errors decrease. In Tables 3 and 4, results are compared with other methods [10,31–33] in terms of the absolute errors. In this paper, J and r are the criteria for the discretization and the degree of polynomials used as a basis, respectively. Taking $r = 7$ and $J = 2$, the results of Tables 3 and 4 indicate that the proposed method solves this equation better than others [10,31–33]. Furthermore, we reported the exact and numerical solution by Figure 5.

Table 2. Effect of the refinement level J and multiplicity parameter r on L_2-error for Example 3.

	r	$J = 2$	$J = 3$	$J = 4$	$J = 5$
	3	4.20×10^{-5}	5.25×10^{-6}	6.56×10^{-7}	8.20×10^{-8}
u_1	4	4.04×10^{-7}	2.53×10^{-8}	1.58×10^{-9}	9.86×10^{-11}
	5	8.31×10^{-9}	2.60×10^{-10}	8.12×10^{-12}	2.54×10^{-13}
	3	2.50×10^{-5}	3.21×10^{-6}	4.01×10^{-7}	5.02×10^{-8}
u_2	4	6.62×10^{-7}	4.14×10^{-8}	2.59×10^{-9}	1.62×10^{-10}
	5	5.06×10^{-9}	1.58×10^{-10}	4.94×10^{-12}	1.54×10^{-13}

Table 3. The comparison between absolute errors of Example 3 for u_1.

	HPM [31]	**Method Based upon Discretization [32]**	**Simpson's 3/8 Rule [10]**	**Bernstein Collocation Method [33]**	**Present Method ($r = 7, J = 2$)**
x	$(n = 5)$	$(h = 200)$	$(h = 0.025)$	$(n = 10)$	
0.1	1.4×10^{-07}	2.4×10^{-05}	3.0×10^{-10}	5.5×10^{-10}	1.0×10^{-12}
0.2	3.5×10^{-06}	9.8×10^{-05}	1.1×10^{-09}	1.7×10^{-10}	1.1×10^{-12}
0.3	5.5×10^{-05}	2.3×10^{-04}	3.6×10^{-09}	2.6×10^{-10}	1.1×10^{-12}
0.4	3.8×10^{-04}	4.1×10^{-04}	6.0×10^{-09}	1.0×10^{-10}	9.6×10^{-13}
0.5	1.6×10^{-03}	6.6×10^{-04}	8.7×10^{-09}	1.1×10^{-10}	3.2×10^{-12}
0.6		9.7×10^{-04}	1.4×10^{-08}	2.5×10^{-10}	8.4×10^{-13}
0.7		1.4×10^{-03}	1.9×10^{-08}	5.8×10^{-10}	9.2×10^{-13}
0.8		1.8×10^{-03}	2.4×10^{-08}	1.0×10^{-08}	9.2×10^{-13}
0.9		2.5×10^{-03}	3.3×10^{-08}	1.0×10^{-07}	5.0×10^{-12}
1.0		3.2×10^{-03}	4.0×10^{-08}	8.2×10^{-07}	6.7×10^{-11}

Table 4. The comparison between absolute errors of Example 3 for u_2.

x	HPM [31] ($n = 5$)	Method Based upon Discretization [32] ($h = 200$)	Simpson's 3/8 Rule [10] ($h = 0.025$)	Bernstein Collocation Method [33] ($n = 10$)	Present Method ($r = 7, J = 2$)
0.1	3.2×10^{-07}	2.5×10^{-04}	5.3×10^{-10}	6.5×10^{-11}	1.2×10^{-13}
0.2	1.1×10^{-05}	5.0×10^{-04}	3.0×10^{-10}	1.3×10^{-10}	1.6×10^{-13}
0.3	1.2×10^{-04}	7.5×10^{-04}	6.0×10^{-10}	6.5×10^{-12}	4.0×10^{-13}
0.4	6.3×10^{-04}	1.0×10^{-04}	2.1×10^{-09}	6.5×10^{-11}	3.9×10^{-13}
0.5	2.2×10^{-03}	1.2×10^{-03}	3.1×10^{-09}	9.0×10^{-11}	1.3×10^{-12}
0.6		1.5×10^{-03}	5.3×10^{-09}	4.2×10^{-11}	6.0×10^{-13}
0.7		1.8×10^{-03}	9.4×10^{-09}	4.2×10^{-11}	6.8×10^{-13}
0.8		2.0×10^{-03}	1.4×10^{-08}	3.6×10^{-09}	8.1×10^{-13}
0.9		2.3×10^{-03}	2.0×10^{-08}	3.5×10^{-08}	4.8×10^{-13}
1.0		2.6×10^{-03}	2.9×10^{-08}	2.9×10^{-07}	4.2×10^{-11}

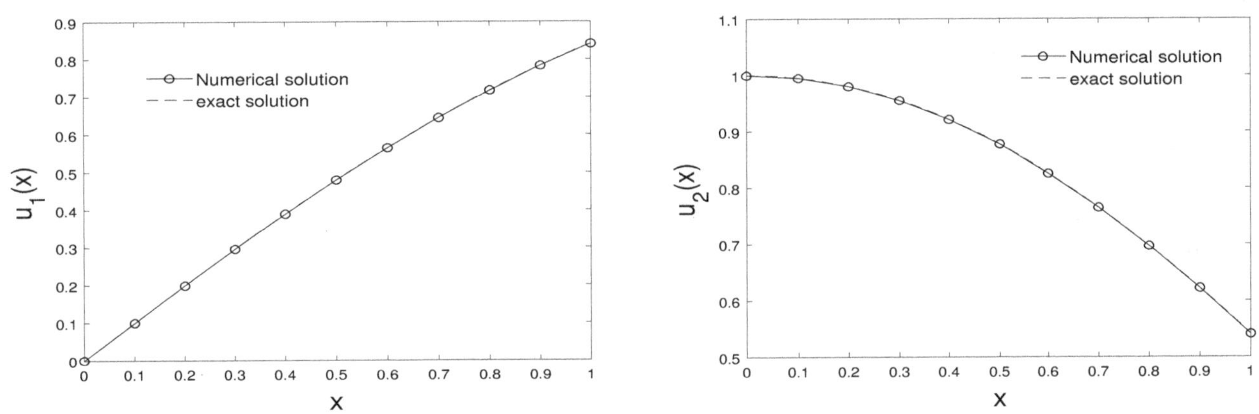

Figure 5. Plot of the exact and numerical solution taking $r = 6$ and $J = 2$ for Example 3.

5. Conclusions

In this paper, we proposed the multi-wavelets Galerkin method to solve the linear and nonlinear Volterra integral equation. The convergence analysis and numerical simulations indicate that the proposed method gives a satisfactory approximation to the exact solution. Thresholding can be used to increase sparsity for a lower computational cost, without affecting the error in L^2. Using the interpolation property of Alpert's multi-wavelets, the proposed method turns out to be fast and very competitive against state-of-the-art techniques. The main advantages of the proposed method are the lower computational cost and lower complexity.

Author Contributions: Conceptualization, H.V.L. and H.B.J.; methodology, software, H.B.J. and S.T.; validation, formal analysis, H.B.J.; writing–original draft preparation, investigation, funding acquisition, H.V.L., H.B.J. and S.T.; writing–review and editing, H.V.L., H.B.J. and S.T. All authors have read and agreed to the published version of the manuscript.

References

1. Jerri, A. *Introduction to Integral Equations with Applications*; Wiley: New York, NY, USA, 1999.
2. Li, Y.; Xu, R.; Lin, J. Global dynamics for a class of infection-age model with nonlinear incidence. *Nonlinear Anal. Model. Control* **2018**, *24*, 47–72. [CrossRef]
3. Nordbo, A.; Wyller, J.; Einevoll, G.T. Neural network firing-rate models on integral form. *Biol. Cybern.* **2007**, *97*, 195–209. [CrossRef] [PubMed]

4. Porter, D.; Stirling, D.S. *Integral Equations: A Practical Treatment from Spectral Theory to Applications*; Cambridge University Press: Cambridge, UK, 2004.
5. Aminikhah, H.; Biazar, J. A new analytical method for solving systems of Volterra integral equations. *Int. J. Comput. Math.* **2010**, *87*, 1142–1157. [CrossRef]
6. Biazar, J.; Babolian, E.; Islam, R. Solution of a system of Volterra integral equations of the first kind by Adomian method. *Appl. Math. Comput.* **2003**, *139*, 249–258.
7. Saray, B.N. An efficient algorithm for solving Volterra integro-differential equations based on Alpert's multi-wavelets Galerkin method. *J. Comput. Appl. Math.* **2019**, *348*, 453–465. [CrossRef]
8. Golbabai, A.; Mammadova, M.; Seifollahi, S. Solving a system of nonlinear integral equations by an RBF network. *Comput. Math. Appl.* **2009**, *57*, 1651–1658.
9. Saberi-Nadjafi, J.; Tamamgar, M. Modified homotopy perturbation method for solving the system of Volterra integral equations. *Int. Nonlinear Sci. Numer. Simul.* **2008**, *9*, 409–413. [CrossRef]
10. Kılıçman, A.; Kargaran Dehkordi, L.; Tavassoli Kajani, M. Numerical Solution of Nonlinear Volterra Integral Equations System Using Simpson's 3/8 Rule. *Math. Prob. Eng.* **2012**, *2012*, 1–16. [CrossRef]
11. Aguilar, M.; Brunner, H. Collocation methods for second-order Volterra integrodifferential equations. *Appl. Numer. Math.* **1988**, *4*, 455–470. [CrossRef]
12. Górska, K.; Horzela, A. The Volterra type equations related to the non-Debye relaxation. *Commun. Nonlinear Sci.* **2020**, *85*, 105246. [CrossRef]
13. Alpert, B.; Beylkin, G.; Gines, D.; Vozovoi, L. Adaptive solution of partial differential equations in multiwavelet bases. *J. Comput. Phys.* **2002**, *182*, 149–190. [CrossRef]
14. Saray, B.N.; Lakestani, M.; Razzaghi, M. Sparse representation of system of Fredholm integro-differential equations by using Alpert multiwavelets. *Comput. Math. Math. Phys.* **2015**, *55*, 1468–1483. [CrossRef]
15. Alpert, B.; Beylkin, G.; Coifman, R.R.; Rokhlin, V. Wavelet-like bases for the fast solution of second-kind integral equations. *SIAM Sci. Stat. Comput.* **1993**, *14*, 159–184. [CrossRef]
16. Mallat, S.G. *A Wavelet Tour of Signal Processing*; Academic Press: Cambridge, MA, USA, 1999.
17. Meyer, Y. *Wavelets and Operators*; Cambridge University Press: Cambridge, UK, 1993.
18. Dehghana, M.; Saray, B.N.; Lakestani, M. Three methods based on the interpolation scaling functions and the mixed collocation finite difference schemes for the numerical solution of the nonlinear generalized Burgers–Huxley equation. *Math. Comput. Model.* **2012**, *55*, 1129–1142. [CrossRef]
19. Saray, B.N.; Lakestani, M.; Cattani, C. Evaluation of mixed Crank–Nicolson scheme and Tau method for the solution of Klein–Gordon equation. *Appl. Math. Comput.* **2018**, *331*, 169–181.
20. Shahriari, M.; Saray, B.N.; Lakestani, M.; Manafia, J. Numerical treatment of the Benjamin-Bona-Mahony equation using Alpert multiwavelets. *Eur. Phys. J. Plus.* **2018**, *133*, 1–12. [CrossRef]
21. Shamsi, M.; Razzaghi, M. Numerical solution of the controlled duffing oscillator by the interpolating scaling functions, Electrmagn. *Waves Appl.* **2004**, *18*, 691–705.
22. Shamsi, M.; Razzaghi, M. Solution of Hallen's integral equation using multiwavelets. *Comput. Phys. Comm.* **2005**, *168*, 187–197. [CrossRef]
23. Seyedi, S.; Saray, B.N.; Nobari, M. Using interpolation scaling functions based on Galerkin method for solving non-Newtonian fluid flow between two vertical flat plates. *Appl. Math. Comput.* **2015**, *269*, 488–496.
24. Saray, B.N.; Manafian, J. Sparse representation of delay differential equation of pantograph type using multiwavelets Galerkin method. *Eng. Comput.* **2018**, *35*, 887–903. [CrossRef]
25. Hovhannisyan, N.; Müller, S.; Schäfer, R. Adaptive multiresolution discontinuous Galerkin schemes for conservation laws. *Math. Comput.* **2014**, *83*, 113–151. [CrossRef]
26. Atkinson KE. *The Numerical Solution of Integral Equations of the Second Kind*; Cambridge University Press: Cambridge, UK, 1997.
27. Lancaster, P.; Farahat, H.K. Norms on direct sums and Tensor products. *Math. Comput.* **1972**, *26*, 401–414. [CrossRef]
28. Goswami, J.C.; Chan, A.K.; Chui, C.K. On solving first-kind integral equations using wavelets on bounded integval. *IEEE Trans. Antennas Propag.* **1995**, *43*, 614–622. [CrossRef]
29. Delves, L.M.; Mohamed, J.L. *Computational Methods for Integral Equations*; Cambridge University Press: New York, NY, USA, 1988.
30. Babolian, E.; Biazar, J. Solution of a system of non-linear Volterra integral equations of the second kind. *Far East J. Math. Sci.* **2000**, 2, 935–945.

31. Biazar, J.; Ghazvini, H. Hes homotopy perturbation method for solving systems of Volterra integral equations of the second kind. *Chaos Soliton Fract.* **2009**, *39*, 770–777. [CrossRef]
32. Yaghouti, M.R. A numerical method for solving a system of Volterra integral equations. *World Appl. Program.* **2012**, *2*, 18–33.
33. Davaeifar, S.; Rashidinia, J. Approximate Solution of System of Nonlinear Volterra Integro-Differential Equations by Using Bernstein Collocation Method. *Int. J. Math. Model. Comput.* **2017**, *7*, 79–91.

10

Construction of Different Types Analytic Solutions for the Zhiber-Shabat Equation

Asıf Yokus [1,†], Hülya Durur [2,†], Hijaz Ahmad [3] and Shao-Wen Yao [4,*]

1 Department of Actuary, Faculty of Science, Firat University, Elazig 23200, Turkey; asfyokus@yahoo.com
2 Department of Computer Engineering, Faculty of Engineering, Ardahan University, Ardahan 75000, Turkey; hulyadurur@ardahan.edu.tr
3 Department of Basic Sciences, University of Engineering and Technology, Peshawar 25000, Pakistan; hijaz555@gmail.com
4 School of Mathematics and Information Science, Henan Polytechnic University, Jiaozuo 454000, China
* Correspondence: yaoshaowen@hpu.edu.cn
† These authors contributed equally to this work.

Abstract: In this paper, a new solution process of $(1/G')$-expansion and $(G'/G,1/G)$-expansion methods has been proposed for the analytic solution of the Zhiber-Shabat (Z-S) equation. Rather than the classical $(G'/G,1/G)$-expansion method, a solution function in different formats has been produced with the help of the proposed process. New complex rational, hyperbolic, rational and trigonometric types solutions of the Z-S equation have been constructed. By giving arbitrary values to the constants in the obtained solutions, it can help to add physical meaning to the traveling wave solutions, whereas traveling wave has an important place in applied sciences and illuminates many physical phenomena. 3D, 2D and contour graphs are displayed to show the stationary wave or the state of the wave at any moment with the values given to these constants. Conditions that guarantee the existence of traveling wave solutions are given. Comparison of $(G'/G,1/G)$-expansion method and $(1/G')$-expansion method, which are important instruments in the analytical solution, has been made. In addition, the advantages and disadvantages of these two methods have been discussed. These methods are reliable and efficient methods to obtain analytic solutions of nonlinear evolution equations (NLEEs).

Keywords: $(1/G')$-expansion method; the Zhiber-Shabat equation; $(G'/G,1/G)$-expansion method; traveling wave solutions; exact solutions

1. Introduction

The analysis of analytic solutions of nonlinear evolution equations (NLEEs) plays a significant role in the study of nonlinear physical phenomena. Various techniques have been tried to obtained analytic solutions, such as the sine–cosine method [1], extended sinh-Gordon equation expansion method [2,3], (G'/G)-expansion method [4,5], improved Bernoulli sub-equation function method [6], variational iteration algorithm-II [7–9], sub equation method [10], collocation method [11,12], $(1/G')$-expansion method [13–15], first integral method [16], adomian decomposition methods [17–19], hirota bilinear method [20], modified variational iteration algorithms [21–24], homotopy perturbation method [25], residual power series method [26], $(G'/G,1/G)$-expansion method [27] and so on [28–38].

In this study, our main purpose is to obtain the traveling wave solutions of the evolution equations in nonlinear dynamics. As it is known, scientific studies take place gradually. The first step is to examine a physical event, the second step is to model the event, the third step is to produce the solution of the model and the fourth step is to load the produced solution into physical meaning. In this article, to produce the solution in the third stage and to prepare for the fourth stage. As it is known, in soliton theory, it will be much more valuable if solitons gain physical meaning. For example, today, the pandemic patients that affect the world may represent a stationary wave on a graph consisting of numerical data related to parameters such as number of patients and number of tests. Employees on this subject can relate to the solutions we will offer in this study. We consider the Zhiber-Shabat (Z-S) equation [39]

$$u_{xt} + pe^{u} + qe^{-u} + re^{-2u} = 0, \tag{1}$$

where p, r, q are arbitrary constants. When $r = 0,\ q \neq 0$, Equation (1) gives the well-known sinh–Gordon equation, while, $r \neq 0,\ q = 0$, gives the Dodd–Bullough–Mikhailov (DBM) equation. However, for $p = 0,\ q = -1,\ r = -1$, Equation (1) reduced to the Tzitzeica–Dodd–Bullough (TDB) equation, while for $r = q = 0$, gives the Liouville equation. These equations play an effective role in various scientific applications such as fluid dynamics, solid state physics, nonlinear optics and chemical kinetics. When the analytical solution of Equation (1) is found, the solutions of the sinh–Gordon, DBM, TDB and the Liouville equations can also be obtained.

Many researchers have investigated the Z-S equations and discussed its applications in different field of science and engineering. Some of these investigations are as follows: various types of solution for the Z-S equation are obtained [40] by using qualitative theory of polynomial differential system, while qualitative behavior and exact travelling wave solutions of the Z-S equation are obtained in [41]. Analytic solutions of the Z-S equation are obtained using the $(-\phi(\xi))$-expansion method [42], exponential rational function method [43], while exact solutions of it are obtained using bifurcation theory and method of phase portraits analysis [44].

In the current work, we are interested in constructing exact solutions of the Zhiber-Shabat (Z-S) equation using $(1/G')$-expansion method and $(G'/G, 1/G)$-expansion method. The solutions of the equation have not been studied with either method. In this study, both to provide the literature with the solution produced by these methods and to discuss the advantages and disadvantages of the methods.

2. (1/G′)-Expansion Method

Consider a general form of NLEEs as

$$S\left(u, \frac{\partial u}{\partial t}, \frac{\partial u}{\partial x}, \frac{\partial^2 u}{\partial x^2}, \ldots\right) = 0. \tag{2}$$

Let $u = u(x,t) = u(\xi),\ \xi = x - wt,\ w \neq 0$. where w is a constant speed of the wave. After using transformation, it can be converted into the following nonlinear ODE for $u(\xi)$:

$$\varrho\left(u, u', u'', u''', \ldots\right) = 0. \tag{3}$$

The solution of Equation (3) is assumed to have the form

$$u(\xi) = a_0 + \sum_{i=1}^{m} a_i \left(\frac{1}{G'}\right)^{i}, \tag{4}$$

where $a_i,\ i = 0, 1, \ldots, m$ are constants and $G = G(\xi)$ provides the following second order IODE

$$G'' + \lambda G' + \mu = 0, \tag{5}$$

where, λ and μ are constants to be determined after,

$$\frac{1}{G'} = \frac{\lambda}{-\mu + \lambda A\left(\mathrm{Cosh}(\xi\lambda) - \mathrm{Sinh}(\xi\lambda)\right)}, \tag{6}$$

where A is an integral constant. If the desired derivatives of the Equation (4) are calculated and substituting in the Equation (3), a polynomial with the argument $(1/G')$ is attained. An algebraic equation system is created by equalizing the coefficients of this polynomial to zero. These equations are solved with the help of the package program and put into place in the default Equation (3) solution function. Finally, the solutions of Equation (1) are obtained.

3. $(G'/G, 1/G)$-Expansion Method

Consider the following general form of NLEEs

$$Z\left(u, u_t, u_x, u_y, u_z, u_{tt}, u_{xx}, \ldots\right) = 0. \tag{7}$$

If $u\left(x,t\right) = u\left(\xi\right), \quad \xi = x - wt$ where w is a constant, when transmutation is applied to Equation (7), it becomes a NLODE and this equation may be written as:

$$z\left(u', u'', u''', \ldots\right) = 0. \tag{8}$$

Complexity can be reduced by integrating Equation (8). By the nature of this method, $G\left(\xi\right)$ is a quadratic function ODE solution,

$$G''\left(\xi\right) + \lambda G\left(\xi\right) = \mu. \tag{9}$$

Furthermore, to provide operational aesthetics as $\phi = \phi\left(\xi\right) = G'/G$ and $\psi = \psi\left(\xi\right) = \frac{1}{G(\xi)}$. We may write derivatives of functions defined here;

$$\phi' = -\phi^2 + \mu\psi - \lambda, \; \psi' = -\phi\psi. \tag{10}$$

We can offer the behavior of solution function Equation (9) according to the state of λ, taking into account the equations given by the Equation (10).

i: If $\lambda < 0$

$$G\left(\xi\right) = c_1 \sinh\left(\sqrt{-\lambda}\xi\right) + c_2 \cosh\left(\sqrt{-\lambda}\xi\right) + \frac{\mu}{\lambda}, \tag{11}$$

whereas c_1 and c_2 are reel numbers. Considering Equation (11);

$$\psi^2 = \frac{-\lambda}{\lambda^2\sigma + \mu^2}\left(\phi^2 - 2\mu\psi + \lambda\right), \; \sigma = c_1^2 - c_2^2, \tag{12}$$

Equation (12) is written.

ii: If $\lambda > 0$

$$G\left(\xi\right) = c_1 \sin\left(\sqrt{\lambda}\xi\right) + c_2 \cos\left(\sqrt{\lambda}\xi\right) + \frac{\mu}{\lambda}, \tag{13}$$

here c_1 and c_2 are reel numbers. Considering Equation (13), there is following equation;

$$\psi^2 = \frac{\lambda}{\lambda^2\sigma - \mu^2}\left(\phi^2 - 2\mu\psi + \lambda\right), \; \sigma = c_1^2 + c_2^2, \tag{14}$$

iii: If $\lambda = 0$

$$G\left(\xi\right) = \frac{\mu}{2}\xi^2 + c_1\xi + c_2, \tag{15}$$

here c_1 and c_2 are reel numbers. Considering Equation (15), there is following equation;

$$\psi^2 = \frac{1}{c_1^2 - 2\mu c_2}\left(\phi^2 - 2\mu\psi\right). \tag{16}$$

In terms of ϕ and ψ polynomials, solution of Equation (8) is ;

$$U(\xi) = a_0 + \sum_{i=1}^{n} \left(a_i\phi^i + b_i\psi^i\right). \tag{17}$$

In this study, we reorganized the solution function in classical $(G'/G, 1/G)$-expansion method as Equation (17) with the logic of solution functions of (G'/G) and $(1/G')$-expansion methods. This logic is considered together with the classical $(G'/G, 1/G)$-expansion method and the method can be developed in future studies and different solutions can be offered.

Wherein, $a_i \ \ (i = 0, 1, \ldots, m)$ and $b_i \ \ (i = 1, \ldots, m)$ counts then are constants to be determined. m is a positive equilibrium term that may be attained by comparing the maximum order derivative and the maximum order nonlinear term in Equation (8). If Equation (17) is written in Equation (8) with Equations (10), (12), (14) or (16), a polynomial function associated with ϕ and ψ is written. Each term coefficient of $\phi^i\psi^j \ \ (i = 0, 1, \ldots, m) \ \ (j = 1, \ldots, m)$ of the attained polynomial functions are equated to zero and a system algebraic equations is attained for $a_i, b_i, w, \mu, c_1, c_2$ and $\lambda \ \ (i = 0, 1, \ldots, m)$. The required coefficients are obtained by solving the algebraic equation with the help of computer package programs. These coefficients found are written in Equation (17) and $u(\xi)$ solution function of Equation (8) is obtained and if $\xi = x - wt$ transmutation is employed in reverse order, we will attain analytic solution $u(x, t)$ of Equation (7).

4. Solutions of The (Z-S) Equation Using $(1/G')$-Expansion Method

We consider Equation (1). Using transmutation $u(x, t) = u(\xi)$, $\xi = x - wt$, we obtain

$$-wu_{\xi\xi} + pe^u + qe^{-u} + re^{-2u} = 0, \tag{18}$$

where w is the wave speed. To implement this method, we use transmutation $u = \ln v$ and $v = V(\xi)$, Equation (18) becomes

$$-w\left(VV'' - (V')^2\right) + pV^3 + qV + r = 0. \tag{19}$$

In Equation (19), we find balancing term $m = 2$ and in Equation (4), the following situation is obtained:

$$V = a_0 + a_1\left(\frac{1}{G'}\right) + a_2\left(\frac{1}{G'}\right)^2, \ a_2 \neq 0, \tag{20}$$

where a_0, a_1, a_2 unknown constants to be determined later. Replacing Equation (20) into Equation (19) and the coefficients of the algebraic Equation (1) are equal to zero, we can establish the following algebraic equation systems

$$\begin{aligned}
&\left(\tfrac{1}{G'[\xi]}\right)^0 : r + qa_0 + pa_0^3 = 0,\\
&\left(\tfrac{1}{G'[\xi]}\right)^1 : qa_1 - w\lambda^2 a_0 a_1 + 3pa_0^2 a_1 = 0,\\
&\left(\tfrac{1}{G'[\xi]}\right)^2 : -3w\lambda\mu a_0 a_1 + 3pa_0 a_1^2 + qa_2 - 4w\lambda^2 a_0 a_2 + 3pa_0^2 a_2 = 0,\\
&\left(\tfrac{1}{G'[\xi]}\right)^3 : -2w\mu^2 a_0 a_1 - w\lambda\mu a_1^2 + pa_1^3 - 10w\lambda\mu a_0 a_2 - w\lambda^2 a_1 a_2 + 6pa_0 a_1 a_2 = 0,\\
&\left(\tfrac{1}{G'[\xi]}\right)^4 : -w\mu^2 a_1^2 - 6w\mu^2 a_0 a_2 - 5w\lambda\mu a_1 a_2 + 3pa_1^2 a_2 + 3pa_0 a_2^2 = 0,\\
&\left(\tfrac{1}{G'[\xi]}\right)^5 : -4w\mu^2 a_1 a_2 - 2w\lambda\mu a_2^2 + 3pa_1 a_2^2 = 0,\\
&\left(\tfrac{1}{G'[\xi]}\right)^6 : -2w\mu^2 a_2^2 + pa_2^3 = 0.
\end{aligned} \tag{21}$$

Case I.

$$\mu = \frac{p\lambda a_0 a_1}{2\left(q + 3pa_0^2\right)}, \ a_2 = \frac{pa_0 a_1^2}{2\left(q + 3pa_0^2\right)}, \ w = \frac{q + 3pa_0^2}{\lambda^2 a_0}, \ r = -qa_0 - pa_0^3, \tag{22}$$

considering Equation (6), substituting Equation (22) into Equation (20), the following solution is attained

$$V = \left[\begin{array}{l} a_0 + \dfrac{p a_0 a_1^2}{2\left(q+3pa_0^2\right)\left(A\cosh\left[\lambda\left(x-wt\right)\right] - A\sinh\left[\lambda\left(x-wt\right)\right] - \frac{pa_0a_1}{2\left(q+3pa_0^2\right)}\right)^2} \\ + \dfrac{a_1}{A\cosh\left[\lambda\left(x-wt\right)\right] - A\sinh\left[\lambda\left(x-wt\right)\right] - \frac{pa_0a_1}{2\left(q+3pa_0^2\right)}} \end{array} \right]. \tag{23}$$

In addition, if Equation (23) is written instead of $u = \ln v$ transformation, the analytical solution of Equation (1) is as follows,

$$u_1(x,t) = \ln \left[\begin{array}{l} a_0 + \dfrac{p a_0 a_1^2}{2\left(q+3pa_0^2\right)\left(A\cosh\left[\lambda\left(x-wt\right)\right] - A\sinh\left[\lambda\left(x-wt\right)\right] - \frac{pa_0a_1}{2\left(q+3pa_0^2\right)}\right)^2} \\ + \dfrac{a_1}{A\cosh\left[\lambda\left(x-wt\right)\right] - A\sinh\left[\lambda\left(x-wt\right)\right] - \frac{pa_0a_1}{2\left(q+3pa_0^2\right)}} \end{array} \right]. \tag{24}$$

The hyperbolic traveling wave solution of Equation (24) produced from the $(1/G')$-expansion method is as in Figure 1.

Case II.

$$\begin{array}{l} a_0 = \dfrac{pa_1^2 + \sqrt{p}\sqrt{pa_1^4 - 48qa_2^2}}{12pa_2}, \quad w = \dfrac{pa_2}{2\mu^2}, \quad \lambda = \dfrac{\mu a_1}{a_2}, \\ r = \dfrac{-\frac{pa_1^6}{a_2} - \frac{\sqrt{p}a_1^4\sqrt{pa_1^4-48qa_2^2}}{a_2} - \frac{24qa_2\sqrt{pa_1^4-48qa_2^2}}{\sqrt{p}}}{432a_2^2}, \end{array} \tag{25}$$

considering Equation (6), replacing Equation (25) into Equation (20), the following solution is attained

$$V = \frac{1}{12}\left[\frac{\sqrt{pa_1^4 - 48qa_2^2}}{\sqrt{p}a_2} + a_1^2 \left(\begin{array}{l} \dfrac{12}{Ae^{a_1\left(\frac{pt}{2\mu} - \frac{x\mu}{a_2}\right)}a_1 - a_2} + \dfrac{1}{a_2} \\ + \dfrac{12a_2}{\left(-Ae^{a_1\left(\frac{pt}{2\mu} - \frac{x\mu}{a_2}\right)}a_1 + a_2\right)^2} \end{array} \right) \right]. \tag{26}$$

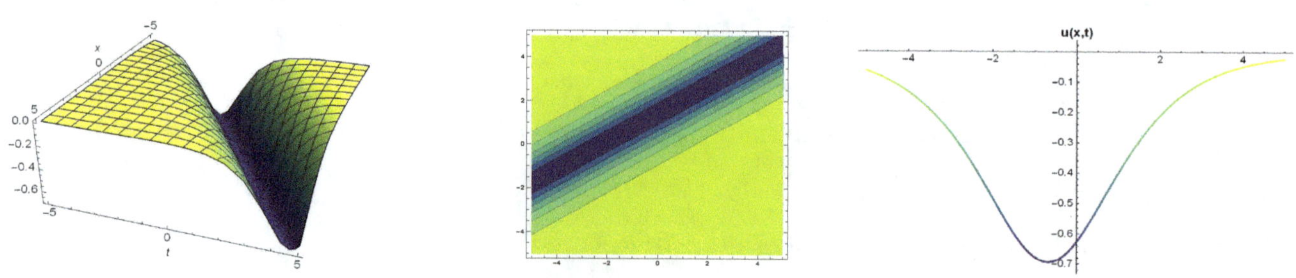

Figure 1. 3D, contour and 2D graphs respectively for $p = -1,\ \lambda = -0.8,\ A = -3,\ a_0 = -1,\ a_1 = -1,\ q = 2$ values of Equation (24).

In addition, if Equation (26) is written instead of $u = \ln v$ transformation, the analytical solution of Equation (1) is as follows,

$$u_2(x,t) = \ln\left[\frac{1}{12}\left[\frac{\sqrt{pa_1^4 - 48qa_2^2}}{\sqrt{pa_2}} + a_1^2\left(\begin{array}{c}\dfrac{12}{Ae^{a_1\left(\frac{pt}{2\mu}-\frac{x\mu}{a_2}\right)}a_1 - a_2} + \dfrac{1}{a_2}\\ +\dfrac{12a_2}{\left(-Ae^{a_1\left(\frac{pt}{2\mu}-\frac{x\mu}{a_2}\right)}a_1 + a_2\right)^2}\end{array}\right)\right]\right]. \tag{27}$$

The analytic solution of Equation (27) produced from the $(1/G)$-expansion method is as in Figure 2.

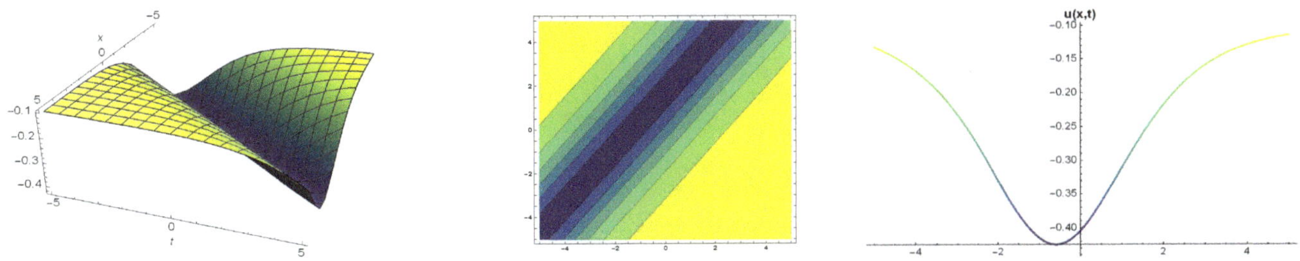

Figure 2. 3D, contour and 2D graphs respectively for $p = -1,\ \mu = -0.8,\ A = -3,\ a_2 = -1,\ a_1 = -1,\ q = 2$ values of Equation (27).

5. Solutions of The (Z-S) Equation Using $(G'/G, 1/G)$-Expansion Method

We consider Equation (1). Using transmutation $u(x,t) = u(\xi),\ \xi = x - wt, w \neq 0$, we get

$$-wu_{\xi\xi} + pe^{u} + qe^{-u} + re^{-2u} = 0, \tag{28}$$

To apply this method, we use transmutation $u = \ln v$ and $v = V(\xi)$, Equation (28) becomes

$$-w\left(VV'' - (V')^2\right) + pV^3 + qV + r = 0. \tag{29}$$

In Equation (29), we find balancing term $m = 2$ and in Equation (10), the following situation is obtained

$$u(\xi) = a_0 + a_1\phi[\xi] + b_1\psi[\xi] + a_2\phi[\xi]^2 + b_2\psi[\xi]^2, \tag{30}$$

where a_0, a_1, a_2, b_1, b_2 constants to be determined are unknown. Replacing Equation (30) into Equation (29) and the coefficients of the algebraic Equation (1) are equal to zero, we can establish the following algebraic equation systems

$$\begin{aligned}
Cons\ :\quad & r+qa_0+pa_0^3+w\lambda^2a_1^2-\frac{w\lambda^2\mu^2a_1^2}{\mu^2+\lambda^2\sigma}-2w\lambda^2a_0a_2+\frac{2w\lambda^2\mu^2a_0a_2}{\mu^2+\lambda^2\sigma}-\frac{w\lambda^2\mu a_0b_1}{\mu^2+\lambda^2\sigma}\\
&+\frac{4w\lambda^3\mu^3a_2b_1}{(\mu^2+\lambda^2\sigma)^2}-\frac{4w\lambda^3\mu a_2b_1}{\mu^2+\lambda^2\sigma}-\frac{2w\lambda^3\mu^2b_1^2}{(\mu^2+\lambda^2\sigma)^2}+\frac{w\lambda^3b_1^2}{\mu^2+\lambda^2\sigma}-\frac{3p\lambda^2a_0b_1^2}{\mu^2+\lambda^2\sigma}\\
&-\frac{2p\lambda^3\mu b_1^3}{(\mu^2+\lambda^2\sigma)^2}-\frac{q\lambda^2b_2}{\mu^2+\lambda^2\sigma}-\frac{4w\lambda^3\mu^2a_0b_2}{(\mu^2+\lambda^2\sigma)^2}+\frac{2w\lambda^3a_0b_2}{\mu^2+\lambda^2\sigma}-\frac{3p\lambda^2a_0^2b_2}{\mu^2+\lambda^2\sigma}\\
&-\frac{8w\lambda^4\mu^2a_2b_2}{(\mu^2+\lambda^2\sigma)^2}+\frac{2w\lambda^4a_2b_2}{\mu^2+\lambda^2\sigma}+\frac{6w\lambda^4\mu b_1b_2}{(\mu^2+\lambda^2\sigma)^2}-\frac{12p\lambda^3\mu a_0b_1b_2}{(\mu^2+\lambda^2\sigma)^2}=0,\\
\phi\,[\xi]\ :\quad & qa_1-2w\lambda a_0a_1+3pa_0^2a_1+2w\lambda^2a_1a_2-\frac{2w\lambda^2\mu^2a_1a_2}{\mu^2+\lambda^2\sigma}-\frac{2w\lambda^2\mu a_1b_1}{\mu^2+\lambda^2\sigma}\\
&-\frac{3p\lambda^2a_1b_1^2}{\mu^2+\lambda^2\sigma}-\frac{2w\lambda^3\mu^2a_1b_2}{(\mu^2+\lambda^2\sigma)^2}-\frac{6p\lambda^2a_0a_1b_2}{\mu^2+\lambda^2\sigma}-\frac{12p\lambda^3\mu a_1b_1b_2}{(\mu^2+\lambda^2\sigma)^2}=0,
\end{aligned}\tag{31}$$

$$\begin{aligned}
(\phi\,[\xi])^2:\quad & -\frac{w\lambda\mu^2a_1^2}{\mu^2+\lambda^2\sigma}+3pa_0a_1^2+qa_2-8w\lambda a_0a_2+\frac{2w\lambda\mu^2a_0a_2}{\mu^2+\lambda^2\sigma}+3pa_0^2a_2+2w\lambda^2a_2^2\\
&-\frac{2w\lambda^2\mu^2a_2^2}{\mu^2+\lambda^2\sigma}-\frac{w\lambda\mu a_0b_1}{\mu^2+\lambda^2\sigma}+\frac{4w\lambda^2\mu^3a_2b_1}{(\mu^2+\lambda^2\sigma)^2}-\frac{11w\lambda^2\mu a_2b_1}{\mu^2+\lambda^2\sigma}-\frac{2w\lambda^2\mu^2b_1^2}{(\mu^2+\lambda^2\sigma)^2}\\
&+\frac{2w\lambda^2b_1^2}{\mu^2+\lambda^2\sigma}-\frac{3p\lambda a_0b_1^2}{\mu^2+\lambda^2\sigma}-\frac{3p\lambda^2a_2b_1^2}{\mu^2+\lambda^2\sigma}-\frac{2p\lambda^2\mu b_1^3}{(\mu^2+\lambda^2\sigma)^2}-\frac{q\lambda b_2}{\mu^2+\lambda^2\sigma}\\
&-\frac{4w\lambda^2\mu^2a_0b_2}{(\mu^2+\lambda^2\sigma)^2}+\frac{8w\lambda^2a_0b_2}{\mu^2+\lambda^2\sigma}-\frac{3p\lambda a_0^2b_2}{\mu^2+\lambda^2\sigma}-\frac{3p\lambda^2a_1^2b_2}{\mu^2+\lambda^2\sigma}--\frac{16w\lambda^3\mu^2a_2b_2}{(\mu^2+\lambda^2\sigma)^2}\\
&+\frac{4w\lambda^3a_2b_2}{\mu^2+\lambda^2\sigma}-\frac{6p\lambda^2a_0a_2b_2}{\mu^2+\lambda^2\sigma}+\frac{14w\lambda^3\mu b_1b_2}{(\mu^2+\lambda^2\sigma)^2}-\frac{12p\lambda^2\mu a_0b_1b_2}{(\mu^2+\lambda^2\sigma)^2}-\frac{12p\lambda^3\mu a_2b_1b_2}{(\mu^2+\lambda^2\sigma)^2}=0,\\
(\phi\,[\xi])^3:\ & -2wa_0a_1+pa_1^3-2w\lambda a_1a_2-\frac{2w\lambda\mu^2a_1a_2}{\mu^2+\lambda^2\sigma}+6pa_0a_1a_2-\frac{2w\lambda\mu a_1b_1}{\mu^2+\lambda^2\sigma}-\frac{3p\lambda a_1b_1^2}{\mu^2+\lambda^2\sigma}\\
&-\frac{2w\lambda^2\mu^2a_1b_2}{(\mu^2+\lambda^2\sigma)^2}+\frac{4w\lambda^2a_1b_2}{\mu^2+\lambda^2\sigma}-\frac{6p\lambda a_0a_1b_2}{\mu^2+\lambda^2\sigma}-\frac{6p\lambda^2a_1a_2b_2}{\mu^2+\lambda^2\sigma}-\frac{12p\lambda^2\mu a_1b_1b_2}{(\mu^2+\lambda^2\sigma)^2}=0,\\
(\phi\,[\xi])^4:\quad & -wa_1^2-6wa_0a_2+3pa_1^2a_2-\frac{2w\lambda\mu^2a_2^2}{\mu^2+\lambda^2\sigma}+3pa_0a_2^2-\frac{7w\lambda\mu a_2b_1}{\mu^2+\lambda^2\sigma}+\frac{w\lambda b_1^2}{\mu^2+\lambda^2\sigma}\\
&-\frac{3p\lambda a_2b_1^2}{\mu^2+\lambda^2\sigma}+\frac{6w\lambda a_0b_2}{\mu^2+\lambda^2\sigma}-\frac{3p\lambda a_1^2b_2}{\mu^2+\lambda^2\sigma}-\frac{8w\lambda^2\mu^2a_2b_2}{(\mu^2+\lambda^2\sigma)^2}+\frac{6w\lambda^2a_2b_2}{\mu^2+\lambda^2\sigma}\\
&-\frac{6p\lambda a_0a_2b_2}{\mu^2+\lambda^2\sigma}-\frac{3p\lambda^2a_2^2b_2}{\mu^2+\lambda^2\sigma}+\frac{8w\lambda^2\mu b_1b_2}{(\mu^2+\lambda^2\sigma)^2}-\frac{12p\lambda^2\mu a_2b_1b_2}{(\mu^2+\lambda^2\sigma)^2}=0,\\
(\phi\,[\xi])^5:\quad & -4wa_1a_2+3pa_1a_2^2+\tfrac{4w\lambda a_1b_2}{\mu^2+\lambda^2\sigma}-\tfrac{6p\lambda a_1a_2b_2}{\mu^2+\lambda^2\sigma}=0,\\
(\phi\,[\xi])^6:\ & -2wa_2^2+pa_2^3+\tfrac{4w\lambda a_2b_2}{\mu^2+\lambda^2\sigma}-\tfrac{3p\lambda a_2^2b_2}{\mu^2+\lambda^2\sigma}=0,
\end{aligned}$$

$$\psi[\xi] : \quad -2w\lambda\mu a_1^2 + \frac{2w\lambda\mu^3 a_1^2}{\mu^2+\lambda^2\sigma} + 4w\lambda\mu a_0 a_2 - \frac{4w\lambda\mu^3 a_0 a_2}{\mu^2+\lambda^2\sigma} + qb_1 - w\lambda a_0 b_1 + \frac{2w\lambda\mu^2 a_0 b_1}{\mu^2+\lambda^2\sigma}$$
$$+3pa_0^2 b_1 - 2w\lambda^2 a_2 b_1 - \frac{8w\lambda^2\mu^4 a_2 b_1}{(\mu^2+\lambda^2\sigma)^2} + \frac{10w\lambda^2\mu^2 a_2 b_1}{\mu^2+\lambda^2\sigma} + \frac{4w\lambda^2\mu^3 b_1^2}{(\mu^2+\lambda^2\sigma)^2} - \frac{3w\lambda^2\mu b_1^2}{\mu^2+\lambda^2\sigma}$$
$$+\frac{6p\lambda\mu a_0 b_1^2}{\mu^2+\lambda^2\sigma} + \frac{4p\lambda^2\mu^2 b_1^3}{(\mu^2+\lambda^2\sigma)^2} - \frac{p\lambda^2 b_1^3}{\mu^2+\lambda^2\sigma} + \frac{2q\lambda\mu b_2}{\mu^2+\lambda^2\sigma} + \frac{8w\lambda^2\mu^3 a_0 b_2}{(\mu^2+\lambda^2\sigma)^2} - \frac{6w\lambda^2\mu a_0 b_2}{\mu^2+\lambda^2\sigma}$$
$$+\frac{6p\lambda\mu a_0^2 b_2}{\mu^2+\lambda^2\sigma} + \frac{16w\lambda^3\mu^3 a_2 b_2}{(\mu^2+\lambda^2\sigma)^2} - \frac{8w\lambda^3\mu a_2 b_2}{\mu^2+\lambda^2\sigma} - \frac{12w\lambda^3\mu^2 b_1 b_2}{(\mu^2+\lambda^2\sigma)^2} + \frac{3w\lambda^3 b_1 b_2}{\mu^2+\lambda^2\sigma}$$
$$+\frac{24p\lambda^2\mu^2 a_0 b_1 b_2}{(\mu^2+\lambda^2\sigma)^2} - \frac{6p\lambda^2 a_0 b_1 b_2}{\mu^2+\lambda^2\sigma} = 0,$$

$$\phi[\xi]\psi[\xi] : \quad 3w\mu a_0 a_1 - 4w\lambda\mu a_1 a_2 + \frac{4w\lambda\mu^3 a_1 a_2}{\mu^2+\lambda^2\sigma} - w\lambda a_1 b_1 + \frac{4w\lambda\mu^2 a_1 b_1}{\mu^2+\lambda^2\sigma} + 6pa_0 a_1 b_1$$
$$+\frac{6p\lambda\mu a_1 b_1^2}{\mu^2+\lambda^2\sigma} + \frac{4w\lambda^2\mu^3 a_1 b_2}{(\mu^2+\lambda^2\sigma)^2} - \frac{w\lambda^2\mu a_1 b_2}{\mu^2+\lambda^2\sigma} + \frac{12p\lambda\mu a_0 a_1 b_2}{\mu^2+\lambda^2\sigma} + \frac{24p\lambda^2\mu^2 a_1 b_1 b_2}{(\mu^2+\lambda^2\sigma)^2} - \frac{6p\lambda^2 a_1 b_1 b_2}{\mu^2+\lambda^2\sigma} = 0,$$

$$\phi[\xi]^2\psi[\xi] : \quad w\mu a_1^2 + 10w\mu a_0 a_2 - 4w\lambda\mu a_2^2 + \frac{4w\lambda\mu^3 a_2^2}{\mu^2+\lambda^2\sigma} - 2wa_0 b_1 + 3pa_1^2 b_1 - 5w\lambda a_2 b_1$$
$$+\frac{16w\lambda\mu^2 a_2 b_1}{\mu^2+\lambda^2\sigma} + 6pa_0 a_2 b_1 - \frac{3w\lambda\mu b_1^2}{\mu^2+\lambda^2\sigma} + \frac{6p\lambda\mu a_2 b_1^2}{\mu^2+\lambda^2\sigma} - \frac{p\lambda b_1^3}{\mu^2+\lambda^2\sigma} - \frac{14w\lambda\mu a_0 b_2}{\mu^2+\lambda^2\sigma}$$
$$+\frac{6p\lambda\mu a_1^2 b_2}{\mu^2+\lambda^2\sigma} + \frac{16w\lambda^2\mu^3 a_2 b_2}{(\mu^2+\lambda^2\sigma)^2} - \frac{12w\lambda^2\mu a_2 b_2}{\mu^2+\lambda^2\sigma} + \frac{12p\lambda\mu a_0 a_2 b_2}{\mu^2+\lambda^2\sigma} - \frac{16w\lambda^2\mu^2 b_1 b_2}{(\mu^2+\lambda^2\sigma)^2}$$
$$+\frac{7w\lambda^2 b_1 b_2}{\mu^2+\lambda^2\sigma} - \frac{6p\lambda a_0 b_1 b_2}{\mu^2+\lambda^2\sigma} + \frac{24p\lambda^2\mu^2 a_2 b_1 b_2}{(\mu^2+\lambda^2\sigma)^2} - \frac{6p\lambda^2 a_2 b_1 b_2}{\mu^2+\lambda^2\sigma} = 0,$$

$$\phi[\xi]^3\psi[\xi] : 5w\mu a_1 a_2 - 2wa_1 b_1 + 6pa_1 a_2 b_1 - \frac{9w\lambda\mu a_1 b_2}{\mu^2+\lambda^2\sigma} + \frac{12p\lambda\mu a_1 a_2 b_2}{\mu^2+\lambda^2\sigma} - \frac{6p\lambda a_1 b_1 b_2}{\mu^2+\lambda^2\sigma} = 0,$$

$$\phi[\xi]^4\psi[\xi] : 2w\mu a_2^2 - 4wa_2 b_1 + 3pa_2^2 b_1 - \frac{12w\lambda\mu a_2 b_2}{\mu^2+\lambda^2\sigma}$$
$$+\frac{6p\lambda\mu a_2^2 b_2}{\mu^2+\lambda^2\sigma} + \frac{4w\lambda b_1 b_2}{\mu^2+\lambda^2\sigma} - \frac{6p\lambda a_2 b_1 b_2}{\mu^2+\lambda^2\sigma} = 0,$$

$$\psi[\xi]^4 : \quad -2w\mu^2 a_2 b_2 + 3w\mu b_1 b_2 + 3pb_1^2 b_2 - 2w\lambda b_2^2 + 3pa_0 b_2^2 = 0,$$
$$\phi[\xi]\psi[\xi]^4 : \quad 3pa_1 b_2^2 = 0,$$
$$\phi[\xi]^2\psi[\xi]^4 : \quad -2wb_2^2 + 3pa_2 b_2^2 = 0,$$
$$\psi[\xi]^5 : \quad 2w\mu b_2^2 + 3pb_1 b_2^2 = 0,$$
$$\psi[\xi]^6 : \quad pb_2^3 = 0,$$

Our aim with the computer package program was reaching the solutions of system (31) and we attained the following situations.

If $\lambda < 0$,

Case I:

$$a_0 = \frac{4pw\lambda - \sqrt{-3p^3 q + 4p^2 w^2 \lambda^2}}{3p^2},\ a_1 = 0,\ b_1 = 0,\ a_2 = \frac{2w}{p},\ b_2 = 0,\ \mu = 0,$$
$$r = \frac{1}{27p^2}2\left(16w^3\lambda^3 + 3q\sqrt{p^2(-3pq + 4w^2\lambda^2)} + \frac{8w^2\lambda^2\sqrt{p^2(-3pq+4w^2\lambda^2)}}{p}\right), \tag{32}$$

considering Equation (6), replacing Equation (32) into Equation (30), the following solution is attained

$$V=\frac{4pw\lambda-\sqrt{-3p^3q+4p^2w^2\lambda^2}}{3p^2}+\frac{2w\left(c_2\sqrt{-\lambda}\cosh\left[(-tw+x)\sqrt{-\lambda}\right]+c_1\sqrt{-\lambda}\sinh\left[(-tw+x)\sqrt{-\lambda}\right]\right)^2}{p\left(c_1\cosh\left[(-tw+x)\sqrt{-\lambda}\right]+c_2\sinh\left[(-tw+x)\sqrt{-\lambda}\right]\right)^2}. \tag{33}$$

In addition, if Equation (33) is written instead of $u=\ln v$ transformation, the hyperbolic traveling wave solution of Equation (1) is as follows,

$$u_1(x,t)=\ln\left[\frac{4pw\lambda-\sqrt{-3p^3q+4p^2w^2\lambda^2}}{3p^2}+\frac{2w\left(c_2\sqrt{-\lambda}\cosh\left[(-tw+x)\sqrt{-\lambda}\right]+c_1\sqrt{-\lambda}\sinh\left[(-tw+x)\sqrt{-\lambda}\right]\right)^2}{p\left(c_1\cosh\left[(-tw+x)\sqrt{-\lambda}\right]+c_2\sinh\left[(-tw+x)\sqrt{-\lambda}\right]\right)^2}\right]. \tag{34}$$

The hyperbolic traveling wave solution of Equation (34) produced from the $(G'/G,1/G)$-expansion method is as in Figure 3.

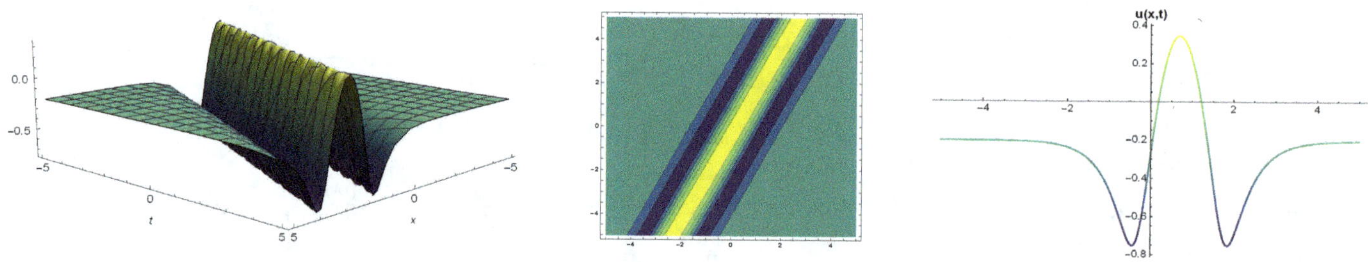

Figure 3. 3D, contour and 2D graphs respectively for $p=-0.5,\ \lambda=-1,\ c_2=-1,\ c_1=5,\ q=-1,\ w=0.5$ values of Equation (34).

Case II:

$$a_0=-\frac{3r}{4q},\ a_1=0,\ b_1=0,\ a_2=\frac{9r}{4q\lambda},\ b_2=0,\ \mu=0,\ w=-\frac{q^2}{12r\lambda},\ p=-\frac{2q^3}{27r^2}, \tag{35}$$

considering Equation (6), replacing Equation (35) into Equation (30), the following solution is attained

$$V=-\frac{3r}{4q}+\frac{9r\left(c_2\sqrt{-\lambda}\cosh\left[\left(x+\frac{q^2t}{12r\lambda}\right)\sqrt{-\lambda}\right]+c_1\sqrt{-\lambda}\sinh\left[\left(x+\frac{q^2t}{12r\lambda}\right)\sqrt{-\lambda}\right]\right)^2}{4q\lambda\left(c_1\cosh\left[\left(x+\frac{q^2t}{12r\lambda}\right)\sqrt{-\lambda}\right]+c_2\sinh\left[\left(x+\frac{q^2t}{12r\lambda}\right)\sqrt{-\lambda}\right]\right)^2}, \tag{36}$$

In addition, if Equation (36) is written instead of $u=\ln v$ transformation, the hyperbolic traveling wave solution of Equation (1) is as follows,

$$u_2(x,t)=\ln\left[-\frac{3r}{4q}+\frac{9r\left(c_2\sqrt{-\lambda}\cosh\left[\left(x+\frac{q^2t}{12r\lambda}\right)\sqrt{-\lambda}\right]+c_1\sqrt{-\lambda}\sinh\left[\left(x+\frac{q^2t}{12r\lambda}\right)\sqrt{-\lambda}\right]\right)^2}{4q\lambda\left(c_1\cosh\left[\left(x+\frac{q^2t}{12r\lambda}\right)\sqrt{-\lambda}\right]+c_2\sinh\left[\left(x+\frac{q^2t}{12r\lambda}\right)\sqrt{-\lambda}\right]\right)^2}\right]. \tag{37}$$

The hyperbolic traveling wave solution of Equation (37) produced from the $(G'/G,1/G)$-expansion method is as in Figure 4.

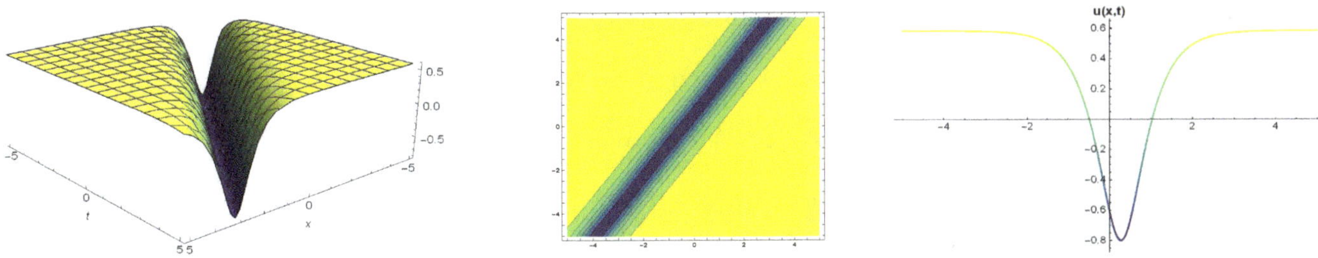

Figure 4. 3D, contour and 2D graphs respectively for $\lambda = -1,\ c_2 = 2, c_1 = 5,\ q = -5,\ r = 3$ values of Equation (37).

If $\lambda > 0$,
Case III:

$$a_0 = \frac{2p\lambda a_2 - \sqrt{-3pq + p^2\lambda^2 a_2^2}}{3p},\ a_1 = 0,\ b_1 = 0,\ b_2 = 0,\ \mu = 0,\ w = \frac{pa_2}{2},$$
$$r = \frac{2}{27}\left(2p\lambda^3 a_2^3 + \frac{3q\sqrt{p\left(-3q + p\lambda^2 a_2^2\right)}}{p} + 2\lambda^2 a_2^2\sqrt{p\left(-3q + p\lambda^2 a_2^2\right)}\right), \tag{38}$$

considering Equation (6), replacing Equation (38) into Equation (30), the following solution is attained

$$V = \frac{\left(c_2\sqrt{\lambda}\cos\left[\sqrt{\lambda}\left(x - \frac{1}{2}pta_2\right)\right] - c_1\sqrt{\lambda}\sin\left[\sqrt{\lambda}\left(x - \frac{1}{2}pta_2\right)\right]\right)^2 a_2}{\left(c_1\cos\left[\sqrt{\lambda}\left(x - \frac{1}{2}pta_2\right)\right] + c_2\sin\left[\sqrt{\lambda}\left(x - \frac{1}{2}pta_2\right)\right]\right)^2} + \frac{2p\lambda a_2 - \sqrt{-3pq + p^2\lambda^2 a_2^2}}{3p}. \tag{39}$$

In addition, if Equation (39) is written instead of $u = \ln v$ transformation, the trigonometric traveling wave solution of Equation (1) is as follows,

$$u_3(x,t) = \ln\left[\frac{\left(c_2\sqrt{\lambda}\cos\left[\sqrt{\lambda}\left(x - \frac{1}{2}pta_2\right)\right] - c_1\sqrt{\lambda}\sin\left[\sqrt{\lambda}\left(x - \frac{1}{2}pta_2\right)\right]\right)^2 a_2}{\left(c_1\cos\left[\sqrt{\lambda}\left(x - \frac{1}{2}pta_2\right)\right] + c_2\sin\left[\sqrt{\lambda}\left(x - \frac{1}{2}pta_2\right)\right]\right)^2} + \frac{2p\lambda a_2 - \sqrt{-3pq + p^2\lambda^2 a_2^2}}{3p}\right]. \tag{40}$$

The trigonometric traveling wave solution of Equation (40) produced from the $(G'/G, 1/G)$-expansion method is as in Figure 5.

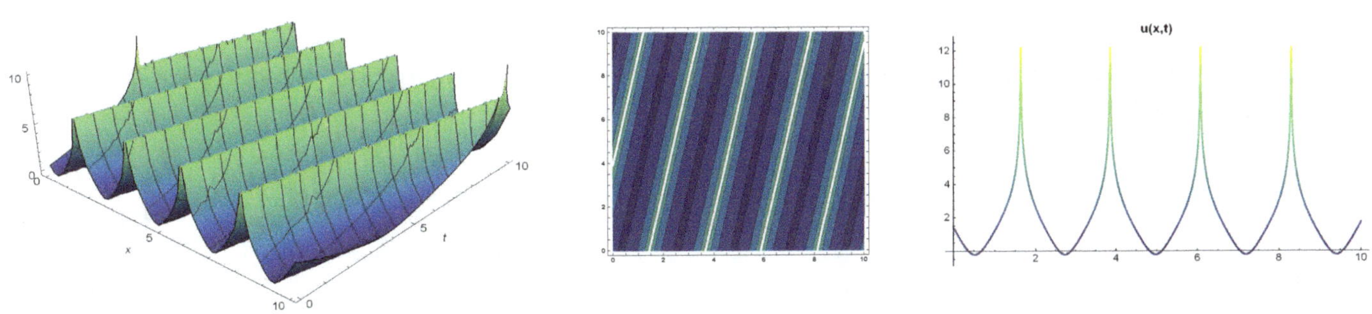

Figure 5. 3D, contour and 2D graphs respectively for $c_2 = -0.5,\ c_1 = -1,\ q = 1,\ a_0 = 0.5,\ p = -0.2,\ \lambda = 2,\ a_2 = -2$ values of Equation (40).

Case IV:

$$a_1 = 0, \quad b_1 = 0, \quad b_2 = 0, \quad \mu = 0, \quad w = \frac{pa_2}{2}, \quad r = 2pa_0(-a_0 + \lambda a_2)^2, \\ q = -3pa_0^2 + 4p\lambda a_0 a_2 - p\lambda^2 a_2^2, \tag{41}$$

considering Equation (6), replacing Equation (41) into Equation (30), the following solution is attained

$$V = a_0 - \lambda a_2 + \frac{\left(c_1{}^2 + c_2{}^2\right) \lambda a_2}{\left(c_1 \cos\left[\sqrt{\lambda}\left(x - \frac{1}{2}pta_2\right)\right] + c_2 \sin\left[\sqrt{\lambda}\left(x - \frac{1}{2}pta_2\right)\right]\right)^2}. \tag{42}$$

In addition, if Equation (42) is written instead of $u = \ln v$ transformation, the analytical solution of Equation (1) is as follows,

$$u_4(x,t) = \ln\left[a_0 - \lambda a_2 + \frac{\left(c_1{}^2 + c_2{}^2\right) \lambda a_2}{\left(c_1 \cos\left[\sqrt{\lambda}\left(x - \frac{1}{2}pta_2\right)\right] + c_2 \sin\left[\sqrt{\lambda}\left(x - \frac{1}{2}pta_2\right)\right]\right)^2}\right]. \tag{43}$$

The trigonometric traveling wave solution of Equation (43) produced from the $(G'/G, 1/G)$-expansion method is as in Figure 6.

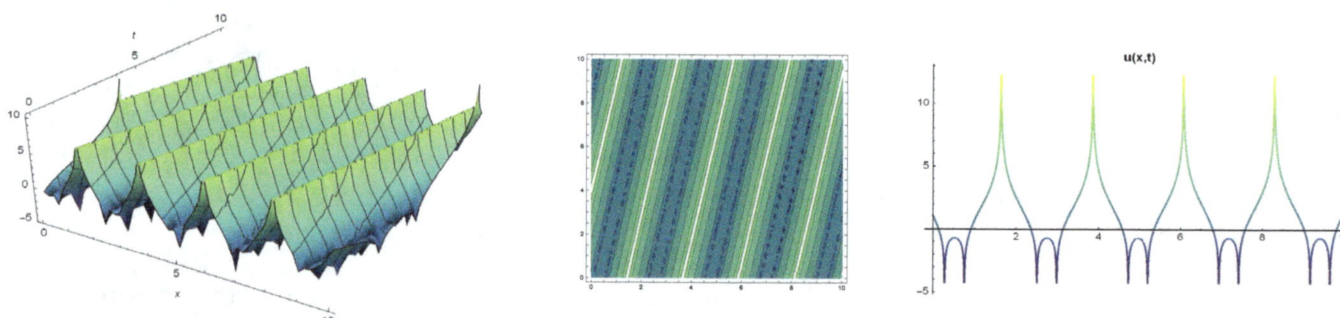

Figure 6. 3D, contour and 2D graphs respectively $c_2 = -0.5,\ c_1 = -1, a_0 = 0.5,\ \ p = -0.2,\ \ \lambda = 2,\ \ a_2 = -2$ values of Equation (43).

If $\lambda = 0$,
Case V:

$$a_0 = \frac{i\sqrt{q}}{\sqrt{3}\sqrt{p}},\ a_1 = 0,\ b_1 = 0,\ a_2 = \frac{2w}{p},\ b_2 = 0,\ \mu = 0,\ r = -\frac{2iq^{3/2}}{3\sqrt{3}\sqrt{p}}, \tag{44}$$

considering Equation (6), replacing Equation (44) into Equation (30), the following solution is attained

$$V = \frac{i\sqrt{q}}{\sqrt{3}\sqrt{p}} + \frac{2c_2{}^2 w}{p(c_1 + c_2(-tw + x))^2}. \tag{45}$$

In addition, if Equation (45) is written instead of $u = \ln v$ transformation, the complex analytical solution of Equation (1) is as follows,

$$u_5(x,t) = \ln\left[\frac{i\sqrt{q}}{\sqrt{3}\sqrt{p}} + \frac{2c_2{}^2 w}{p(c_1 + c_2(-tw + x))^2}\right]. \tag{46}$$

The complex analytical solution of Equation (46) produced from the $(G'/G, 1/G)$-expansion method is as in Figures 7 and 8.

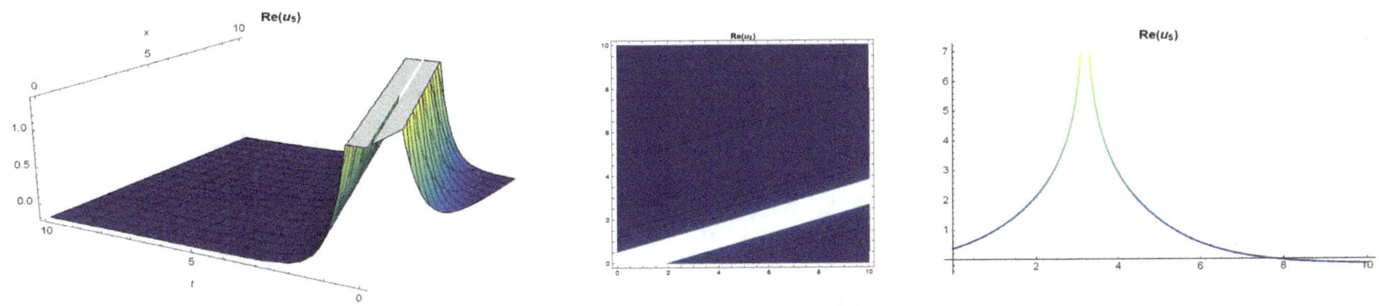

Figure 7. The real part of the 3D, contour and 2D graphics respectively for $c_2 = 5,\ c_1 = -1,\ q = -1,\ p = -0.5,\ w = 3$ values of Equation (46).

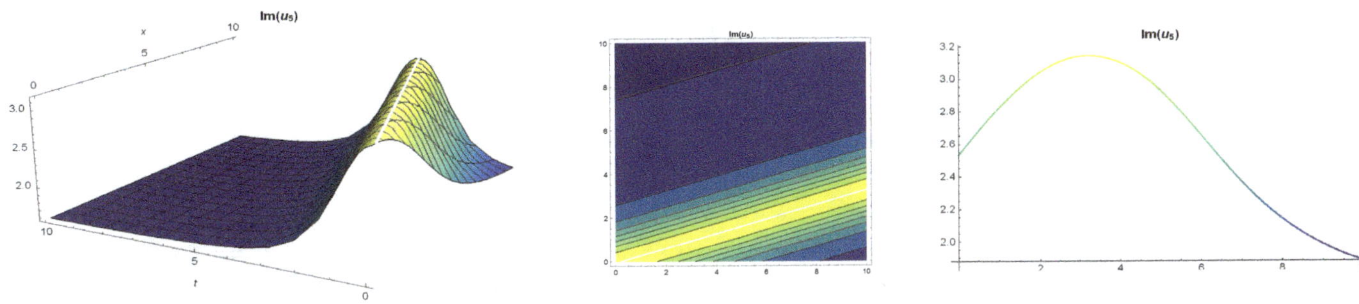

Figure 8. The imaginary part of 3D, contour and 2D graphs respectively for $c_2 = 5,\ c_1 = -1,\ q = -1,\ p = -0.5,\ w = 3$ values of Equation (46).

Case VI:

$$a_1 = 0,\ b_1 = 0,\ b_2 = 0,\ \mu = 0,\ r = 2pa_0^3,\ w = \frac{pa_2}{2},\ q = -3pa_0^2, \tag{47}$$

considering Equation (6), replacing Equation (47) into Equation (30), the following solution is attained

$$V = a_0 + \frac{{c_2}^2 a_2}{\left(c_1 + c_2\left(x - \frac{1}{2}pta_2\right)\right)^2}. \tag{48}$$

In addition, if Equation (48) is written instead of $u = \ln v$ transformation, the analytical solution of Equation (1) is as follows,

$$u_5(x,t) = \ln\left[a_0 + \frac{{c_2}^2 a_2}{\left(c_1 + c_2\left(x - \frac{1}{2}pta_2\right)\right)^2}\right]. \tag{49}$$

The analytical solution of Equation (49) produced from the $(G'/G, 1/G)$-expansion method is as in Figure 9.

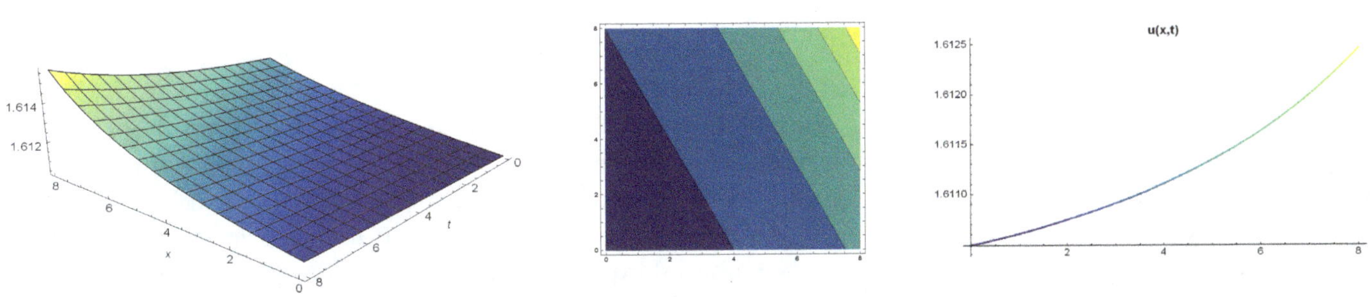

Figure 9. 3D, contour and 2D graphs respectively for $c_2 = -0.1,\ c_1 = 2,\ a_2 = 2,\ p = -0.5,\ a_0 = 5$ values of Equation (49).

6. Results and Discussion

There are various methods for obtaining the exact solution of NLEEs. Analytic solutions of the Z-S equations were successfully constructed by using both methods. When $u = \ln v$ transformation is performed in both methods, solution functions are logarithmic. This is a result of the exponential functions in the structure of the Z-S equation. In different cases of p, q, r coefficients of Z-S equation, this equation is recognized by a different name. The main purpose of this article is to present the solution of Z-S equation and also to solve the equations of the sinh–Gordon, (DBM), (TDB) and the Liouville equations. For example; The Z-S equation for $r = q = 0$ is called the Liouville equation. In this study, if the values $r = q = 0$ in Equation (24) are written, the traveling wave solution of the Liouville equation is obtained as

$$u(x,y) = \ln\left[\begin{array}{l} a_0 + \dfrac{a_1^2}{6a_0\left(A\cosh\left[\lambda\left(x - \frac{3pta_0}{\lambda^2}\right)\right] - A\sinh\left[\lambda\left(x - \frac{3pta_0}{\lambda^2}\right)\right] - \frac{a_1}{6a_0}\right)^2} \\ + \dfrac{a_1}{A\cosh\left[\lambda\left(x - \frac{3pta_0}{\lambda^2}\right)\right] - A\sinh\left[\lambda\left(x - \frac{3pta_0}{\lambda^2}\right)\right] - \frac{a_1}{6a_0}} \end{array}\right], \tag{50}$$

Similarly, the solutions of the sinh–Gordon, DBM, TDB and Liouville equations can be obtained by using both the methods. The V solution described above is presented in hyperbolic form in $(1/G')$-expansion method, and in hyperbolic, trigonometric and rational forms in $(G'/G, 1/G)$-expansion method. In this case, $(G'/G, 1/G)$-expansion method is advantageous in terms of solution. However, in $(G'/G, 1/G)$-expansion method, the process complexity is higher. This can be observed in the system Equation (31), $(1/G')$-expansion method is more advantageous in terms of process. 2D, 3D and contour graphics which we consider will help in traveling wave solutions which have considerable importance in applied sciences, are presented. In order to draw these graphs, real values are given to arbitrary constants in the analytical solution.

This problem contains the properties of many equations. For the different states of the coefficients, to offer the solution of the equation which includes equations with different names, also to offer the solution of the subclass equations. It is known that each equation has different meanings. For example, with the interaction of solitons produced by sinh-Gordon equation, kink and antikink solutions came to the fore. We can make the same comments for the solutions obtained in these studies for. In this case, the equation we dealt with is the umbrella task. It makes the wave solutions valuable because it will carry the properties of the equations under the umbrella. The most important factor that stands out in this study is to take a different solutions from the solution in classical $(G'/G, 1/G)$-expansion method. This results in obtaining different types of traveling wave solutions from the classical method. It also creates a basis for a new study. This is an improved method that can produce different solutions by adding the solution we offer with the Equation (17) to the classical $(G'/G, 1/G)$ solution. Because the Equation (17) presented and the equilibrium term 2 in the equation discussed are different from the solutions offered in the classical $(G'/G, 1/G)$-expansion method. Different types of solutions were obtained with both methods and both methods can be used as important instruments to get traveling wave solution for many different NLEEs.

7. Conclusions

In this article, we have applied the $(1/G')$-expansion and $(G'/G, 1/G)$-expansion methods to derive analytic solutions for the Zhiber-Shabat equation. The solutions obtained are complex rational, hyperbolic, rational and trigonometric type traveling wave solutions. The 2D, 3D and contour graphics of these solutions were presented by giving value to arbitrary parameters. These graphs represent the stationary wave at any given moment. As it is very difficult to obtain the solutions of NLEEs, in this study traveling wave solutions of Zhiber-Shabat equation are presented applying two complex methods using many complex operations and transformations. These are very effective and powerful methods for obtaining analytical solutions and can be used to obtain solutions of many mathematical models representing physical phenomena. The accuracy of the attained solutions has been assured by putting them back into the original equations with the help of the computer package program.

Author Contributions: Conceptualization, A.Y.; data curation, A.Y. and H.D.; formal analysis, H.A.; funding acquisition, S.-W.Y.; investigation, H.D.; methodology, A.Y.; software, H.D.; supervision, A.Y.; writing—original draft, H.A.; writing—review and editing, S.-W.Y. All authors have read and agreed to the published version of the manuscript.

Acknowledgments: The authors thank the Editor-in-Chief and unknown referees for the fruitful comments and significant remarks that helped them in improving the quality and readability of the paper, which led to a significant improvement of the paper.

References

1. Wazwaz, A.M. A sine-cosine method for handlingnonlinear wave equations. *Math. Comput. Model.* **2004**, *40*, 499–508.
2. Baskonus, H.M.; Sulaiman, T.A.; Bulut, H.; Aktürk, T. Investigations of dark, bright, combined dark-bright optical and other soliton solutions in the complex cubic nonlinear Schrödinger equation withδ-potential. *Superlattices Microstruct.* **2018**, *115*, 19–29.
3. Cattani, C.; Sulaiman, T.A.; Baskonus, H.M.; Bulut, H. On the soliton solutions to the Nizhnik-Novikov-Veselov and the Drinfel'd-Sokolov systems. *Opt. Quantum Electron.* **2018**, *50*. doi:10.1007/s11082-018-1406-3.
4. Yokuş, A.; Kaya, D. Traveling wave solutions of some nonlinear partial differential equations by using extended-expansion method. *İstanbul Ticaret Üniversitesi Fen Bilimleri Dergisi* **2015**, *28*, 85–92.
5. Durur, H. Different types analytic solutions of the (1+1)-dimensional resonant nonlinear Schrödinger's equation using (G'/G)-expansion method. *Mod. Phys. Lett. B* **2020**, *34*. doi:10.1142/S0217984920500360.
6. Bulut, H.; Yel, G.; Başkonuş, H.M. An application of improved Bernoulli sub-equation function method to the nonlinear time-fractional burgers equation. *Turk. J. Math. Comput. Sci.* **2016**, *5*, 1–7.
7. Ahmad, H.; Seadawy, A.R.; Khan, T.A. Numerical solution of Korteweg-de Vries-Burgers equation by the modified variational Iteration algorithm-II arising in shallow water waves. *Phys. Scr.* **2020**. doi:10.1088/1402-4896/ab6070.
8. Ahmad, H.; Seadawy, A.R.; Khan, T.A. Study on Numerical Solution of Dispersive Water Wave Phenomena by Using a Reliable Modification of Variational Iteration Algorithm. *Math. Comput. Simul.* **2020**. doi:10.1016/j.matcom.2020.04.005.
9. Ahmad, H.; Khan, T.A.; Cesarano, C. Numerical Solutions of Coupled Burgers' Equations. *Axioms* **2019**, *8*, 119.
10. Durur, H.; Taşbozan, O.; Kurt, A.; Şenol, M. New Wave Solutions of Time Fractional Kadomtsev-Petviashvili Equation Arising In the Evolution of Nonlinear Long Waves of Small Amplitude. *Erzincan Univ. J. Inst. Sci. Technol.* **2019**, *12*, 807–815.
11. Aziz, I.; Šarler, B. The numerical solution of second-order boundary-value problems by collocation method with the Haar wavelets. *Math. Comput. Model.* **2010**, *52*, 1577–1590.
12. Nawaz, M.; Ahmad, I.; Ahmad, H. A radial basis function collocation method for space-dependent inverse heat problems. *J. Appl. Comput. Mech.* **2020**, doi:10.22055/jacm.2020.32999.2123.
13. Yokuş, A.; Durur, H. Complex hyperbolic traveling wave solutions of Kuramoto-Sivashinsky equation using $(1/G')$ expansion method for nonlinear dynamic theory. *J. BalıKesir Univ. Inst. Sci. Technol.* **2010**, *21*, 590–599.
14. Durur, H.; Yokuş, A. Analytical solutions of Kolmogorov–Petrovskii–Piskunov equation.*Balıkesir Üniversitesi Fen Bilimleri Enstitüsü Dergisi* **2020**, *22*, 628–636.
15. Yokuş, A.; Durur, H.; Ahmad, H. Hyperbolic type solutions for the couple Boiti-Leon-Pempinelli system. *Facta Univ. Ser. Math. Inform.* **2020**, *35*, 523–531.
16. Darvishi, M.; Arbabi, S.; Najafi, M.; Wazwaz, A. Traveling wave solutions of a (2 + 1)-dimensional Zakharov-like equation by the first integral method and the tanh method. *Optik* **2016**, *127*, 6312–6321.

17. Kaya, D.; Yokus, A. A numerical comparison of partial solutions in the decomposition method for linear and nonlinear partial differential equations. *Math. Comput. Simul.* **2002**, *60*, 507–512.
18. Kaya, D.; Yokus, A. A decomposition method for finding solitary and periodic solutions for a coupled higher-dimensional Burgers equations. *Appl. Math. Comput.* **2005**, *164*, 857–864.
19. Yavuz, M.; Özdemir, N. A quantitative approach to fractional option pricing problems with decomposition series. *Konuralp J. Math.* **2018**, *6*, 102–109.
20. Jin-Ming, Z.; Yao-Ming, Z. The Hirota bilinear method for the coupled Burgers equation and the high-order Boussinesq—Burgers equation. *Chin. Phys. B* **2011**, *20*. doi:10.1088/1674-1056/20/1/010205.
21. Ahmad, H. Variational iteration method with an auxiliary parameter for solving differential equations of the fifth order. *Nonlinear Sci. Lett. A* **2018**, *9*, 27–35.
22. Ahmad, H.; Khan, T.A. Variational iteration algorithm-I with an auxiliary parameter for wave-like vibration equations. *J. Low Freq. Noise Vib. Act. Control* **2019**, *38*, 1113–1124.
23. Ahmad, H.; Khan, T.A. Variational iteration algorithm I with an auxiliary parameter for the solution of differential equations of motion for simple and damped mass–spring systems. *Noise Vib. Worldw.* **2020**, *51*, 12–20.
24. Ahmad, H.; Seadawy, A.R.; Khan. T.A.; Thounthong, P. Analytic Approximate Solutions for Some Nonlinear Parabolic Dynamical Wave Equations. *J. Taibah Univ. Sci.* **2020**, *14*, 346–358.
25. Yang, X.J.; Srivastava, H.M.; Cattani, C. Local fractional homotopy perturbation method for solving fractal partial differential equations arising in mathematical physics. *Rom. Rep. Phys.* **2015**, *67*, 752–761.
26. Durur, H.; Şenol, M.; Kurt, A.; Taşbozan, O. Zaman-Kesirli Kadomtsev-Petviashvili Denkleminin Conformable Türev ile Yaklaşık Çözümleri. *Erzincan Univ. J. Inst. Sci. Technol.* **2019**, *12*, 796–806.
27. Yokus, A.; Kuzu, B.; Demiroğlu, U. Investigation of solitary wave solutions for the (3 + 1)-dimensional Zakharov–Kuznetsov equation. *Int. J. Mod. Phys. B* **2019**, *33*. doi:10.1142/S0217979219503508.
28. Ricceri, B. A Class of Equations with Three Solutions. *Mathematics* **2020**, *8*, 478.
29. Treanţă, S. On the Kernel of a Polynomial of Scalar Derivations. *Mathematics* **2020**, *8*, 515.
30. Treanţă, S.; Vârsan C. Weak small controls and approximations associated with controllable affine control systems. *J. Differ. Equ.* **2013**, *255*, 1867–1882.
31. Ahmad, H.; Khan, T.; Stanimirovic, P.; Ahmad, I. Modified Variational Iteration Technique for the Numerical Solution of Fifth Order KdV Type Equations. *J. Appl. Comput. Mech.* **2020**. doi:10.22055/jacm.2020.33305.2197.
32. Doroftei, M.; Treanta, S. Higher order hyperbolic equations involving a finite set of derivations. *Balk. J. Geom. Its Appl.* **2012**, *17*, 22–33.
33. Treanţă, S. Gradient Structures Associated with a Polynomial Differential Equation. *Mathematics* **2020**, *8*, 535.
34. Ahmad, H.; Khan, T.A.; Yao, S. Numerical solution of second order Painlevé differential equation. *J. Math. Comput. Sci.* **2020**, *21*, 150–157.
35. Ahmad, H.; Rafiq, M.; Cesarano, C.; Durur, H. Variational Iteration Algorithm-I with an Auxiliary Parameter for Solving Boundary Value Problems. *Earthline J. Math. Sci.* **2020**, *3*, 229–247.
36. Kaya, D.; Yokuş, A.; Demiroğlu, U. Comparison of Exact and Numerical Solutions for the Sharma–Tasso–Olver Equation. In *Numerical Solutions of Realistic Nonlinear Phenomena*; Springer: Cham, Switzerland, 2020; pp. 53–65.
37. Kurt, A.; Tasbozan, O.; Durur, H. The Exact Solutions of Conformable Fractional Partial Differential Equations Using New Sub Equation Method. *Fundam. J. Math. Appl.* **2019**, *2*, 173–179.
38. Ali, K.K.; Yilmazer, R.; Yokus, A.; Bulut, H. Analytical solutions for the (3 + 1)-dimensional nonlinear extended quantum Zakharov–Kuznetsov equation in plasma physics. *Phys. A Stat. Mech. Its Appl.* **2020**, *548*. doi:10.1016/j.physa.2020.124327.
39. Borhanifar, A.; Moghanlu, A.Z. Application of the (G'/G)-expansion method for the Zhiber-Shabat equation and other related equations. *Math. Comput. Model.* **2011**, *54*, 2109–2116.
40. Tang, Y.; Xu, W.; Shen, J.; Gao, L. Bifurcations of traveling wave solutions for Zhiber-Shabat equation. *Nonlinear Anal. Theory Methods Appl.* **2007**, *67*, 648–656.
41. Chen, A.; Huang, W.; Li, J. Qualitative behavior and exact travelling wave solutions of the Zhiber-Shabat equation. *J. Comput. Appl. Math.* **2009**, *230*, 559–569.
42. Hafez, M.G.; Kauser, M.A.; Akter, M.T. Some New Exact Traveling Wave Solutions for the Zhiber-Shabat Equation. *J. Adv. Math. Comput. Sci.* **2014**, 2582–2593, doi:10.9734/BJMCS/2014/11563.

43. Tala-Tebue, E.; Djoufack, Z.I.; Tsobgni-Fozap, D.C.; Kenfack-Jiotsa, A.; Kapche-Tagne, F.; Kofané, T.C. Traveling wave solutions along microtubules and in the Zhiber-Shabat equation. *Chin. J. Phys.* **2017**, *55*, 939–946.
44. He, B.; Long, Y.; Rui, W. New exact bounded travelling wave solutions for the Zhiber-Shabat equation. *Nonlinear Anal. Theory Methods Appl.* **2009**, *71*, 1636–1648.

11

Efficient Computation of Highly Oscillatory Fourier Transforms with Nearly Singular Amplitudes over Rectangle Domains

Zhen Yang and Junjie Ma *

School of Mathematics and Statistics, Guizhou University, Guiyang 550025, Guizhou, China
* Correspondence: jjma@gzu.edu.cn

Abstract: In this paper, we consider fast and high-order algorithms for calculation of highly oscillatory and nearly singular integrals. Based on operators with regard to Chebyshev polynomials, we propose a class of spectral efficient Levin quadrature for oscillatory integrals over rectangle domains, and give detailed convergence analysis. Furthermore, with the help of adaptive mesh refinement, we are able to develop an efficient algorithm to compute highly oscillatory and nearly singular integrals. In contrast to existing methods, approximations derived from the new approach do not suffer from high oscillatory and singularity. Finally, several numerical experiments are included to illustrate the performance of given quadrature rules.

Keywords: highly oscillatory integral; Chebyshev polynomial; nearly singular; Levin quadrature rule; adaptive mesh refinement

---

1. Introduction

Highly oscillatory integrals frequently arise in acoustic scattering [1], computational physical optics [2], computational electromagnetics [3], and related fields. Generally, dramatically changing integrands make classical approximations perform poor. Therefore, studies on numerical calculation of highly oscillatory integrals have attracted much attention during the past few decades, and a variety of contributions has been made, for example, Filon-type quadrature [4,5], numerical steepest descent method [6], Levin method [7], and so on.

When the phase is nonlinear, researchers usually resort to Levin-type quadrature, which originates from David Levin's pioneering work in [7]. By transforming the oscillatory integration problem into a special ordinary differential equation, one could get an efficient approximation to the generalized Fourier transform with the help of collocation methods. Afterwards, Levin analyzed the convergence rate of the innovative approach in [8]. Analogous to Filon-type quadrature, the Levin-type method based on Hermite interpolation was developed by Olver in [9]. Application of Hermite interpolation definitely increased the convergence rate of the numerical method with respect to the frequency. In [10], Li et al. proposed a stable and high-order Levin quadrature rule by employing the spectral Chebyshev collocation method and truncated singular value decomposition technique. Multiquadric radial basis functions were applied to Levin's equation and an innovative composite Levin method was presented in [11]. Numerical tests manifested that such kind of algorithms was able to deal with stationary problems. Sparse solvers for Levin's equation in one-dimension were constructed by employing recurrence of Chebyshev polynomials in [12,13]. Meanwhile, a class of preconditioners was proposed by the second author to deal with the ill-conditioned linear system in [13]. Molabahrami studied the Galerkin method for Levin's equation and developed the Galerkin–Levin method for oscillatory integrals in [14].

Levin-type quadrature rules were extended to solving singular problems in the past several years. In [15], Wang and Xiang employed the technique of singularity separation and transformed Levin's equation into coupled non-singular ordinary differential equations. By solving the transformed equations numerically, they obtained an efficient Levin quadrature rule for weakly singular integrals with highly oscillatory Fourier kernels. Recently, the second author proposed the fractional Jacobi-Galerkin–Levin quadrature by investigating fractional Jacobi approximations in [16]. Through properly choosing weighted Jacobi polynomials, the discretized Levin's equation was turned into a sparse linear system. It had been verified that the convergence rate of this kind of Levin quadrature rules could be analyzed by studying coefficients of the fractional Jacobi expansion of the error function. In [17], a multi-resolution quadrature rule was applied to deal with the singularity, and the modified Levin quadrature rule coupled with the multi-quadric radial basis function was developed to calculate oscillatory integrals with Bessel and Airy kernels.

Levin's quadrature rule also plays an important role in solving multi-dimensional problems. By introducing a multivariate ordinary differential equation, Levin found a non-oscillatory approximation to the integrand in [7], which led to an efficient algorithm for computing oscillatory integrals over rectangular regions. In [18], Li et al. devised a class of spectral Levin methods for multi-dimensional integrals by utilizing the Chebyshev differential matrix and delaminating quadrature rule. An innovative procedure for multivariate highly oscillatory integrals was devised by employing multi-resolution analysis in [19]. Meanwhile, the meshless approximation was obtained by truncated singular value decomposition. In [20], the second author studied a fast algorithm for Hermite differential matrix by the barycentric formula. With the help of delaminating quadrature, the spectral Levin-type method for calculation of highly oscillatory integrals over rectangular regions was constructed.

Although researchers have made much contribution to numerical calculation of highly oscillatory integrals, little attention has been paid to the computation of nearly singular and highly oscillatory integrals, for example,

$$I[F,G_1,G_2,\omega] = \int_{-1}^{1}\int_{-1}^{1} \frac{F(x,y)}{(x-a)^2+(x-b)^2+\epsilon^2} e^{i\omega(G_1(x)+G_2(y))}dxdy. \qquad (1)$$

In this paper, we are concerned with efficient computation of Integral (1), and partly fill in the gap in this field. Moreover, we suppose that $F(x,y)$ is analytic with respect to both variables, $G_1(x)$ and $G_2(y)$ are sufficiently smooth functions without stationary points, and the frequency parameter $\omega \gg 1$, $(a,b) \in \mathbb{R}^2$, and $|\epsilon| \ll 1$.

A large frequency parameter ω implies that integrands of Integral (1) are highly oscillatory, and classical quadrature rules suffer from the computational cost. In Table 1, we list numerical results computed by the classical delaminating quadrature rule coupled with Clenshaw–Curtis quadrature (CCQ), where the quadrature nodes are fixed 16. Referenced values are computed by CHEBFUN toolbox (see [21]). CHEBFUN, which approximate functions by Chebyshev interplant, was firstly developed in 2004. Due to the fast and high-order approximation to the integrand, numerical integration methods in CHEBFUN usually provide efficient numerical approaches for univariate and multivariate integrals. Hence, the 2D quadrature method in CHEBFUN is chosen as a benchmark. It can be seen from Table 1 that, as ω goes to infinity, CCQ diverges from the referenced values when we do not add quadrature nodes.

Table 1. Numerical results of classical delaminating quadrature rules for highly oscillatory multivariate integrals $\int_{-1}^{1}\int_{-1}^{1}\cos(x+y)e^{i\omega(x+y)}dxdy$.

	CCQ	Referenced Value
$\omega = 10$	0.020725079138303	0.020722222756906
$\omega = 40$	0.325821665438573	0.001251371233532
$\omega = 160$	0.084546798935177	0.000107898607873
$\omega = 640$	0.098831710673239	0.000004494735846
$\omega = 2560$	0.213379710792315	0.000000211475406

In contrast to oscillatory integrals arising in existing studies, when the point (a, b) in Integral (1) is close to or falls in the integration domain and ϵ is particularly small, the integrand attains its peak value around (a, b) and decays dramatically away from such a critical point. In general, the point (a, b) is called the nearly singular point. Plenty of additional quadrature nodes have to be used if we want to make the numerical formula retain a tolerance error.

There also exist several contributions to tackle the nearly singular problem. The sinh transformation is deemed one of the most important tools. For nearly singular moments arising in Laplace's equation, Johnston et al. proposed the sinh transformation in [22]. Occorsio and Serafini considered two kinds of cubature rules for nearly singular and highly oscillatory integrals in [23]. With the help of 2D-dilation technique, Occorsio and Serafini were able to relax the fast changing integrand and applied Gauss–Jacobi quadrature to the transformed integral. Numerical experiments verified that such an approximation procedure greatly increased the numerical performance of Gauss quadrature.

The remaining parts are organized as follows. In the second section, we review some results with regard to the calculation of Chebyshev series and present the convergence property of Chebyshev interplant and series. In Section 3, we first extend the idea in [13] to two-dimensional oscillatory integrals. Compared with existing Levin quadrature, the new approach has an advantage in computational time. Then, noting that there is little convergence analysis of 2D Levin quadrature rules, we try to fill the gap through examining the modified Levin equation. Finally, we present an innovative composite Levin quadrature rule for solving nearly singular and highly oscillatory problems. In contrast to existing numerical integration methods, the proposed composite method does not suffer from high oscillation and nearly singular amplitudes. Numerical tests included in Section 4 are conducted to verify the efficiency of the proposed approach, and some remarks are concluded in Section 5.

2. Auxillary Tools

In this section, we first revisit auxillary operators with regard to Chebyshev series, which help develop numerical algorithms for computation of two-dimensional oscillatory integrals. Then, error bounds for coefficients of Chebyshev series and Clenshaw–Curtis interplant are introduced.

When the given function $f(x)$ is analytic in a sufficiently large domain containing $[-1, 1]$, one can compute its Chebyshev series by (see [24])

$$f(x) = \sum_{n=0}^{\infty} f_n T_n(x),$$

with

$$f_0 = \frac{1}{\pi}\int_{-1}^{1}\frac{f(s)}{\sqrt{1-s^2}}ds, f_n = \frac{2}{\pi}\int_{-1}^{1}\frac{f(s)T_n(s)}{\sqrt{1-s^2}}ds, \; n = 1, 2, \cdots.$$

Here, $T_n(x)$ denotes the first-kind Chebyshev polynomial of order n. Noting the relation between the first- and second-kind Chebyshev polynomials $T_n(x), U_n(x)$ (see [24])

$$\frac{d}{dx}T_n(x) = \begin{cases} nU_{n-1}(x), & n \geq 1, \\ 0, & n = 0, \end{cases} \tag{2}$$

and

$$T_n(x) = \begin{cases} U_0(x), & n = 0, \\ \frac{1}{2}U_1(x), & n = 1, \\ \frac{1}{2}(U_n(x) - U_{n-2}(x)), & n \geq 2, \end{cases} \tag{3}$$

we are able to compute

$$f'(x) = \sum_{n=0}^{\infty} f'_n T_n(x) = \sum_{n=0}^{\infty}{}' \left(\sum_{k=0}^{\infty} (2n + 4k + 2) f_{n+2k+1} \right) T_n(x),$$

which implies Chebyshev coefficients of the derivative can be represented by

$$\begin{pmatrix} f'_0 \\ f'_1 \\ f'_2 \\ f'_3 \\ \vdots \end{pmatrix} = \begin{pmatrix} 0 & 1 & 0 & 3 & 0 & 5 & \cdots \\ 0 & 0 & 4 & 0 & 8 & 0 & \ddots \\ 0 & 0 & 0 & 6 & 0 & 10 & \ddots \\ 0 & 0 & 0 & 0 & 8 & 0 & \ddots \\ \vdots & \ddots & \ddots & \ddots & \ddots & \ddots & \ddots \end{pmatrix} \begin{pmatrix} f_0 \\ f_1 \\ f_2 \\ f_3 \\ \vdots \end{pmatrix} = \mathcal{D} \begin{pmatrix} f_0 \\ f_1 \\ f_2 \\ f_3 \\ \vdots \end{pmatrix}. \tag{4}$$

Secondly, suppose that there exists a sufficiently smooth function

$$a(x) = \sum_{n=0}^{\infty} a_n T_n(x).$$

Noting the identity (see [24])

$$T_m(x)T_n(x) = \frac{1}{2}(T_{m+n}(x) + T_{|m-n|}(x)),$$

we can compute the product $a(x)f(x)$ by

$$a(x)f(x) = \sum_{n=0}^{\infty} c_n T_n(x),$$

where $\mathbf{a} = [a_0, a_1, \cdots]^T$, and coefficients $\{c_n\}_{n=0}^{\infty}$ are defined by

$$\begin{pmatrix} c_0 \\ c_1 \\ c_2 \\ \vdots \end{pmatrix} = \frac{1}{2}\left(\begin{pmatrix} 2a_0 & a_1 & a_2 & a_3 & \cdots \\ a_1 & 2a_0 & a_1 & a_2 & \ddots \\ a_2 & a_1 & 2a_0 & a_1 & \ddots \\ \vdots & \ddots & \ddots & \ddots & \ddots \end{pmatrix} + \begin{pmatrix} 0 & 0 & 0 & 0 & \cdots \\ a_1 & a_2 & a_3 & a_4 & \ddots \\ a_2 & a_3 & a_4 & a_5 & \ddots \\ \vdots & \ddots & \ddots & \ddots & \ddots \end{pmatrix}\right)\begin{pmatrix} f_0 \\ f_1 \\ f_2 \\ \vdots \end{pmatrix}$$
$$= \mathcal{M}[\mathbf{a}]\begin{pmatrix} f_0 \\ f_1 \\ f_2 \\ \vdots \end{pmatrix}. \tag{5}$$

Operators $\mathcal{D}, \mathcal{M}[\mathbf{a}]$ have been verified to be efficient tools for discretizing Levin's equation. For more details, one can refer to [13].

On the other hand, when $f(x)$ is analytic with $|f(x)| \leq M$ in the region bounded by the Bernstein ellipse with the radius $\rho > 1$, we have for every $n \neq 0$ (see [25])

$$|f_n| \leq 2M\rho^{-n}.$$

Noting that

$$f'_n = \sum_{k=0}^{\infty}(2n+4k+2)f_{n+2k+1},$$

we can compute

$$|f'_n| \leq \sum_{k=0}^{\infty}(2n+4k+2)|f_{n+2k+1}| \leq \sum_{k=0}^{\infty}(2n+4k+2)2M\rho^{-n-2k-1} \leq 4M\rho^{-n-1}\left((n+1)\sum_{k=0}^{\infty}\rho^{-2k} + 2\sum_{k=0}^{\infty}k\rho^{-2k}\right).$$

Employing

$$\sum_{k=0}^{\infty}\rho^{-2k} = \frac{\rho^2}{\rho^2-1}, \quad \sum_{k=0}^{\infty}k\rho^{-2k} = \frac{\rho^2}{(\rho^2-1)^2}$$

leads to

$$|f'_n| \leq 4M\rho^{-n-1}\left((n+1)\frac{\rho^2}{\rho^2-1} + 2\frac{\rho^2}{(\rho^2-1)^2}\right) \leq 4M\rho^{-n-1}\frac{\rho(\rho-1)(n+1)+2}{(\rho-1)^2} \leq 4M(n+1)\frac{\rho^{-n+1}}{(\rho-1)^2}.$$

Furthermore, according to [25, Theorems 2.1, 2.4], we have

$$\|f - p_N\|_\infty \leq 4M\frac{\rho^{-N}}{\rho-1} \tag{6}$$

and

$$\|f' - p'_N\|_\infty \leq 4M(N+1)^2\frac{\rho^{-N+2}}{(\rho-1)^3}. \tag{7}$$

Here, $p_N(x)$ denotes the interplant of $f(x)$ at Clenshaw–Curtis nodes or the truncated Chebyshev series of $f(x)$.

3. Main Results

This section is devoted to investigating fast algorithms for calculation of Integral (1). To begin with, let us consider the computation of oscillatory integral without nearly singular integrands, that is,

$$\hat{I}[F, G_1, G_2, \omega] = \int_{-1}^{1}\int_{-1}^{1} F(x,y)e^{i\omega(G_1(x)+G_2(y))}dxdy. \tag{8}$$

Here, $F(x,y), G_1(x), G_2(y)$ are smooth functions with sufficiently large analytic regions, and $G_1(x), G_2(y)$ do not have stationary points in the complex plane.

Consider the inner integral

$$H(y) = \int_{-1}^{1} F(x,y)e^{i\omega G_1(x)}dx. \tag{9}$$

For fixed $y_j = \cos\frac{j}{N}\pi, j = 0, 1, \cdots, N$, we are restricted to finding a function $P_j(x)$ satisfying

$$P_j'(x) + i\omega G_1'(x)P_j(x) = F(x, y_j). \tag{10}$$

Noting that $G_1'(x)$ never vanishes over the interval $[-1,1]$, we can get the modified Levin equation,

$$\frac{P_j'(x)}{G_1'(x)} + i\omega P_j(x) = \frac{F(x,y_j)}{G_1'(x)}. \tag{11}$$

Let

$$P_j(x) = \sum_{n=0}^{\infty} p_n^j T_n(x),\ 1/G_1'(x) = \sum_{n=0}^{\infty} g_n^1 T_n(x),\ F(x,y_j) = \sum_{n=0}^{\infty} f_n^j T_n(x).$$

With the help of operators $\mathcal{D}, \mathcal{M}[\mathbf{a}]$, we rewrite modified Levin's Equation (11) as

$$\mathcal{M}[\mathbf{G}_1]\mathcal{D}\mathbf{P}_j + i\omega\mathbf{P}_j = \mathcal{M}[\mathbf{G}_1]\mathbf{F}_j, \tag{12}$$

where

$$x_j = \cos\frac{j}{N}\pi,\ \mathbf{P}_j = \begin{pmatrix} p_0^j \\ p_1^j \\ \vdots \end{pmatrix},\ \mathbf{G}_1 = \begin{pmatrix} g_0^1 \\ g_1^1 \\ \vdots \end{pmatrix},\ \mathbf{F}_j = \begin{pmatrix} f_0^j \\ f_1^j \\ \vdots \end{pmatrix}.$$

Solving Equation (12) by the truncation method [26] gives the unknown coefficients p_n^j, $n = 0, 1, \cdots, N$, and we can get approximations to $P_j(\pm 1)$ by Clenshaw algorithm,

$$P_j(\pm 1) \approx \frac{b_0(\pm 1) - b_2(\pm 1)}{2}$$

with

$$\begin{cases} b_{N+1}(\pm 1) = 0, b_N(\pm 1) = p_N^{j,N}, \\ b_k(\pm 1) = (\pm 2) \times b_{k+1}(\pm 1) + p_k^{j,N}, \end{cases}$$

where $p_k^{j,N}$ denotes the approximation to p_k^j. Hence, the inner integral (9) is computed by

$$\int_{-1}^{1} F(x,y_j)e^{i\omega G_1(x)}dx \approx e^{i\omega G_1(1)}\frac{b_0(1) - b_2(1)}{2} - e^{i\omega G_1(-1)}\frac{b_0(-1) - b_2(-1)}{2}. \tag{13}$$

Since $H_N(y_j)$, the approximation to $H(y)$ at Clenshaw–Curtis nodes, has been obtained, we are able to construct the polynomial $H_N(y)$ by

$$H_N(y) = \sum_{j=0}^{N} H_N(y_j)L_j(y) = \sum_{n=0}^{N} h_n T_n(y),$$

where $L_j(y)$ denotes Lagrange basis with respect to Clenshaw–Curtis nodes and h_n can be computed by fast Fourier transform. Letting

$$Q(y) = \sum_{n=0}^{\infty} q_n T_n(y)$$

denote the function satisfying

$$\mathcal{M}[\mathbf{G}_2]\mathcal{D}\mathbf{Q} + \mathrm{i}\omega\mathbf{Q} = \mathcal{M}[\mathbf{G}_2]\mathbf{H}_N, \tag{14}$$

where

$$1/G_2'(y) = \sum_{n=0}^{\infty} g_n^2 T_n(y), H_N(y) = \sum_{n=0}^{N} h_n T_n(y),$$

and

$$\mathbf{Q} = \begin{pmatrix} q_0 \\ q_1 \\ \vdots \end{pmatrix}, \mathbf{G}_2 = \begin{pmatrix} g_0^2 \\ g_1^2 \\ \vdots \end{pmatrix}, \mathbf{H}_N = \begin{pmatrix} h_0 \\ h_1 \\ \vdots \end{pmatrix},$$

we are able to approximate $q_0, \cdots, q_N$ by the truncation method again. Computing $a_0(\pm 1), a_2(\pm 1)$ by

$$\begin{cases} a_{N+1}(\pm 1) = 0, a_N(\pm 1) = q_N^N, \\ a_k(\pm 1) = (\pm 2) \times a_{k+1}(\pm 1) + q_k^N, \end{cases}$$

where q_k^N denotes the approximation to q_k, we arrive at 2D spectral coefficient Levin quadrature for Integral (8)

$$\hat{I}[F, G_1, G_2, \omega] \approx \hat{I}_N[F, G_1, G_2, \omega] := \mathrm{e}^{\mathrm{i}\omega G_2(1)} \frac{a_0(1) - a_2(1)}{2} - \mathrm{e}^{\mathrm{i}\omega G_2(-1)} \frac{a_0(-1) - a_2(-1)}{2}. \tag{15}$$

In [27], Xiang established the relation between Filon and Levin quadrature rules in the case of the phase $g(x) = 1$ and analyzed the convergence property of Levin quadrature. Instead of resorting to Filon quadrature, we consider the convergence rate of the above spectral coefficient Levin method with respect to quadrature nodes and frequency in the case of nonlinear oscillators through examining the decaying rate of the coefficients.

For any fixed $y \in [-1,1]$, $F(x,y)$ turns to the univariate function with regard to x. Let $M_H, M_F(y), M_{G1}, M_{G2}$ denote the maximum of $H(y), F(x,y), 1/G_1'(x), 1/G_2'(y)$ within their corresponding Bernstein ellipse with radiuses $\rho_H, \rho_F(y), \rho_{G1}, \rho_{G2}$, respectively. Furthermore, denoting

$$\rho_F = \inf_{y\in[-1,1]}\{\rho_F(y)\}, M_F := \sup_{y\in[-1,1]}\{M_F(y)\}, M_2 := \sup_{y\in[-1,1]}\{G_2''(y)\}, m_2 := \inf_{y\in[-1,1]}\{G_2'(y)\},$$

we summarize the convergence analysis in the following theorem.

Theorem 1. *Suppose*

- $F(x,y), 1/G_1'(y)$, *and* $1/G_2'(y)$ *are analytic within corresponding Bernstein ellipses;*
- $G_1(x)$, $G_2(y)$ *are smooth and bounded over* $[-1,1]$*;*
- *The analytic radiuses* P_F, P_{G1} *satisfy* $P_F < P_{G1}$.

Then, for sufficiently large ω, we have

$$
\begin{aligned}
&\left|\hat{I}[F,G_1,G_2,\omega]-\hat{I}_N[F,G_1,G_2,\omega]\right|\\
\leq &C\left(\frac{(N+1)^2}{\omega}\frac{\rho_H^{-N+2}}{(\rho_H-1)^3}+\frac{(N+2)^3\log(N+1)}{\omega}\frac{\rho_F^{-N+4}}{(\rho_F-1)^6}+\frac{(N+2)^3}{\omega}\frac{\rho_H^{-N+4}}{(\rho_H-1)^6}\right),
\end{aligned}
$$

where the constant C does not depend on ω, N.

Proof. Let $\hat{H}_N(y)=\sum_{j=0}^{N}H(y_j)L_j(y)$ denote the interplant of $H(y)$ at Clenshaw–Curtis points. A direct calculation implies the quadrature error can be decomposed into

$$
\begin{aligned}
&\hat{I}[F,G_1,G_2,\omega]-\hat{I}_N[F,G_1,G_2,\omega]\\
=&\int_{-1}^{1}H(y)e^{i\omega G_2(y)}dy-\left(e^{i\omega G_2(1)}\frac{a_0(1)-a_2(1)}{2}-e^{i\omega G_2(-1)}\frac{a_0(-1)-a_2(-1)}{2}\right)\\
=&\int_{-1}^{1}\int_{-1}^{1}F(x,y)e^{i\omega(G_1(x)+G_2(y))}dxdy-\int_{-1}^{1}\hat{H}_N(y)e^{i\omega G_2(y)}dy\\
&+\int_{-1}^{1}\sum_{j=0}^{N}H(y_j)L_j(y)e^{i\omega G_2(y)}dy-\int_{-1}^{1}\sum_{j=0}^{N}\left(P_j(1)e^{i\omega G_1(1)}-P_j(-1)e^{i\omega G_1(-1)}\right)L_j(y)e^{i\omega G_2(y)}dy\\
&+\int_{-1}^{1}H_N(y)e^{i\omega G_2(y)}dy-\left(Q(1)e^{i\omega G_2(1)}-Q(-1)e^{i\omega G_2(-1)}\right)\\
=&E_1+E_2+E_3.
\end{aligned}
$$

Here,

$$
\begin{aligned}
E_1:=&\int_{-1}^{1}\int_{-1}^{1}F(x,y)e^{i\omega(G_1(x)+G_2(y))}dxdy-\int_{-1}^{1}\hat{H}_N(y)e^{i\omega G_2(y)}dy,\\
E_2:=&\int_{-1}^{1}\sum_{j=0}^{N}H(y_j)L_j(y)e^{i\omega G_2(y)}dy-\int_{-1}^{1}\sum_{j=0}^{N}\left(P_j(1)e^{i\omega G_1(1)}-P_j(-1)e^{i\omega G_1(-1)}\right)L_j(y)e^{i\omega G_2(y)}dy,\\
E_3:=&\int_{-1}^{1}H_N(y)e^{i\omega G_2(y)}dy-\left(Q(1)e^{i\omega G_2(1)}-Q(-1)e^{i\omega G_2(-1)}\right),
\end{aligned}
$$

In the remaining work, we give estimates for E_1, E_2, E_3 with respect to the increasing truncation term N and frequency ω.

For E_1, note that $H(y)$ is bounded within its Bernstein ellipse by

$$
|H(y)|=\left|\int_{-1}^{1}F(x,y)e^{i\omega G_1(x)}dx\right|\leq\int_{-1}^{1}|F(x,y)||e^{i\omega G_1(x)}|dx\leq 2M_F. \tag{16}
$$

As a result, we have according to integration by parts

$$
\begin{aligned}
|E_1| =& \left|\int_{-1}^{1} (H(y) - \hat{H}_N(y)) e^{i\omega G_2(y)} dy\right| \\
\leq& \left|\frac{H(y) - \hat{H}_N(y)}{i\omega G_2'(y)} e^{i\omega G_2(y)}\Big|_{y=-1}^{y=1}\right| + \left|\frac{1}{i\omega}\int_{-1}^{1}\left(\frac{H(y)-\hat{H}_N(y)}{G_2'(y)}\right)' e^{i\omega G_2(y)} dy\right| \\
=& \left|\frac{H(y) - \hat{H}_N(y)}{i\omega G_2'(y)} e^{i\omega G_2(y)}\Big|_{y=-1}^{y=1}\right| + \left|\frac{1}{i\omega}\int_{-1}^{1}\frac{(H'(y)-\hat{H}_N'(y))G_2'(y) - (H(y)-\hat{H}_N(y))G_2''(y)}{(G_2'(y))^2} e^{i\omega G_2(y)} dy\right| \\
\leq& \frac{2\|H(y)-\hat{H}_N(y)\|_\infty}{\omega m_2} + \frac{2\|H'(y)-\hat{H}_N'(y)\|_\infty}{\omega m_2} + \frac{2M_2\|H(y)-\hat{H}_N(y)\|_\infty}{\omega m_2^2} \\
\leq& \frac{8M_F}{m_2\omega}\frac{\rho_H^{-N}}{\rho_H - 1} + \frac{16M_F}{m_2\omega}\frac{(N+1)^2\rho_H^{-N+2}}{(\rho_H-1)^3} + \frac{8M_F M_2}{m_2^2\omega}\frac{\rho_H^{-N}}{\rho_H-1} \\
\leq& \frac{8M_F}{m_2\omega}\left(1+\frac{M_2}{m_2}\right)\left(\frac{\rho_H^{-N}}{\rho_H-1} + \frac{(N+1)^2\rho_H^{-N+2}}{(\rho_H-1)^3}\right) \\
\leq& \frac{8M_F}{m_2\omega}\left(1+\frac{M_2}{m_2}\right)\frac{\rho_H^{-N}(\rho_H-1)^2 + (N+1)^2\rho_H^{-N+2}}{(\rho_H-1)^3} \\
\leq& \frac{8M_F}{m_2\omega}\left(1+\frac{M_2}{m_2}\right)\frac{2(N+1)^2\rho_H^{-N+2}}{(\rho_H-1)^3} \\
=& C_1\frac{(N+1)^2}{\omega}\frac{\rho_H^{-N+2}}{(\rho_H-1)^3},
\end{aligned}
\tag{17}
$$

where $C_1 := \frac{16M_F}{m_2}\left(1+\frac{M_2}{m_2}\right)$.

For E_2, since

$$\hat{H}_N(y) = \sum_{j=0}^{N} H(y_j)L_j(y), H_N(y) = \sum_{j=0}^{N} H_N(y_j)L_j(y), H_N(y_j) = P_j(1)e^{i\omega G_1(1)} - P_j(-1)e^{i\omega G_1(-1)},$$

letting

$$E_{2,j} := H(y_j) - H_N(y_j),$$

we obtain

$$E_2 = \sum_{j=0}^{N} E_{2,j}\int_{-1}^{1} L_j(y)e^{i\omega G_2(y)}dy.$$

Furthermore, letting

$$\mathrm{Err}_{2,j}(x) := \frac{F(x,y_j) - P_j'(x) - i\omega G_1'(x)P_j(x)}{G_1'(x)},$$

we have

$$E_{2,j} = \int_{-1}^{1} \mathrm{Err}_{2,j}(x)G_1'(x)e^{i\omega G_1(x)}dx.$$

A direct calculation as is done in the estimation procedure for E_1 results in

$$|E_{2,j}| \leq \frac{2\|\mathrm{Err}_{2,j}(x)\|_\infty}{\omega} + \frac{2\|\mathrm{Err}_{2,j}'(x)\|_\infty}{\omega}. \tag{18}$$

Then, let us consider the decaying rate of coefficients of Chebyshev expansions of $\mathrm{Err}_{2,j}(x)$, which helps to analyze $\|\mathrm{Err}_{2,j}(x)\|_\infty$ and $\|\mathrm{Err}_{2,j}'(x)\|_\infty$. In fact, the truncation technique implies

$$\mathcal{M}_N[\mathbf{G}_1]\mathcal{D}_N\mathbf{P}_{j,N} + i\omega\mathbf{P}_{j,N} = \mathcal{M}_N[\mathbf{G}_1]\mathbf{F}_{j,N} \tag{19}$$

with

$$\mathbf{P}_{j,N}=\begin{pmatrix} p_0^{j,N} \\ p_1^{j,N} \\ \vdots \\ p_N^{j,N} \end{pmatrix},\ \mathbf{F}_{j,N}=\begin{pmatrix} f_0^j \\ f_1^j \\ \vdots \\ f_N^j \end{pmatrix},$$

and $p_k^{j,N}$ denotes the approximation to p_k^j in Equation (12). For sufficiently large $\omega > N$, it follows that $\frac{1}{\omega}\|\mathcal{M}_N[\mathbf{G}_1]\mathcal{D}_N\|_\infty < 1$. By Neumann's lemma, we have

$$\mathbf{P}_{j,N}=\frac{1}{\mathrm{i}\omega}\left(\sum_{n=0}^{\infty}\left(-\frac{1}{\mathrm{i}\omega}\right)^n \mathcal{M}_N^n[\mathbf{G}_1]\mathcal{D}_N^n\right)\mathcal{M}_N[\mathbf{G}_1]\mathbf{F}_{j,N}. \tag{20}$$

Denoting the maximum of $\left(\sum_{n=0}^{\infty}\left(-\frac{1}{\mathrm{i}\omega}\right)^n \mathcal{M}_N^n[\mathbf{G}_1]\mathcal{D}_N^n\right)\mathcal{M}_N[\mathbf{G}_1]$ by S_N, we notice that

$$|p_n^{j,N}| \le \frac{2S_N M_F}{\omega}\rho_F^{-n},\ |dp_n^{j,N}| \le \frac{4S_N M_F}{\omega}(n+1)\frac{\rho_F^{-n+1}}{(\rho_F-1)^2} \le 4S_N M_F\frac{\rho_F^{-n+1}}{(\rho_F-1)^2}.$$

The Chebyshev coefficients of $\mathrm{Err}_{2,j}(x)$ can be computed by

$$c_n^{2,j}=\int_{-1}^{1}\frac{\mathrm{Err}_{2,j}(x)T_n(x)}{\sqrt{1-x^2}}dx,\ n=0,1,\cdots.$$

Noting the construction technique in the modified spectral Levin coefficient method, we get $c_n^{2,j}=0$ for $n=0,1,\cdots,N$. On the other hand, for $n \ge N+1$, it follows that

$$|c_n^{2,j}| \le \left|\int_{-1}^{1}\frac{F(x,y_j)T_n(x)}{G_1'(x)\sqrt{1-x^2}}dx\right| + \left|\int_{-1}^{1}\frac{P_j'(x)T_n(x)}{G_1'(x)\sqrt{1-x^2}}dx\right|.$$

It is noted that the first term in the right-hand side of the above equation is the coefficient of $\frac{F(x,y_j)}{G_1'(x)}$ and the second term is that of $\frac{P_j'(x)}{G_1'(x)}$, where we denote coefficients to be c_n^{FG}, c_n^{PG}, respectively. Recalling the product operator in Equation (5), we have

$$|c_n^{FG}| \le \frac{3}{2}\left(|g_n^1||f_0^j| + |g_{n-1}^1||f_1^j| + \cdots + |g_0^1||f_n^j|\right) \le 6M_{G_1}M_F(n+1)\rho_F^{-n},$$

and

$$|c_n^{PG}| \le \frac{3}{2}\left(|g_n^1||dp_0^{j,N}| + |g_{n-1}^1||dp_1^{j,N}| + \cdots + |g_0^1||dp_n^{j,N}|\right) \le 12M_{G_1}M_F S_N(n+1)\frac{\rho_F^{-n+1}}{(\rho_F-1)^2}.$$

Therefore, it follows

$$\begin{aligned}|c_n^{2,j}| \le& 6M_{G_1}M_F(n+1)\rho_F^{-n} + 12M_{G_1}M_F S_N(n+1)\frac{\rho_F^{-n+1}}{(\rho_F-1)^2}\\ \le& 6M_{G_1}M_F(n+1)\frac{\rho_F^{-n+1}}{(\rho_F-1)^2} + 12M_{G_1}M_F S_N(n+1)\frac{\rho_F^{-n+1}}{(\rho_F-1)^2}\\ \le& C'(n+1)\frac{\rho_F^{-n+1}}{(\rho_F-1)^2}.\end{aligned} \tag{21}$$

Here, $C' := 2\max\{6M_{G_1}M_F, 12M_{G_1}M_FS_N\}$. Hence, $\|\mathrm{Err}_{2,j}(x)\|_\infty$ and $\|\mathrm{Err}'_{2,j}(x)\|_\infty$ can be bounded by

$$\|\mathrm{Err}_{2,j}(x)\|_\infty \leq \sum_{n=N+1}^{\infty} |c_n^{2,j}| \leq \frac{C'}{(\rho_F-1)^2}\sum_{n=N+1}^{\infty}(n+1)\rho_F^{-n+1} \leq C'\frac{(N+2)\rho_F^{-N+2}}{(\rho_F-1)^4},$$

and

$$\|\mathrm{Err}'_{2,j}(x)\|_\infty \leq 2\sum_{n=N+1}^{\infty}|c_n^{2,j}|\|T'_n(x)\|_\infty \leq 2\sum_{n=N+1}^{\infty}C'(n+1)\frac{\rho_F^{-n+1}}{(\rho_F-1)^2}n^2 \leq C'(N+2)^3\frac{\rho_F^{-N+4}}{(\rho_F-1)^6}.$$

As a result, it follows that

$$|E_{2,j}| \leq \frac{2}{\omega}(\|\mathrm{Err}_{2,j}(x)\|_\infty + \|\mathrm{Err}'_{2,j}(x)\|_\infty) \leq \frac{8C'}{\omega}(N+2)^3\frac{\rho_F^{-N+4}}{(\rho_F-1)^6}.$$

Now, we arrive at the fact

$$|E_2| \leq \sum_{j=0}^{N}|E_{2,j}|\left|\int_{-1}^{1}L_j(y)e^{i\omega G_2(y)}dy\right| \leq C_2\frac{(N+2)^3\log(N+1)}{\omega}\frac{\rho_F^{-N+4}}{(\rho_F-1)^6}, \tag{22}$$

where $C_2 := \dfrac{64C'}{\pi}$.

The estimation procedure for E_3 is similar to that of $E_{2,j}$. We ignore details and give the conclusion directly

$$|E_3| \leq C_3\frac{(N+2)^3}{\omega}\frac{\rho_H^{-N+4}}{(\rho_H-1)^6}, \tag{23}$$

where C_3 does not depend on N and ω.

To sum up, we arrive at the following error bound by combining Equations (17), (22), and (23),

$$\begin{aligned}
&\left|\hat{I}[F,G_1,G_2,\omega] - \hat{I}_N[F,G_1,G_2,\omega]\right| \\
\leq&|E_1|+|E_2|+|E_3| \\
\leq&\frac{C_1(N+1)^2}{\omega}\frac{\rho_H^{-N+2}}{(\rho_H-1)^3} + C_2\frac{(N+2)^3\log(N+1)}{\omega}\frac{\rho_F^{-N+4}}{(\rho_F-1)^6} + C_3\frac{(N+2)^3}{\omega}\frac{\rho_H^{-N+4}}{(\rho_H-1)^6} \\
\leq&C\left(\frac{(N+1)^2}{\omega}\frac{\rho_H^{-N+2}}{(\rho_H-1)^3} + \frac{(N+2)^3\log(N+1)}{\omega}\frac{\rho_F^{-N+4}}{(\rho_F-1)^6} + \frac{(N+2)^3}{\omega}\frac{\rho_H^{-N+4}}{(\rho_H-1)^6}\right),
\end{aligned} \tag{24}$$

with $C = \max\{C_1, C_2, C_3\}$. It is easily seen that the constant C does not depend on N and ω. This completes the proof. □

Finally, let us turn to the construction of the composite quadrature rule for calculation of Integral (1). It is observed in the above theorem that, when the radiuses ρ_F, ρ_H are close to 1, the error bound would expand dramatically. Therefore, an efficient quadrature rule has to guarantee the fact that the integrand has a relatively large analytic radius over the integration domain. To make this judgment be satisfied, we choose a non-uniform grid instead of partitioning the integration region uniformly.

To begin with, the singular point z^* is projected into the plane containing the integration region and we get the projection point z. In the case of the projected point z falling into the integration domain (Case I), the first box is determined by the distance between z^* and z. We construct a square with its center being z and its side length being $2\|z^* - z\|$. Then, the side length of level-2 box's with the center z is set to be $2^2\|z^* - z\|$. To devise the composite quadrature rule, we first select level-1 box as a subdomain. Noting that the remaining domain is not a rectangle, we partition it into four subdomains, that is, Box21, Box22, Box23, and Box24 (see Figure 1). In general, the side length of level-l box's with the center z is set to be $2^l \|z^* - z\|$, and the integration subdomain is constructed similarly, which finally results in a nonuniform grid (see Figure 2).

When the projected point falls out of the integration domain, for example, it is around the side (Case II) or vertex (Case III), we implement a similar partition procedure like that in Case I. The final partition grid is shown in Figures 3 and 4. Applying 2D spectral coefficient Levin quadrature rule in the subdomain leads to a class of composite 2D spectral coefficient Levin quadrature. It is noted that such kind of partition techniques guarantee the fact that the distance between the singular and the integration interval is no less than 2 when we map the integration domain into $[-1,1] \times [-1,1]$.

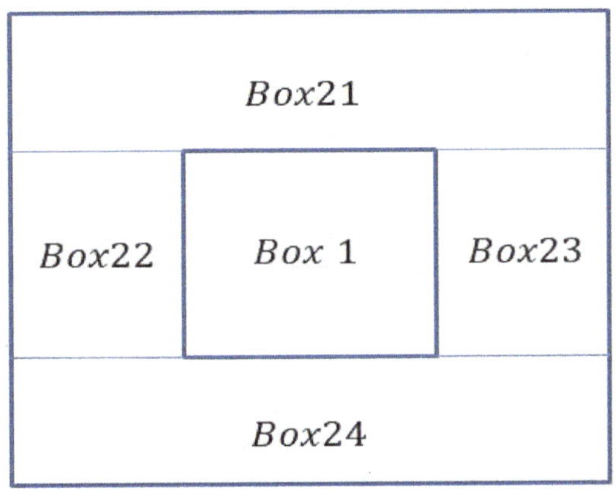

Figure 1. The integration subdomains for level-2 box.

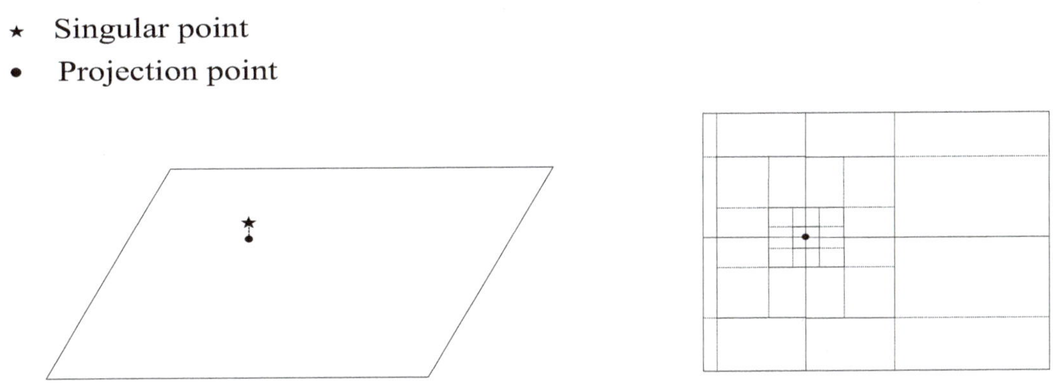

Figure 2. The partition method in Case I (left: location of the singular point z^* and projection point z. right: the nonuniform grid).

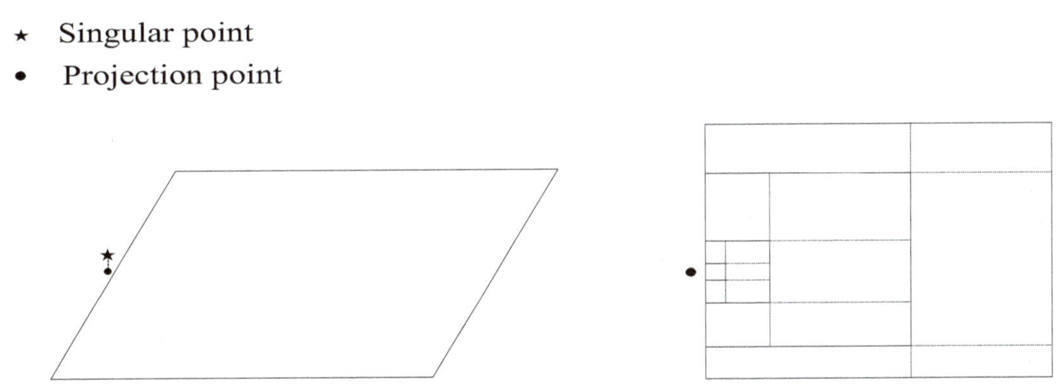

Figure 3. The partition method in Case II (left: location of the singular point z^* and projection point z. right: the nonuniform grid).

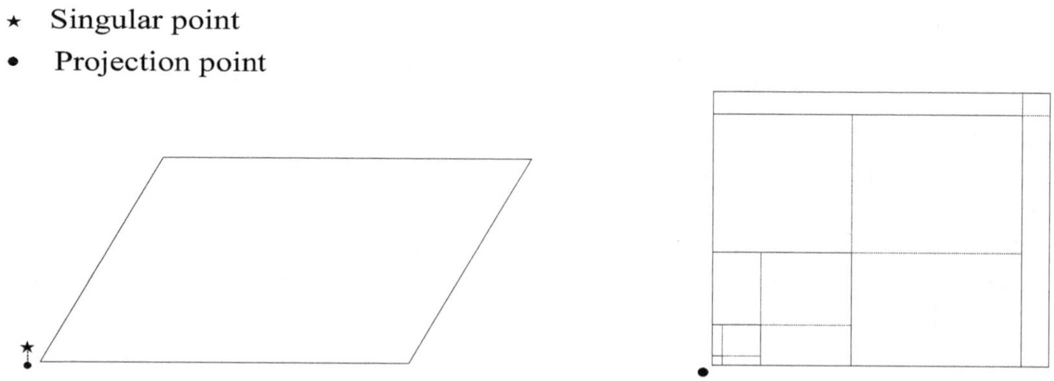

Figure 4. The partition method in Case III (left: location of the singular point z^* and projection point z. right: the nonuniform grid).

4. Numerical Experiments

This section is devoted to illustrating the numerical performance of 2D spectral coefficient Levin quadrature (2DSC-Levin) and composite 2D spectral coefficient Levin quadrature (C2DSC-Levin) given in Section 3.

Example 1. *Let us consider the computation of the oscillatory integral*

$$\int_{-1}^{1}\int_{-1}^{1}\cos(x+y)\mathrm{e}^{\mathrm{i}\omega(x+y)}dxdy.$$

The phase $x+y$ has no stationary points within the domain $[-1,1]\times[-1,1]$, and the amplitude $\cos(x+y)$ is analytic with respect to both variables. Therefore, it is expected that the new approach has an exponential convergence rate.

It is noted that approximation results derived from classical cubature usually do not make sense when $\omega \gg 1$. We first employ CCQ and give the computational results in Table 2, where both quantity of quadrature nodes N and oscillation parameter ω are variables.

Table 2. Absolute errors of CCQ for Example 1.

	$\omega = 200$	$\omega = 500$	$\omega = 2000$	$\omega = 5000$	$\omega = 10{,}000$
$N = 200$	2.2×10^{-16}	3.5×10^{-2}	8.8×10^{-2}	6.2×10^{-4}	1.6×10^{-2}
$N = 400$	1.6×10^{-16}	1.9×10^{-16}	1.0×10^{-3}	2.5×10^{-2}	8.1×10^{-3}
$N = 800$	2.0×10^{-17}	4.2×10^{-17}	7.2×10^{-3}	1.5×10^{-4}	1.7×10^{-2}
$N = 1600$	4.4×10^{-17}	8.8×10^{-18}	8.2×10^{-17}	1.4×10^{-3}	4.9×10^{-4}

Although plenty of quadrature nodes have been used in the above example, computed results are not satisfactory especially in the case of high oscillation. Now, we list approximated results of 2DSC-Levin in Table 3, where the referenced exact value is computed by the CHEBFUN toolbox again. It can be seen from Table 3 that absolute errors do not increase as the frequency ω enlarges, which implies that the new method is robust to high oscillation. On the other hand, when we raise the truncation term of Chebyshev series, the absolute error decays fast. In Figure 5, we give tendencies of absolute errors with respect to increasing frequencies and compare the computational time of 2DSC-Levin and referenced algorithm in CHEBFUN. It can be found that the consumed time of 2DSC-Levin does not vary as the frequency ω enlarges, whereas that of CHEBFUN's 2D quadrature (2D-Cheb) increases dramatically. Since 2D-Cheb is a class of self-adaptive algorithms, it has to increase quadrature nodes when the frequency goes to infinity to retain a tolerance error, which results in the dramatically growing curve. However, approximations derived from 2DSC-Levin do not suffer from high oscillation according to Theorem 1. Hence, there is no need to raise quadrature nodes of 2DSC-Levin in high oscillation, and we do not witness an obvious change in Figure 5.

Table 3. Absolute errors of 2DSC-Levin for Example 1.

	$\omega = 200$	$\omega = 500$	$\omega = 2000$	$\omega = 5000$	$\omega = 10{,}000$
$N = 6$	1.7×10^{-10}	6.0×10^{-11}	7.3×10^{-13}	2.1×10^{-14}	1.8×10^{-13}
$N = 8$	7.1×10^{-13}	2.0×10^{-13}	2.6×10^{-15}	8.4×10^{-17}	8.0×10^{-16}
$N = 10$	1.9×10^{-15}	4.3×10^{-16}	7.6×10^{-18}	3.5×10^{-17}	1.8×10^{-16}
$N = 12$	9.2×10^{-18}	4.4×10^{-18}	1.3×10^{-18}	3.5×10^{-17}	1.8×10^{-16}
$N = 16$	5.9×10^{-18}	3.7×10^{-18}	1.3×10^{-18}	3.5×10^{-17}	1.8×10^{-16}

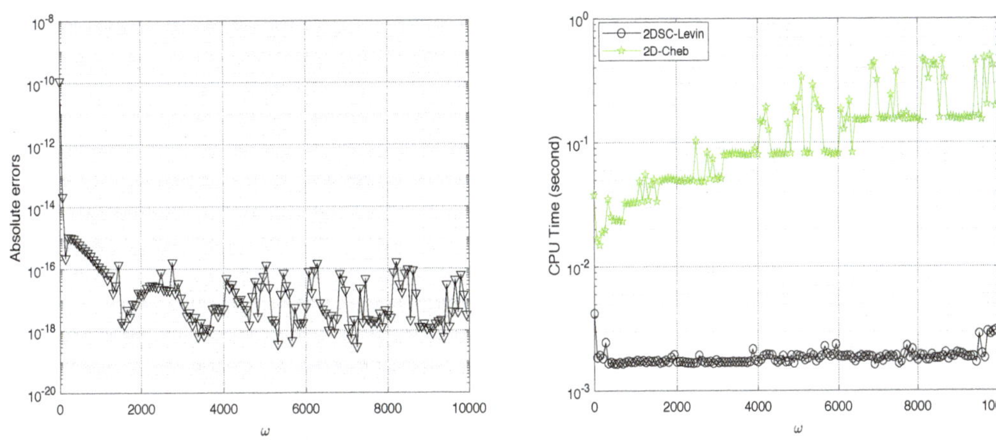

Figure 5. Comparison between 2DSC-Levin and 2D-Cheb in Example 1, ω is a variable (left: absolute errors, right: CPU time).

We also employ 2D-Cheb–Levin quadrature given in [18] to give a comparison. Computed results are shown in Table 4.

Table 4. Absolute errors of 2D-Cheb–Levin quadrature for Example 1.

	$\omega = 200$	$\omega = 500$	$\omega = 2000$	$\omega = 5000$	$\omega = 10{,}000$
$N = 6$	1.8×10^{-10}	8.7×10^{-12}	1.2×10^{-13}	3.5×10^{-15}	6.5×10^{-16}
$N = 8$	1.4×10^{-12}	5.7×10^{-14}	8.4×10^{-16}	4.2×10^{-17}	1.8×10^{-16}
$N = 10$	5.8×10^{-15}	2.0×10^{-16}	1.9×10^{-18}	3.5×10^{-17}	1.8×10^{-16}
$N = 12$	1.1×10^{-17}	4.2×10^{-18}	1.3×10^{-18}	3.5×10^{-17}	1.8×10^{-16}
$N = 16$	6.2×10^{-18}	3.7×10^{-18}	1.3×10^{-18}	3.5×10^{-17}	1.8×10^{-16}

Comparison between Tables 3 and 4 illustrates that the accuracy of 2DSC-Levin and 2D-Cheb–Levin quadrature is similar. However, 2DSC-Levin does a little better than 2D-Cheb–Levin quadrature when CPU time is considered. Since both approaches consist of approximations to a series of one-dimensional integrals, we show the consuming time of both approaches for computing the final highly oscillatory integrals in Table 5, where it can be seen that 2DSC-Levin is slightly faster than 2D-Cheb–Levin quadrature, which is partly due to the sparse structure of the discretizatized modified Levin equation.

Table 5. Comparison of CPU time for 2DSC-Levin and 2D-Cheb–Levin quadrature for fixed $\omega = 1000$.

	2DSC-Levin	**2D-Cheb–Levin Quadrature**
$N = 32$	0.000822 s	0.001703 s
$N = 64$	0.002251 s	0.005618 s
$N = 128$	0.009882 s	0.017755 s
$N = 256$	0.054631 s	0.088747 s

For computation of univariate oscillatory integrals, Levin quadrature does well in solving problems with complicate phases. In the following example, we consider a highly oscillatory integral with nonlinear oscillators over $[0,1] \times [0,1]$.

Example 2. *Let us consider the computation of the oscillatory integral*

$$\int_0^1 \int_0^1 \frac{1}{x^2 + y^2 + 15} e^{i\omega(x^2+x+y^2+y)} dxdy.$$

The amplitude $\frac{1}{x^2 + y^2 + 15}$ *is no longer an entire function, and the inverse of the phase function* $x^2 + x + y^2 + y$ *can not be calculated directly.*

We show absolute errors and CPU time of 2DSC-Levin in Table 6 and Figure 6, respectively. Due to the fact that the amplitude in Example 2 has a limited analytic radius, 2DSC-Levin converges to the machine precision much more slowly than that in Example 1. However, noting the decaying curve in the left part of Figure 6 manifests that 2DSC-Levin has the property that the higher the oscillation, the better the approximation, which also coincides with the theoretical estimate in Theorem 1. Hence, 2DSC-Levin is feasible for calculation of highly oscillatory integrals over rectangle regions when the oscillator $g(x,y)$ is nonlinear. In addition, it is interesting that the curve of CPU time of 2DSC-Levin has a jump at about $\omega = 5500$ in the right part of Figure 6. Such a phenomenon may originate from the fact that the Levin equation can be solved more efficiently as the frequency becomes larger. However, this is still a conjecture and we need more theoretical investigation in the future work.

Table 6. Absolute errors of 2DSC-Levin for Example 2.

	$\omega = 200$	$\omega = 500$	$\omega = 2000$	$\omega = 5000$	$\omega = 10{,}000$
$N = 6$	7.3×10^{-10}	8.3×10^{-11}	3.8×10^{-12}	2.7×10^{-13}	2.2×10^{-13}
$N = 8$	6.3×10^{-11}	5.5×10^{-12}	2.9×10^{-13}	2.0×10^{-14}	1.6×10^{-14}
$N = 10$	5.3×10^{-12}	3.6×10^{-13}	2.2×10^{-14}	1.5×10^{-15}	1.1×10^{-15}
$N = 12$	4.2×10^{-13}	2.2×10^{-14}	1.7×10^{-15}	1.2×10^{-16}	8.0×10^{-17}
$N = 16$	2.2×10^{-15}	6.4×10^{-17}	9.2×10^{-18}	8.6×10^{-19}	3.1×10^{-19}

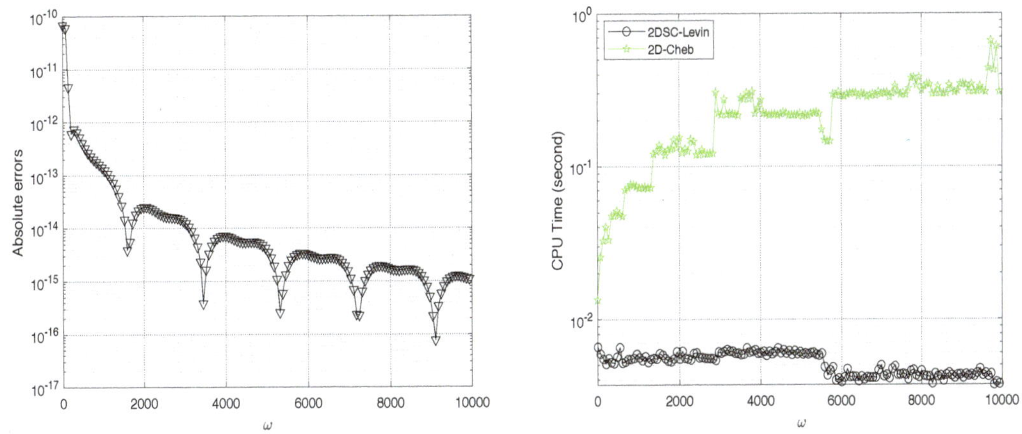

Figure 6. Comparison between 2DSC-Levin and 2D-Cheb in Example 2, ω is a variable (left: absolute errors, right: CPU time).

To illustrate the effectiveness of the composite 2D spectral coefficient Levin quadrature rule (C2DSC Levin), we give a comparison among the new approach, 2D sinh transformation (JJE) in [22], and 2D dilation quadrature (2D-d) in [23].

For Integral (1), the sinh transformation is defined by

$$x = a + \epsilon \sinh(\mu_1 u - \eta_1), y = b + \epsilon \sinh(\mu_2 u - \eta_2),$$

where

$$\mu_1 = \frac{1}{2}\left(\operatorname{arcsinh}\left(\frac{1+a}{\epsilon}\right) + \operatorname{arcsinh}\left(\frac{1-a}{\epsilon}\right)\right),$$
$$\mu_2 = \frac{1}{2}\left(\operatorname{arcsinh}\left(\frac{1+a}{\epsilon}\right) - \operatorname{arcsinh}\left(\frac{1-a}{\epsilon}\right)\right).$$

Since the transformed integrand is no longer nearly singular, a direct 2D Gauss cubature can be applied in practical computation.However, it should be noted that, although JJE can efficiently deal with nearly singular problems, it generally suffers from the highly oscillatory integrands.

In [23], Occorsio and Serafini proposed 2D-d for the integral

$$\mathcal{I}(F,\omega) = \int_D F(\mathbf{x})\mathbf{K}(\mathbf{x},\omega)d\mathbf{x}, \tag{25}$$

where $D := [-1,1] \times [-1,1]$, $\mathbf{x} = (x,y)$, and

$$F(\mathbf{x}) = \frac{1}{(x-a)^2 + (y-b)^2 + \epsilon^2}, \mathbf{K}(\mathbf{x},\omega) = e^{i\omega(G_1(x)+G_2(y))}.$$

Letting $\omega_1 = \sqrt{|\omega|}, x = \dfrac{\eta}{\omega_1}, y = \dfrac{\theta}{\omega_1}$, we have

$$\mathcal{I}(F,\omega) = \omega_1^2 \int_{[-\omega_1,\omega_1]^2} F\left(\frac{\eta}{\omega_1}, \frac{\theta}{\omega_1}\right) \mathbf{K}\left(\frac{\eta}{\omega_1}, \frac{\theta}{\omega_1}, \omega\right) d\eta d\theta.$$

Properly choosing $d \in \mathbb{R}^+$ and $S = \dfrac{2\omega_1}{d} \in \mathbb{N}$ results in

$$\mathcal{I}(F,\omega) = \omega \sum_{i=1}^{S}\sum_{j=1}^{S} \int_{D_{i,j}} F\left(\frac{\eta}{\omega_1}, \frac{\theta}{\omega_1}\right) \mathbf{K}\left(\frac{\eta}{\omega_1}, \frac{\theta}{\omega_1}, \omega\right) d\eta d\theta.$$

Here, $D_{i,j} := [-\omega_1 + (i-1)d, -\omega_1 + id] \times [-\omega_1 + (j-1)d, -\omega_1 + jd]$. Employing the transformed Gauss–Jacobi quadrature to the moment integral gives 2D-dilation quadrature.

Example 3. *Consider the computation of*

$$\int_{-1}^{1}\int_{-1}^{1} \frac{\sin(xy)}{(x+0.5)^2 + (y-0.5)^2 + 0.09} e^{i\omega(x+y)} dxdy.$$

It is noted that the integrand will reach its peak value at $(-0.5, 0.5)$, and dramatically decrease away from such a critical point.

We list computed absolute errors of JJE, 2D-d and C2DSC-Levin in Tables 7–9. As the number of quadrature nodes N increases, all of algorithms converge fast to the referenced value. When the integrand does not change rapidly, JJE provides the best numerical approximation. However, as the frequency ω enlarges, JJE and 2D-d suffer from high oscillation, while C2DSC-Levin is still able to maintain a relatively high-order approximation. It should be noted that the dilation parameter d in 2D-d is restricted to make the number of quadrature nodes coincide with the other two methods, and a slightly modified choice as is done in [23] may make 2D-d be able to deal with some highly oscillatory problems. The corresponding results are shown in Figure 7. Numerical results in this figure indicate that the absolute error derived from JJE increases dramatically when ω goes beyond 500, while the error of 2D-d rises slowly. It is also found that both absolute errors and computational time of the new approach do not suffer from the varying frequency ω. Hence, C2DSC-Levin is the most effective tool for computing oscillatory and nearly singular integrals.

Table 7. Absolute errors of JJE for Example 3.

	$N = 28$	$N = 56$	$N = 84$	$N = 112$	$N = 140$	$N = 168$
$\omega = 10$	2.5×10^{-16}	4.9×10^{-16}	5.9×10^{-16}	2.3×10^{-16}	1.5×10^{-15}	2.3×10^{-16}
$\omega = 20$	4.8×10^{-11}	9.0×10^{-16}	8.9×10^{-16}	9.2×10^{-16}	8.6×10^{-16}	9.0×10^{-16}
$\omega = 40$	2.6×10^{-4}	2.6×10^{-16}	8.3×10^{-17}	7.8×10^{-17}	1.2×10^{-16}	1.3×10^{-16}
$\omega = 80$	1.1×10^{-2}	2.3×10^{-5}	1.2×10^{-16}	3.5×10^{-17}	8.1×10^{-17}	6.1×10^{-17}
$\omega = 160$	1.5×10^{-2}	8.5×10^{-3}	8.6×10^{-3}	5.9×10^{-7}	2.5×10^{-17}	3.0×10^{-17}

Table 8. Absolute errors of 2D-d for Example 3.

	$N = 32$	$N = 64$	$N = 96$	$N = 128$	$N = 160$	$N = 192$
$\omega = 10$	1.6×10^{-2}	6.1×10^{-4}	1.9×10^{-6}	4.2×10^{-8}	6.4×10^{-10}	4.3×10^{-12}
$\omega = 20$	2.7×10^{-1}	5.7×10^{-3}	4.1×10^{-5}	7.8×10^{-8}	1.9×10^{-9}	4.7×10^{-11}
$\omega = 40$	6.9×10^{-1}	7.7×10^{-1}	1.6×10^{-2}	1.1×10^{-4}	3.3×10^{-7}	1.6×10^{-9}
$\omega = 80$	7.2×10^{-2}	6.7×10^{-2}	5.8×10^{-2}	1.9×10^{-1}	2.9×10^{-2}	4.5×10^{-4}
$\omega = 160$	6.8×10^{-2}	3.1×10^{-2}	8.3×10^{-2}	2.3×10^{-2}	6.1×10^{-3}	2.8×10^{-2}

Table 9. Absolute errors of C2DSC-Levin for Example 3.

	$N = 28$	$N = 56$	$N = 84$	$N = 112$	$N = 140$	$N = 168$
$\omega = 10$	7.3×10^{-4}	6.7×10^{-4}	9.0×10^{-5}	4.4×10^{-6}	1.1×10^{-8}	3.4×10^{-10}
$\omega = 20$	7.2×10^{-4}	1.5×10^{-6}	3.4×10^{-6}	1.5×10^{-6}	3.2×10^{-8}	5.0×10^{-9}
$\omega = 40$	1.2×10^{-4}	2.4×10^{-5}	5.9×10^{-7}	3.8×10^{-7}	2.0×10^{-7}	2.0×10^{-8}
$\omega = 80$	8.8×10^{-6}	9.2×10^{-7}	4.6×10^{-7}	4.3×10^{-8}	2.3×10^{-9}	1.1×10^{-10}
$\omega = 160$	5.5×10^{-6}	5.3×10^{-7}	2.1×10^{-8}	9.3×10^{-9}	3.6×10^{-9}	5.5×10^{-10}

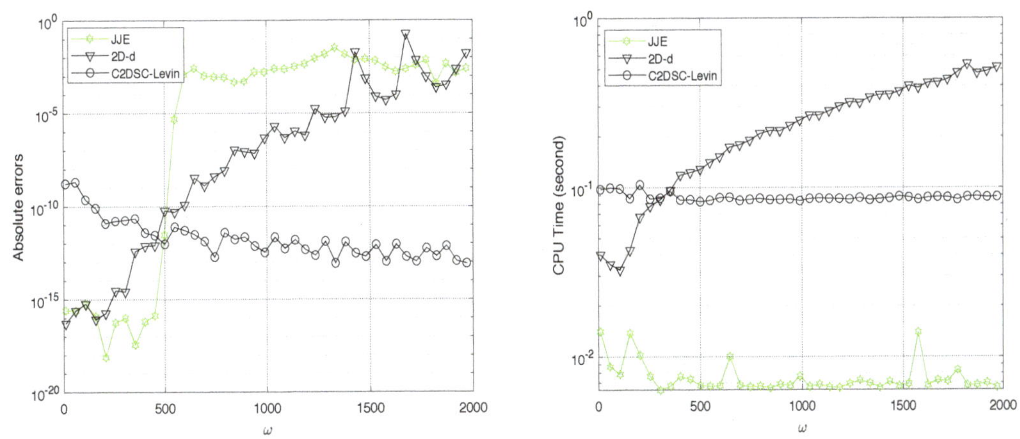

Figure 7. Comparison of JJE, 2D-d and C2DSC-Levin in Example 3, ω is a variable (left: absolute errors, right: CPU time).

Although JJE is efficient for solving some nearly singular problems, it may fail when $b = 0$ in the integrand.

Example 4. *Let us consider the computation of the following integral*

$$\int_0^1 \int_0^1 \frac{1}{(x+0.02)^2 + (y+0.02)^2} e^{i\omega(x^3+3x+y^2+6y)} dxdy,$$

It is noted that JJE does not work in this case.

Absolute errors derived from 2D-d and C2DSC-Levin are listed in Tables 10 and 11, respectively. It can be found that 2D-d provides more accurate approximation than that of C2DSC-Levin in the relatively low frequency. Nevertheless, the absolute error of C2DSC-Levin never increases in the high frequency while its 2D-d counterpart enlarges. Furthermore, although 2D-d will provide a more accurate approximation if we employ the choice of the dilation parameter considered in [23], it still cannot beat C2DSC-Levin in the high frequency when both absolute errors and computational time are considered (see Figure 8).

Table 10. Absolute errors of 2D-d for Example 4.

	$N = 64$	$N = 128$	$N = 192$	$N = 256$	$N = 320$	$N = 384$
$\omega = 10$	3.0×10^{-1}	1.3×10^{-4}	6.5×10^{-7}	2.9×10^{-9}	2.1×10^{-11}	5.5×10^{-14}
$\omega = 20$	4.1×10^{-1}	7.0×10^{-2}	1.1×10^{-4}	8.9×10^{-8}	5.3×10^{-11}	4.3×10^{-13}
$\omega = 40$	9.4×10^{-1}	3.9×10^{-1}	9.0×10^{-2}	2.2×10^{-2}	1.1×10^{-3}	8.2×10^{-6}
$\omega = 80$	3.7×10^{-1}	1.6×10^{-1}	4.6×10^{-1}	1.3×10^{-1}	5.0×10^{-2}	3.5×10^{-2}
$\omega = 160$	1.3×10^{0}	1.9×10^{-1}	2.6×10^{-1}	1.8×10^{-1}	2.0×10^{-1}	6.3×10^{-2}

Table 11. Absolute errors of C2DSC-Levin for Example 4.

	$N = 64$	$N = 128$	$N = 192$	$N = 256$	$N = 320$	$N = 384$
$\omega = 10$	9.5×10^{-5}	8.4×10^{-8}	3.3×10^{-11}	1.5×10^{-12}	1.8×10^{-13}	4.8×10^{-13}
$\omega = 20$	7.2×10^{-5}	5.6×10^{-8}	5.1×10^{-12}	3.9×10^{-13}	6.8×10^{-15}	1.5×10^{-13}
$\omega = 40$	6.8×10^{-6}	1.3×10^{-8}	4.9×10^{-12}	3.3×10^{-14}	1.8×10^{-13}	6.4×10^{-13}
$\omega = 80$	1.1×10^{-5}	1.3×10^{-8}	9.4×10^{-12}	1.2×10^{-14}	3.7×10^{-12}	5.5×10^{-13}
$\omega = 160$	1.6×10^{-6}	2.6×10^{-9}	2.2×10^{-12}	4.4×10^{-15}	3.8×10^{-15}	2.6×10^{-13}

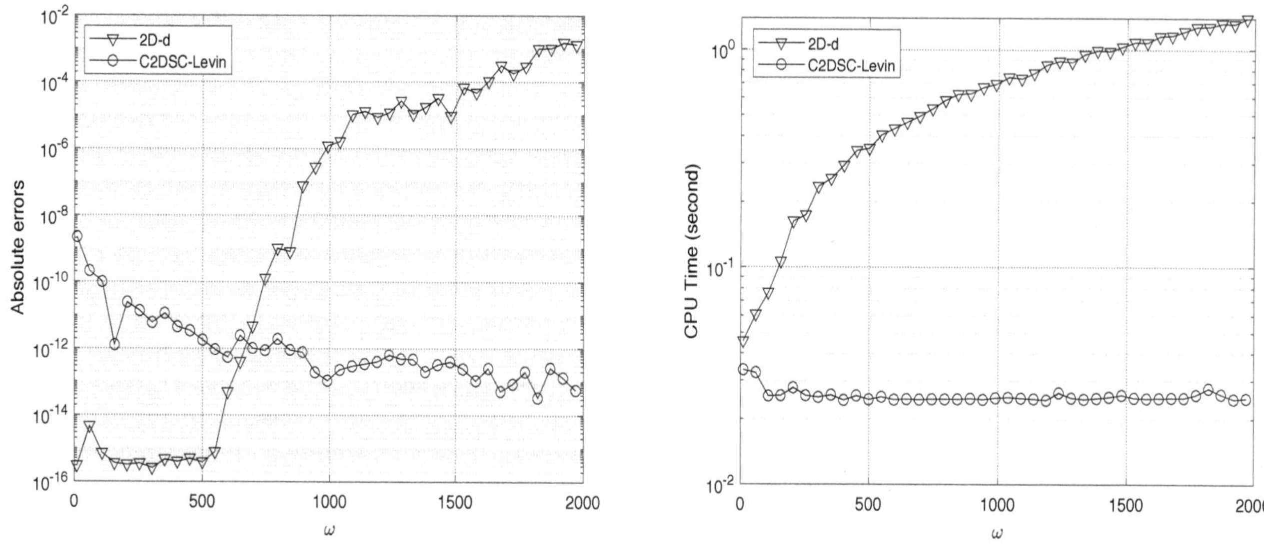

Figure 8. Comparison of 2D-d and C2DSC-Levin in Example 4, ω is a variable (left: absolute errors, right: CPU time).

5. Conclusions

In this paper, we have presented the modified spectral coefficient Levin quadrature for calculation of highly oscillatory integrals over rectangle regions and established its convergence rate with respect to the truncation term and oscillation parameter. Furthermore, by considering numerical calculation of moments over a non-uniform mesh, we derive the composite Levin quadrature. Numerical experiments indicate that the non-uniform partition technique greatly reduces the nearly singular problem. Recently, sharp bounds for coefficients of multivariate Gegenbauer expansion of analytic functions have been studied in [28], which definitely opens a door for our ongoing work about convergence analysis of Levin quadrature in high dimensional hypercube.

On the other hand, studies on the asymptotic and oscillatory behavior of solutions to highly oscillatory integral and differential equations have attracted much attention during the past decades [29–32]. It is noted that computation and numerical analysis of oscillatory integrals provide efficient tools for such kinds of studies and investigation of application of the proposed approaches to oscillatory equations is also necessary in the future work.

Author Contributions: Z.Y. and J.M. conceived and designed the experiments; Z.Y. performed the experiments; Z.Y. and J.M. analyzed the data; J.M. contributed reagents/materials/analysis tools; Z.Y. and J.M. wrote the paper. All authors have read and agreed to the published version of the manuscript.

Acknowledgments: The authors thank referees for their helpful suggestions.

Abbreviations

The following abbreviations are used in this manuscript:

CCQ	Clenshaw–Curtis quadrature
2DSC-Levin	2D spectral coefficient Levin quadrature
C2DSC-Levin	Composite 2D spectral coefficient Levin quadrature
2D-Cheb	CHEBFUN's 2D quadrature
JJE	cubature with 2D sinh transformation in [21]
2D-d	2D dilation quadrature

References

1. Chandler-Wilde, S.N.; Graham, I.G.; Langdon, S.; Spence, E.A. Numerical-asymptotic boundary integral methods in high-frequency acoustic scattering. *Acta Numer.* **2012**, *21*, 89–305.
2. Wu, Y.; Jiang, L.; Chew, W. An efficient method for computing highly oscillatory physical optics integral. *Prog. Electromagn. Res.* **2012**, *127*, 211–257.
3. Ma, J. Fast and high-precision calculation of earth return mutual impedance between conductors over a multilayered soil. *COMPEL Int. J. Comput. Math. Electr. Electron. Eng.* **2018**, *37*, 1214–1227.
4. Iserles, A.; Nørsett, S.P. Efficient quadrature of highly oscillatory integrals using derivatives. *Proc. R. Soc. A Math. Phys. Eng. Sci.* **2005**, *46*, 1383–1399.
5. Xiang, S. Efficient Filon-type methods for $\int_a^b f(x)e^{i\omega g(x)}dx$. *Numer. Math.* **2007**, *105*, 633–658.
6. Huybrechs, D.; Vandewalle, S. On the evaluation of highly oscillatory integrals by analytic continuation. *SIAM J. Numer. Anal.* **2006**, *44*, 1026–1048.
7. Levin, D. Procedures for computing one- and two-dimensional integrals of functions with rapid irregular oscillations. *Math. Comput.* **1982**, *38*, 531–538.
8. Levin, D. Analysis of a collocation method for integrating rapidly oscillatory functions. *J. Comput. Appl. Math.* **1997**, *78*, 131–138.
9. Olver, S. Moment-free numerical integration of highly oscillatory functions. *IMA J. Numer. Anal.* **2006,** *26*, 213–227.
10. Li, J.; Wang, X.; Wang, T. A universal solution to one-dimensional oscillatory integrals. *Sci. China Ser. F Inf. Sci.* **2008**, *51*, 1614–1622.
11. Zaman, S. New quadrature rules for highly oscillatory integrals with stationary points. *J. Comput. Appl. Math.* **2015**, *278*, 75–89.
12. Hasegawa, T.; Sugiura, H. A user-friendly method for computing indefinite integrals of oscillatory functions. *J. Comput. Appl. Math.* **2017**, *315*, 126–141.
13. Ma, J.; Liu, H. A well-conditioned Levin method for calculation of highly oscillatory integrals and its application. *J. Comput. Appl. Math.* **2018**, *342*, 451–462.
14. Molabahrami, A. Galerkin Levin method for highly oscillatory integrals. *J. Comput. Appl. Math.* **2017,** *321*, 499–507.
15. Wang, Y.; Xiang, S. Levin methods for highly oscillatory integrals with singularities. *Sci. China Math.* **2020,** doi:10.1007/s11425-018-1626-x.
16. Ma, J.; Liu, H. A sparse fractional Jacobi-Galerkin–Levin quadrature rule for highly oscillatory integrals. *Appl. Math. Comput.* **2020**, *367*, 124775.
17. Zaman, S.; Hussain, I. Approximation of highly oscillatory integrals containing special functions. *J. Comput. Appl. Math.* **2020**, *365*, 112372.
18. Li, J.; Wang, X.; Wang, T.; Shen, C. Delaminating quadrature method for multi-dimensional highly oscillatory integrals. *Appl. Math. Comput.* **2009**, *209*, 327–338.
19. Siraj-ul-Islam and S. Zaman. Numerical methods for multivariate highly oscillatory integrals. *Int. J. Comput. Math.* **2018**, *95*, 1024–1046.
20. Ma, J.; Duan, S. Spectral Levin-type methods for calculation of generalized Fourier transforms. *Comput. Appl. Math.* **2019**, *38*, 1–14.
21. Trefethen, L.N. Chebfun Version 5.7.0. The Chebfun Development Team. 2017. Available online: http://www.maths.ox.ac.uk/chebfun/ (accessed on 31 July 2020).
22. Johnston, B.M.; Johnston, P.R.; Elliott, D. A sinh transformation for evaluating twodimensional nearly singular boundary element integrals. *Int. J. Numer. Methods Eng.* **2007**, *69*, 1460–1479.
23. Occorsio, D.; Serafini, G. Cubature formulae for nearly singular and highly oscillating integrals. *Calcolo* **2018,** *55*, 1–33.
24. Mason, J.C.; Handscomb, D.C. *Chebyshev Polynomials*; Taylor and Francis: Oxfordshire, UK, 2002.
25. Xiang, S.; Chen, X.; Wang, H. Error bounds for approximation in Chebyshev points. *Numer. Math.* **2010,** *116*, 463–491.
26. Olver, S.; Townsend, A. A fast and well-conditioned spectral method. *SIAM Rev.* **2013**, *55*, 462–489.
27. Xiang, S. On the Filon and Levin methods for highly oscillatory integral $\int_a^b f(x)e^{i\omega g(x)}dx$. *J. Comput. Appl. Math.* **2007**, *208*, 434–439.

28. Wang, H.; Zhang, L. Analysis of multivariate Gegenbauer approximation in the hypercube. *Adv. Comput. Math.* **2020**, *46*, 53.
29. Xiang, S.; Brunner, H. Efficient methods for Volterra integral equations with highly oscillatory Bessel kernels. *BIT Numer. Math.* **2013**, *53*, 241–263.
30. Ma, J.; Kang, H. Frequency-explicit convergence analysis of collocation methods for highly oscillatory Volterra integral equations with weak singularities. *Appl. Numer. Math.* **2020**, *151*, 1–12.
31. Bazighifan, O.; Ramos, H. On the asymptotic and oscillatory behavior of the solutions of a class of higher-order differential equations with middle term. *Appl. Math. Lett.* **2020**, *107*, 106431.
32. Bazighifan, O. Kamenev and Philos-types oscillation criteria for fourth-order neutral differential equations. *Adv. Differ. Equ.* **2020**, *2020*, 201.

12

Improved Oscillation Criteria for 2nd-Order Neutral Differential Equations with Distributed Deviating Arguments

Osama Moaaz [1,†], **Rami Ahmad El-Nabulsi** [2,*,†], **Waad Muhsin** [1,†] **and Omar Bazighifan** [3,4,†]

1 Department of Mathematics, Faculty of Science, Mansoura University, Mansoura 35516, Egypt; o_moaaz@mans.edu.eg (O.M.); waed.zarebah@gmail.com (W.M.)
2 Athens Institute for Education and Research, Mathematics and Physics Divisions, 10671 Athens, Greece
3 Department of Mathematics, Faculty of Science, Hadhramout University, 50512 Hadhramout, Yemen; o.bazighifan@gmail.com
4 Department of Mathematics, Faculty of of Education, Seiyun University, 50512 Hadhramout, Yemen
* Correspondence: nabulsiahmadrami@yahoo.fr
† These authors contributed equally to this work.

Abstract: In this study, we establish new sufficient conditions for oscillation of solutions of second-order neutral differential equations with distributed deviating arguments. By employing a refinement of the Riccati transformations and comparison principles, we obtain new oscillation criteria that complement and improve some results reported in the literature. Examples are provided to illustrate the main results.

Keywords: deviating argument; second order; neutral differential equation; oscillation

---

1. Introduction

This study is concerned with creating new oscillation criteria for the second-order non-linear neutral differential equation with distributed deviating arguments

$$\left(r(t)\left(z'(t)\right)^{\alpha}\right)' + \int_a^b q(t,s)\, f(x(\sigma(t,s)))\, \mathrm{d}s = 0, \tag{1}$$

where $t \geq t_0$ and

$$z(t) := x(t) + \int_c^d p(t,s)\, x(\tau(t,s))\, \mathrm{d}s.$$

Throughout this paper, we assume that:

(H_1) α is a quotient of add positive integers;
(H_2) $r \in C(I,(0,\infty))$, $p \in C(I \times [c,d],[0,\infty))$, $q \in C(I \times [a,b],[0,\infty))$, $q(t,s)$ is not zero on any half line $[t_*,\infty) \times [a,b]$, $t_* \geq t_0$, $\int_c^d p(t,s)\,\mathrm{d}s < 1$ and

$$\int_{t_0}^{\infty} r^{-1/\alpha}(s)\, \mathrm{d}s = \infty; \tag{2}$$

(H_3) $\tau, \sigma \in C(I,\mathbb{R})$, $\tau(t,s) \leq t$, $\sigma(t,s) \leq t$ and $\lim_{t\to\infty} \tau(t,s) = \lim_{t\to\infty} \sigma(t,s) = \infty$;
(H_4) $f \in C(\mathbb{R},\mathbb{R})$ and there exists a constant $k > 0$ such that $f(x) \geq kx^{\alpha}$ for $x \neq 0$.

By a solution of (1), we mean a function $x \in C^1([t,\infty),\mathbb{R})$, $t_x \geq t_0$, which has the property $r(t)(z'(t))^{\alpha} \in C^1([t_0,\infty),\mathbb{R})$, and satisfies (1) on $[t_x,\infty)$. We consider only those solutions x of (1) which satisfy $\sup\{|x(t)| : t \geq t_x\} > 0$, for all $t > t_x$. If x is neither eventually positive nor eventually negative, then x is called oscillatory; otherwise it is called non-oscillatory. The equation itself is called oscillatory if all its solutions oscillate.

In a differential equation with neutral delay, the highest-order derivative appears both with and without delay. In addition to the theoretical importance, the qualitative study of neutral equations has great practical importance. In fact, the neutral equations arise in the study of vibrating masses attached to an elastic bar, in problems concerning electric networks containing lossless transmission lines (as in high-speed computers), and in the solution of variational problems with time delays, see [1,2].

Over the past decades, the issue of studying the oscillation properties for delay/neutral differential equations has been a very active research area see [1–19].

For some related works, Sun et al. [13] and Dzurina et al. [5] obtained some oscillation criteria for

$$(r(t)|x'(t)|^{\alpha-1}x'(t))' + q(t)|x[\sigma(t)]|^{\alpha-1}x[\sigma(t)] = 0. \tag{3}$$

Xu et al. [15,16] and Liu et al. [8] extended the results of [5,13] to (3) with neutral term. Sahiner [12] obtained some general oscillation criteria for neutral delay equations

$$\left(r(t)(x(t) + p(t)x(t-\tau_0))'\right)' + q(t)f(x(\sigma(t))) = 0,$$

In [14], Wang established some general oscillation criteria for equation

$$\left(r(t)(x(t) + p(t)x(t-\tau_0))'\right)' + \int_a^b q(t,s)\,x(\sigma(t,s))\,\mathrm{d}s = 0, \tag{4}$$

by using Riccati technique and averaging functions method. Xu and Weng [17] and Zhao and Meng [19], established some oscillation criteria for (4), which complemented and extended the results in [12,14]. In 2011, Baculikova and Dzurina [3] investigated the properties of delayed equations

$$\left(r(t)\left((x(t) + p(t)x(\tau(t)))'\right)^{\alpha}\right)' + q(t)x^{\beta}(\sigma(t)) = 0. \tag{5}$$

They are provided some comparison theorems which compare the second-order (5) with the first-order differential equations.

It is known that the determination of the signs of the derivatives of the solution is necessary and significant effect before studying the oscillation of delay differential equations. The other essential thing is to establish relationships between derivatives of different orders. Depending on improving the relationship between the neutral function z and its first derivative z', we create new and improved criteria for oscillation of solutions of Equation (1). During this study, we use Riccati transformations and comparison principles to obtain the different criteria for oscillation of (1). Examples are provided to illustrate the main results.

2. Preliminary Results

For convenience, we denote that

$$U(t) := \int_a^b q(t,s)\left[1 - \int_c^d p(\sigma(t,s),v)\,\mathrm{d}v\right]^{\alpha} \mathrm{d}s,$$

$$\eta_{t_0}(t) := \int_{t_0}^{t} r^{-1/\alpha}(u)\,\mathrm{d}u,\ \widetilde{\eta}_{t_0}(t) := \eta_{t_0}(t) + \frac{k}{\alpha}\int_{t_0}^{t} \eta_{t_1}(u)\,\eta_{t_0}^{\alpha}(\sigma(u,a))\,U(u)\,\mathrm{d}u,$$

$$\widehat{\eta}(t) := \exp\left(-\alpha\int_{\sigma(t,a)}^{t} \frac{\mathrm{d}u}{\widetilde{\eta}_{t_0}(u)r^{1/\alpha}(u)}\right),$$

$$R(t) = \alpha / (r(t))^{1/\alpha}, \;\; Q(t) := kU(t)\,\widehat{\eta}(t) \;\text{ and }\; G(t) := \int_t^\infty Q(s)\,\mathrm{d}s.$$

The following lemmas mainly help us to prove the main results:

Lemma 1. *Let $g(x) = Ax - Bx^{(\alpha+1)/\alpha}$ where $A, B > 0$ are constants. Then g attains its maximum value on $\mathbb{R}$ at $x^* = (\alpha A / ((\alpha+1)B))^\alpha$ and*

$$\max_{x \in r} g = g(x^*) = \frac{\alpha^\alpha}{(\alpha+1)^{\alpha+1}} \frac{A^{\alpha+1}}{B^\alpha}. \tag{6}$$

Lemma 2. *[3] If x is a positive solution of (1) on $[t_0, \infty)$, then there exists a $t_1 \geq t_0$ such that*

$$z(t) > 0, \; z'(t) > 0, \left(r(t)(z'(t))^\alpha\right)' \leq 0, \tag{7}$$

for $t \geq t_1$.

Lemma 3. *Let x be a positive solution of Equation (1). Then the function z satisfies*

$$\left(r(t)(z'(t))^\alpha\right)' \leq -kU(t)(z(\sigma(t,a)))^\alpha, \tag{8}$$

$$z(t) \geq \widetilde{\eta}_{t_1}(t)\, r^{1/\alpha}(t)\, z'(t) \tag{9}$$

and

$$\left(r(t)(z'(t))^\alpha\right)' \leq -kU(t)\,\widehat{\eta}(t)\, z^\alpha(t). \tag{10}$$

Proof. Assume that there exists a $t_1 \geq t_0$ such that $x(t) > 0$, $x(\tau(t,v)) > 0$ and $x(\sigma(t,s)) > 0$ for $t \geq t_1$, $v \in [c,d]$ and $s \in [a,b]$. From Lemma 2, we have (7) holds. Thus, by definition of $z(t)$, we obtain

$$\begin{aligned} x(t) &= z(t) - \int_c^d p(t,v)\, x(\tau(t,v))\,\mathrm{d}v \\ &\geq z(t) - \int_c^d p(t,v)\, z(\tau(t,v))\,\mathrm{d}v \\ &\geq z(t)\left[1 - \int_c^d p(t,v)\,\mathrm{d}v\right], \end{aligned}$$

which, with (1), implies that

$$\left(r(t)(z'(t))^\alpha\right)' \leq -k \int_a^b q(t,s)\, z^\alpha(\sigma(t,s)) \left[1 - \int_c^d p(\sigma(t,s),v)\,\mathrm{d}v\right]^\alpha \mathrm{d}s.$$

Since $z'(t) > 0$ and $\frac{\partial}{\partial s}\sigma(t,s) > 0$, we obtain $z(\sigma(t,s)) > z(\sigma(t,a))$ and so

$$\left(r(t)(z'(t))^\alpha\right)' \leq -kU(t)\, z^\alpha(\sigma(t,a)).$$

Applying the chain rule and simple computation, it is easy to see that

$$\begin{aligned} \eta_{t_1}(t)\left(r(t)(z'(t))^\alpha\right)' &= \alpha\left(r^{1/\alpha}(t)\, z'(t)\right)^{\alpha-1} \eta_{t_1}(t)\left(r^{1/\alpha}(t)\, z'(t)\right)' \\ &= -\alpha\left(r^{1/\alpha}(t)\, z'(t)\right)^{\alpha-1} \frac{\mathrm{d}}{\mathrm{d}t}\left(z(t) - \eta_{t_1}(t)\, r^{1/\alpha}(t)\, z'(t)\right). \end{aligned} \tag{11}$$

Combining (8) and (11), we obtain

$$\frac{d}{dt}\left(z(t)-\eta_{t_1}(t)\,r^{1/\alpha}(t)\,z'(t)\right)\geq\frac{k}{\alpha}\eta_{t_1}(t)\left(r^{1/\alpha}(t)\,z'(t)\right)^{1-\alpha}U(t)\,z^{\alpha}(\sigma(t,a)).$$

Integrating this inequality from t_1 to t, we have

$$z(t)\geq\eta_{t_1}(t)\,r^{1/\alpha}(t)\,z'(t)+\frac{k}{\alpha}\int_{t_1}^{t}\eta_{t_1}(u)\,U(u)\left(r^{1/\alpha}(u)\,z'(u)\right)^{1-\alpha}z^{\alpha}(\sigma(u,a))\,du. \tag{12}$$

From the monotonicity of $r^{1/\alpha}(t)\,z'(t)$, we have

$$z(t)=z(t_1)+\int_{t_1}^{t}\frac{1}{r^{1/\alpha}(u)}\left(r^{1/\alpha}(u)\,z'(u)\right)du\geq\eta_{t_1}(t)\,r^{1/\alpha}(t)\,z'(t).$$

Thus, from the fact that $\left(r^{1/\alpha}(t)\,z'(t)\right)'\leq 0$, (12) becomes

$$\begin{aligned}
z(t) &\geq \eta_{t_1}(t)\,r^{1/\alpha}(t)\,z'(t)\\
&\quad+\frac{k}{\alpha}\int_{t_1}^{t}\eta_{t_1}(u)\,U(u)\left(r^{1/\alpha}(u)\,z'(u)\right)^{1-\alpha}\eta_{t_1}^{\alpha}(\sigma(u,a))\left[r(\sigma(u,a))\left(z'(\sigma(u,a))\right)^{\alpha}\right]du.\\
&\geq \eta_{t_1}(t)\,r^{1/\alpha}(t)\,z'(t)+\frac{k}{\alpha}\int_{t_1}^{t}\left(r^{1/\alpha}(u)\,z'(u)\right)^{1-\alpha}\eta_{t_1}(u)\,\eta_{t_1}^{\alpha}(\sigma(u,a))\,U(u)\left[r^{1/\alpha}(u)\,z'(u)\right]^{\alpha}du\\
&\geq r^{1/\alpha}(t)\,z'(t)\left[\eta_{t_1}(t)+\frac{k}{\alpha}\int_{t_1}^{t}\eta_{t_1}(u)\,\eta_{t_1}^{\alpha}(\sigma(u,a))\,U(u)\,du\right].\\
&\geq \widetilde{\eta}_{t_1}(t)r^{1/\alpha}(t)\,z'(t),
\end{aligned}$$

or

$$\frac{z'(t)}{z(t)}\leq\frac{1}{\widetilde{\eta}_{t_1}(t)r^{1/\alpha}(t)}.$$

Integrating the latter inequality from $\sigma(t,a)$ to t, we get

$$\frac{z(\sigma(t,a))}{z(t)}\geq\exp\left(-\int_{\sigma(t,a)}^{t}\frac{du}{\widetilde{\eta}_{t_1}(u)r^{1/\alpha}(u)}\right).$$

which with (8), gives

$$\begin{aligned}
\frac{\left(r(t)\left(z'(t)\right)^{\alpha}\right)'}{z^{\alpha}(t)} &\leq -kU(t)\left(\frac{z(\sigma(t,a))}{z(t)}\right)^{\alpha}\\
&\leq -kU(t)\,\widehat{\eta}(t).
\end{aligned}$$

The proof is complete. □

Lemma 4. *Let x be a positive solution of equation (1). If we define the function*

$$\Psi(t)=\phi(t)\,r(t)\left(\frac{z'(t)}{z(t)}\right)^{\alpha}, \tag{13}$$

then

$$\Psi'(t)\leq\frac{\phi_{+}'(t)}{\phi(t)}\Psi(t)-k\phi(t)U(t)\,\widehat{\eta}(t)-\frac{\alpha}{\left(\phi(t)\,r(t)\right)^{1/\alpha}}\Psi^{(\alpha+1)/\alpha}(t). \tag{14}$$

Proof. Assume that there exists a $t_1\geq t_0$ such that $x(t)>0$, $x(\tau(t,v))>0$ and $x(\sigma(t,s))>0$ for $t\geq t_1$, $v\in[c,d]$ and $s\in[a,b]$. From Lemma 3, we have (10) holds. Thus, from the definition of $\Psi(t)$, we obtain $\Psi(t)>0$ for $t\geq t_1$. Differentiating (13), we arrive at

$$\Psi'(t) = \frac{\phi'(t)}{\phi(t)}\Psi(t) + \phi(t)\frac{(r(t)z'(t))'}{z^{\alpha}(t)} - \alpha\phi(t)r(t)\left(\frac{z'(t)}{z(t)}\right)^{\alpha+1}.$$

From (10) and (13), we deduce that

$$\Psi'(t) \leq \frac{\phi'_{+}(t)}{\phi(t)}\Psi(t) - k\phi(t)U(t)\widehat{\eta}(t) - \frac{\alpha}{(\phi(t)r(t))^{1/\alpha}}\Psi^{(\alpha+1)/\alpha}(t).$$

The proof is complete. □

3. Main Results

In this section, we establish the oscillation criteria for the solutions of (1).

Theorem 1. *If the first-order delay differential equation*

$$\omega'(t) + k\widetilde{\eta}_{t_1}^{\alpha}(\sigma(t,a))U(t)\omega(\sigma(t,a)) = 0 \tag{15}$$

is oscillatory, then (1) is oscillatory.

Proof. Suppose the contrary that (1) has a non-oscillatory solution x on $[t_0,\infty)$. Without loss of generality, we assume that there exists a $t_1 \geq t_0$ such that $x(t) > 0$, $x(\tau(t,v)) > 0$ and $x(\sigma(t,s)) > 0$ for $t \geq t_1$, $v \in [c,d]$ and $s \in [a,b]$. From Lemma 3, we have (8) and (9) hold. Using (8) and (9), one can see that $\omega(t) = r(t)(z'(t))^{\alpha}$ is a positive solution of the first order delay differential inequality

$$\omega'(t) + k\widetilde{\eta}_{t_1}^{\alpha}(\sigma(t,a))U(t)\omega(\sigma(t,a)) \leq 0.$$

In view of ([11] Theorem 1), the associated delay equation (15) also has a positive solution, we find a contradiction. The proof is complete. □

Corollary 1. *If*

$$\limsup_{t\to\infty}\int_{\sigma(t,a)}^{t}\widetilde{\eta}_{t_1}^{\alpha}(\sigma(u,a))U(u)\,du > \frac{1}{k}, \ \frac{\partial}{\partial t}\sigma(t,s) \geq 0 \tag{16}$$

or

$$\liminf_{t\to\infty}\int_{\sigma(t,a)}^{t}\widetilde{\eta}_{t_1}^{\alpha}(\sigma(u,a))U(u)\,du > \frac{1}{ke}, \tag{17}$$

then (1) is oscillatory.

Proof. It is well known that (16) or (17) ensures oscillation of (15), see ([7] Theorem 2.1.1). □

Lemma 5. *Assume that σ is strictly increasing with respect to t for all $s \in (a,b)$. Suppose for some $\delta > 0$ that*

$$\liminf_{t\to\infty}\int_{\sigma(t,a)}^{t}\widetilde{\eta}_{t_1}^{\alpha}(\sigma(u,a))U(u)\,du \geq \delta \tag{18}$$

and (1) has an eventually positive solution x. Then,

$$\frac{w(\sigma(t,a))}{w(t)} \geq \theta_n(\delta), \tag{19}$$

for every $n \geq 0$ and t large enough, where $w(t) := r(t)(z'(t))^{\alpha}$,

$$\theta_0(u) := 1 \text{ and } \theta_n(u) := \exp(\rho\theta_{n-1}(u)). \tag{20}$$

Proof. Assume that (1) has a positive solution x on $[t_0, \infty)$. Then, we can expect the existence of a $t_1 \geq t_0$ such that $x(t) > 0$, $x(\tau(t,v)) > 0$ and $x(\sigma(t,s)) > 0$ for $t \geq t_1$, $v \in [c,d]$ and $s \in [a,b]$. Proceeding as in the proof of Theorem 1, we deduce that ω is a positive solution of first order delay differential equation (15). In a similar way to that followed in proof of Lemma 1 in [18], we can prove that (19) holds. □

Theorem 2. *Assume that σ is strictly increasing with respect to t for all $s \in (a,b)$ and (18) holds for some $\delta < 0$. If there exists a function $\varphi \in C^1(I, (0,\infty))$ such that*

$$\limsup_{t\to\infty} \int_{t_1}^{t} \left(k\varphi(u) U(u) - \frac{(\varphi'_+(u))^{\alpha+1} r(\sigma(u,a))}{(\alpha+1)^{\alpha+1} \theta_n(\delta) \varphi^{\alpha}(u) (\sigma'(u,a))^{\alpha}} \right) = \infty, \tag{21}$$

for some sufficiently large $t \geq t_1$and for some $n \geq 0$, where $\theta_n(\delta)$ is defined as (20) and $\varphi'_+(t) = \max\{0, \varphi'(t)\}$, then (1) is oscillatory.

Proof. Suppose the contrary that (1) has a non-oscillatory solution x on $[t_0, \infty)$. Without loss of generality, we assume that there exists a $t_1 \geq t_0$ such that $x(t) > 0$, $x(\tau(t,v)) > 0$ and $x(\sigma(t,s)) > 0$ for $t \geq t_1$, $v \in [c,d]$ and $s \in [a,b]$. From Lemma 3, we have (8) holds. It follows from Lemma 5 that there exists a $t \geq t_1$ large enough such that

$$\frac{z'(\sigma(t,a))}{z'(t)} \geq \left(\frac{\theta_n(\delta) r(t)}{r(\sigma(t))} \right)^{1/\alpha}. \tag{22}$$

Define the function

$$\Phi(t) := \varphi(t) r(t) \left(\frac{z'(t)}{z(\sigma(t,a))} \right)^{\alpha}. \tag{23}$$

Then, $\Phi(t) > 0$ for $t \geq t_1$. Differentiating (23), we get

$$\Phi'(t) = \frac{\varphi'(t)}{\varphi(t)} \Phi(t) + \varphi(t) \frac{(r(t)(z'(t))^{\alpha})'}{z^{\alpha}(\sigma(t,a))} - \alpha\varphi(t) r(t) \left(\frac{z'(t)}{z(\sigma(t,a))} \right)^{\alpha} \left(\frac{z'(\sigma(t))}{z(\sigma(t,a))} \right) \sigma'(t,a).$$

From (8), (22) and (23), we obtain

$$\Phi'(t) \leq -k\varphi(t) U(t) + \frac{\varphi'_+(t)}{\varphi(t)} \Phi(t) - \frac{\alpha\theta_n^{1/\alpha}(\delta) \sigma'(t,a)}{(\varphi(t) r(\sigma(t,a)))^{1/\alpha}} \Phi^{(\alpha+1)/\alpha}(t). \tag{24}$$

Using Lemma 1 with $A = \varphi'_+(t) / \varphi(t)$ and $B = \alpha\theta_n^{1/\alpha}(\delta) / (\varphi(t) r(\sigma(t)))^{-1/\alpha}$, (24) yield

$$\Phi'(t) \leq -k\varphi(t) U(t) + \frac{\varphi'_+(t)^{\alpha+1} r(\sigma(t,a))}{(\alpha+1)^{\alpha+1} \theta_n(\delta) \varphi^{\alpha}(t) (\sigma'(t,a))^{\alpha}}.$$

Integrating this inequality from t_1 to t, we have

$$\int_{t_1}^{t} \left(k\varphi(u) U(u) - \frac{(\varphi'_+(u))^{\alpha+1} r(\sigma(u,a))}{(\alpha+1)^{\alpha+1} \theta_n(\delta) \varphi^{\alpha}(u) (\sigma'(u,a))^{\alpha}} \right) du \leq \Phi(t),$$

then we find a contradiction with condition (21). The proof is complete. □

Theorem 3. *Assume that there exists a function $\phi \in C^1(I, (0,\infty))$ such that*

$$\limsup_{t\to\infty} \int_{t_1}^{t} \left(k\phi(u) U(u) \widehat{\eta}(u) - \frac{r(u) (\phi'_+(u))^{\alpha+1}}{(\alpha+1)^{\alpha+1} \phi^{\alpha}(u)} \right) du = \infty. \tag{25}$$

for some sufficiently large $t \geq t_1$, where $\phi'_+ (t) = \max\{0, \psi'(t)\}$, then (1) is oscillatory.

Proof. Suppose the contrary that (1) has a non-oscillatory solution x on $[t_0, \infty)$. Without loss of generality, we assume that there exists a $t_1 \geq t_0$ such that $x(t) > 0$, $x(\tau(t,v)) > 0$ and $x(\sigma(t,s)) > 0$ for $t \geq t_1$, $v \in [c,d]$ and $s \in [a,b]$. From Lemma 3, we have (8)–(10) hold. Next, using Lemma 4, we arrive at (14). Using Lemma 1 with $A = \phi'_+(t)/\phi(t)$ and $B = \alpha(\phi(t)r(t))^{-1/\alpha}$, (14) becomes

$$\Psi'(t) \leq -k\phi(t)U(t)\widehat{\eta}(t) + \frac{r(t)(\phi'_+(t))^{\alpha+1}}{(\alpha+1)^{\alpha+1}\phi^{\alpha}(t)}.$$

Integrating this inequality from t_1 to t, we have

$$\int_{t_1}^{t} \left(k\phi(u)U(u)\widehat{\eta}(u) - \frac{r(u)\left(\phi'_+(u)^{\alpha+1}\right)}{(\alpha+1)^{\alpha+1}\phi^{\alpha}(u)} \right) du \leq \Psi(t),$$

This is the contrary with condition (25). The proof is complete. □

By different method, we establish new oscillation results for Equation (1).

Theorem 4. *Assume that*

$$\int_{t_0}^{\infty} Q(t)\,dt = \infty, \tag{26}$$

then, Equation (1) is oscillatory.

Proof. Suppose the contrary that (1) has a non-oscillatory solution x on $[t_0, \infty)$. Without loss of generality, we assume that there exists a $t_1 \geq t_0$ such that $x(t) > 0$, $x(\tau(t,v)) > 0$ and $x(\sigma(t,s)) > 0$ for $t \geq t_1$, $v \in [c,d]$ and $s \in [a,b]$. Consider the function Ψ defined as in (13), it follows from Lemma 4 that (14) holds. Set $\phi(t) := 1$, (14) becomes

$$\Psi'(t) + Q(t) + R(t)\Psi^{\frac{\alpha+1}{\alpha}}(t) \leq 0 \tag{27}$$

or

$$\Psi'(t) + Q(t) \leq 0. \tag{28}$$

Integrating (28) from t_3 to t and using (26), we arrive at

$$\Psi(t) \leq \Psi(t_3) - \int_{t_3}^{t} Q(t)\,ds \to \infty \quad \text{as} \quad t \to \infty,$$

which is a contradiction with the fact that $\Psi(t) > 0$ and therefore the proof is complete. □

Definition 1. *Let $\{y_n(t)\}_{n=0}^{\infty}$ be a sequence of functions defined as*

$$y_n(t) = \int_t^{\infty} R(s)\,y_{n-1}^{\frac{\alpha+1}{\alpha}}(s)\,ds + y_0(t), \quad t \geq t_0, \quad n = 1,2,3,\ldots. \tag{29}$$

and

$$y_0(t) = G(t), \quad t \geq t_0,$$

where $y_n(t) \leq y_{n+1}(t)$, $t \geq t_0$.

Lemma 6. *Assume that x is a positive solution of (1). Then $\Psi(t) \geq y_n(t)$ such that $\Psi(t)$ and $y_n(t)$ are defined as in (13) and (29), respectively. Moreover, there exists a positive function $y(t)$ on $[T, \infty)$ such that $\lim_{n\to\infty} y_n(t) = y(t)$ for $t \geq T \geq t_0$ and*

$$y(t) = \int_t^{\infty} R(s)\, y^{\frac{\alpha+1}{\alpha}}(s)\, ds + y_0(t),\ t \geq T. \tag{30}$$

Proof. Let x be a positive solution of (1). Proceeding as in the proof of Theorem 4, we arrive at (27). By integrating (27) from t to t', we obtain

$$\Psi(t') - \Psi(t) + \int_t^{t'} Q(s)\, ds + \int_t^{t'} \Psi^{\frac{\alpha+1}{\alpha}}(s)\, R(s)\, ds \leq 0.$$

This implies

$$\Psi(t') - \Psi(t) + \int_t^{t'} \Psi^{\frac{\alpha+1}{\alpha}}(s)\, R(s)\, ds \leq 0.$$

Then, we conclude that

$$\int_t^{\infty} \Psi^{\frac{\alpha+1}{\alpha}}(s)\, R(s)\, ds < \infty \text{ for } t \geq T, \tag{31}$$

otherwise, $\Psi(t') \leq \Psi(t) - \int_t^{t'} \Psi^{\frac{\alpha+1}{\alpha}}(s)\, R(s) ds \to -\infty$ as $t' \to \infty$, which is a contradiction with $\Psi(t) > 0$. Since $\Psi(t) > 0$ and $\Psi'(t) > 0$, it follows from (27) that

$$\Psi(t) \geq G(t) + \int_t^{\infty} \Psi^{\frac{\alpha+1}{\alpha}}(s)\, R(s)\, ds = y_0(t) + \int_t^{\infty} \Psi^{\frac{\alpha+1}{\alpha}}(s)\, R(s)\, ds, \tag{32}$$

or

$$\Psi(t) \geq G(t) := y_0(t).$$

Hence, $\Psi(t) \geq y_n(t)$, $n = 1, 2, 3, \ldots$. Since $\{y_n(t)\}_{n=0}^{\infty}$ increasing and bounded above, we get that$y_n \to y$ as $n \to \infty$. Using Lebesgue's monotone convergence theorem, we see that (29) turns into (30) as $n \to \infty$. □

Theorem 5. *Assume that*

$$\liminf_{t\to\infty} \frac{1}{y_0(t)} \int_t^{\infty} y_0^{\frac{\alpha+1}{\alpha}}(s)\, R(s)\, ds > \frac{\alpha}{(\alpha+1)^{\frac{\alpha+1}{\alpha}}}, \tag{33}$$

then, (1) is oscillatory.

Proof. Suppose the contrary that (1) has a non-oscillatory solution x on $[t_0, \infty)$. Without loss of generality, we assume that there exists a $t_1 \geq t_0$ such that $x(t) > 0$ for $t \geq t_1$. Proceeding as in the proof of Lemma 6, we arrive at (32). From (32), we find

$$\frac{\Psi(t)}{y_0(t)} \geq 1 + \frac{1}{y_0(t)} \int_t^{\infty} y_0^{\frac{\alpha+1}{\alpha}}(s)\, R(s) \left(\frac{\Psi(s)}{y_0(s)}\right)^{\frac{\alpha+1}{\alpha}} ds. \tag{34}$$

If we consider $\mu = inf_{t\geq T}(\Psi(t)/y_0(t))$, then obviously $\mu \geq 1$. Using (33) and (34), we see that

$$\mu \geq 1 + \alpha \left(\frac{\mu}{\alpha+1}\right)^{\frac{\alpha+1}{\alpha}}$$

or

$$\frac{\mu}{\alpha+1} \geq \frac{1}{\alpha+1} + \frac{\alpha}{\alpha+1}\left(\frac{\mu}{\alpha+1}\right)^{\frac{\alpha+1}{\alpha}},$$

which contradicts the expected value of μ and α, therefore, the proof is complete. □

Theorem 6. *If there exist some $y_n(t)$ such that*

$$\limsup_{t\to\infty} y_n(t)\left(\int_{t_0}^{t} r^{-\frac{1}{\alpha}}(s)\mathrm{d}s\right)^{\alpha} > 1, \tag{35}$$

then, (1) is oscillatory.

Proof. Suppose the contrary that (1) has a non-oscillatory solution x on $[t_0,\infty)$. Without loss of generality, we assume that there exists a $t_1 \geq t_0$ such that $x(t) > 0$ for $t \geq t_1$. Let $\Psi(t)$ defined as in (13). Then,

$$\begin{aligned}
\frac{1}{\Psi(t)} &= \frac{1}{r(t)}\left(\frac{z(t)}{z'(t)}\right)^{\alpha} = \frac{1}{r(t)}\left(\frac{z(T) + \int_T^t r^{-1/\alpha}(s)\, r^{1/\alpha}(s)\, z'(s)\,\mathrm{d}s}{z'(t)}\right)^{\alpha} \\
&\geq \frac{1}{r(t)}\left(\frac{r^{1/\alpha}(t)\, z'(t) \int_T^t r^{-1/\alpha}(s)\,\mathrm{d}s}{z'(t)}\right)^{\alpha} \\
&= \left(\int_T^t r^{-1/\alpha}(s)\,\mathrm{d}s\right)^{\alpha},
\end{aligned} \tag{36}$$

for $t \geq T$. Thus, it follows from (36) that

$$\Psi(t)\left(\int_{t_0}^{t} r^{-1/\alpha}(s)\,\mathrm{d}s\right)^{\alpha} \leq \left(\frac{\int_{t_0}^t r^{-1/\alpha}(s)\,\mathrm{d}s}{\int_T^t r^{-1/\alpha}(s)\,\mathrm{d}s}\right)^{\alpha},$$

and so

$$\limsup_{t\to\infty} \Psi(t)\left(\int_{t_0}^{t} r^{-\frac{1}{\alpha}}(s)\mathrm{d}s\right)^{\alpha} \leq 1,$$

which contradicts (35). The proof is complete. □

Corollary 2. *If there exist some $y_n(t)$ such that either*

$$\int_{t_0}^{\infty} Q(t)\exp\left(\int_{t_0}^{t} y_n^{\frac{1}{\alpha}}(s) R(s)\,\mathrm{d}s\right)\mathrm{d}t = \infty \tag{37}$$

or

$$\int_{t_0}^{\infty} R(t)\, y_n^{\frac{1}{\alpha}}(t) y_0(t)\exp\left(\int_{t_0}^{t} R(s)\, y_n^{\frac{1}{\alpha}}(s)\mathrm{d}s\right)\mathrm{d}t = \infty, \tag{38}$$

then (1) is oscillatory.

Proof. Suppose the contrary that (1) has a non-oscillatory solution x on $[t_0,\infty)$. Without loss of generality, we assume that there exists a $t_1 \geq t_0$ such that $x(t) > 0$ for $t \geq t_1$. From Lemma 6, we get that (30) holds. Using (30), we have

$$\begin{aligned}
y'(t) &= -R(t)\, y^{\frac{\alpha+1}{\alpha}}(t) - Q(t) \\
&\leq -R(t)\, y_n^{\frac{1}{\alpha}}(t)\, y(t) - Q(t).
\end{aligned} \tag{39}$$

Hence,

$$\int_T^t Q(s)\exp\left(\int_T^s y_n^{\frac{1}{\alpha}}(u) R(u)\,\mathrm{d}u\right)\mathrm{d}s \leq y(T) < \infty,$$

which contradicts (37).

Next, let $M(t) = \int_t^\infty R(s) y^{\frac{\alpha+1}{\alpha}}(s)\,ds$. Then, we obtain

$$\begin{aligned} M'(t) &= -R(t) y^{\frac{\alpha+1}{\alpha}}(t) \\ &\leq -R(t) y_n^{\frac{1}{\alpha}}(t) y(t) \\ &= -R(t) y_n^{\frac{1}{\alpha}}(t)(M(t) + y_0(t)). \end{aligned}$$

Therefore, we find

$$\int_T^\infty R(t) y_n^{\frac{1}{\alpha}}(t) y_0(t) \exp\left(\int_T^t R(s) y_n^{\frac{1}{\alpha}}(s) ds\right) dt < \infty,$$

which contradicts (38). The proof is complete. □

4. Examples

Example 1. *Consider the differential equation*

$$\left(\left((x(t) + p_0 x(\tau_0 t))'\right)^\alpha\right)' + \int_\lambda^1 \frac{q_0}{t^{\alpha+1}} x^\alpha(ts)\,ds = 0, \tag{40}$$

where $\lambda, \tau_0 \in (0,1)$. It is easy to verify that

$$U(t) = \frac{q_0}{t^{\alpha+1}}(1-\lambda)[1-p_0]^\alpha,\ \eta_{t_0}(t) = t \text{ and } \widetilde{\eta}_{t_0}(t) = Mt,$$

where

$$M := 1 + \lambda^\alpha \frac{q_0}{\alpha}(1-\lambda)[1-p_0]^\alpha.$$

Using Corollary 1, we see that (40) is oscillatory if

$$\left(M^\alpha \lambda^\alpha q_0 (1-\lambda)[1-p_0]^\alpha\right) \ln \frac{1}{\lambda} > \frac{1}{e}$$

or

$$\alpha(M-1) M^\alpha \ln \frac{1}{\lambda} > \frac{1}{e}. \tag{41}$$

Next, we note that $R(t) = \alpha$,

$$\widehat{\eta}_{t_1}(t) = \lambda^{1/M},\ Q(t) = \frac{N}{t^{\alpha+1}} \lambda^{\alpha/M},\ G(t) = \frac{N\lambda^{\alpha/M}}{\alpha} \frac{1}{t^{\alpha+1}},$$

where $N = q_0(1-p_0)^\alpha(1-\lambda)$. From Theorem 5, (40) is oscillatory if

$$\left(\frac{N}{\alpha}\lambda^{\alpha/M}\right)^{1/\alpha} > \frac{\alpha}{(\alpha+1)^{(\alpha+1)/\alpha}}.$$

Remark 1. *Consider a particular case of (40), namely,*

$$\left(x(t) + \frac{1}{2}x(\tau_0 t)\right)'' + \frac{q_0}{t^2} x(\lambda t) = 0, \tag{42}$$

From the results in Example 1, Equation (42) is oscillatory if

$$\lambda \frac{q_0}{2}\left(1 + \frac{1}{2}\lambda q_0\right) \ln \frac{1}{\lambda} > \frac{1}{e}. \tag{43}$$

Applying Corollary 2 in [3], we see that (42) is oscillatory if

$$q_0 \lambda \ln \frac{1}{2\lambda} > \frac{2}{\mathrm{e}}. \tag{44}$$

Obviously, in the case where $\lambda = 1/3$, conditions (43) and (44) reduce to $q > 1.588$ and $q > 5.443$, respectively. Thus, a new criterion improve some related results in [3].

Example 2. *Consider the differential equation*

$$\left(x(t) + \int_0^1 \frac{1}{2} x\left(\frac{t-x}{3}\right) \mathrm{d}x\right)'' + \int_0^1 \left(\frac{q_0}{t^2}\right) x\left(\frac{t-s}{2}\right) \mathrm{d}s = 0, \tag{45}$$

where $q_0 > 0$. It is easy to verify that

$$U(t) = \frac{q_0}{t^2}, \eta_{t_0}(t) = t$$

and

$$\widetilde{\eta}_{t_0} = t + \frac{q_0}{4} \int_{t_0}^{t} dx = t\left(1 + \frac{q_0}{4}\right).$$

Using Corollary 1, if

$$\frac{q_0}{4}\left(1 + \frac{q_0}{4}\right) \ln 2 > \frac{1}{e},$$

then (45) is oscillatory.

5. Conclusions

The growing interest in the oscillation theory of functional differential equation is due to the many applications of this theory in many fields, see [1,2]. In this work, we used comparison principles and Riccati transformation techniques to obtain new oscillation criteria for neutral differential Equation (1). Our new criteria improved a number of related results [3,4,14]. Further, we extended and generalized the recent works [9,10].

Author Contributions: O.M., W.M., O.B.: Writing original draft, Formal analysis, writing review and editing. R.A.E.-N.: writing review and editing, funding and supervision. All authors have read and agreed to the published version of the manuscript.

Acknowledgments: The authors thank the reviewers for for their useful comments, which led to the improvement of the content of the paper.

References

1. Gyori, I.; Ladas, G. *Oscillation Theory of Delay Differential Equations with Applications*; Oxford University Press: New York, NY, USA, 1991.
2. Hale, J.K. *Theory of Functional Differential Equations*; Springer: New York, NY, USA, 1977.
3. Baculikova, B.; Dzurina, J. Oscillation theorems for second order nonlinear neutral differential equations. *Comput. Math. Appl.* **2011**, *62*, 4472–4478. [CrossRef]
4. Candan, T. Oscillatory behavior of second order nonlinear neutral differential equations with distributed deviating arguments. *Appl. Math. Comput.* **2015**, *262*, 199–203. [CrossRef]
5. Dzurina, J.; Stavroulakis, I.P. Oscillation criteria for second-order delay differential equations. *Appl. Math. Comput.* **2003**, *140*, 445–453. [CrossRef]
6. Elabbasy, E.M.; Hassan, T.S.; Moaaz, O. Oscillation behavior of second order nonlinear neutral differential equations with deviating arguments. *Opuscula Math.* **2012**, *32*, 719–730. [CrossRef]
7. Ladde, G.; Lakshmikantham, V.; Zhang, B. *Oscillation Theory of Differential Equationswith Deviating Arguments*; Marcel Dekker: NewYork, NY, USA, 1987.
8. Liu, L.; Bai, Y. New oscillation criteria for second-order nonlinear neutral delay differential equations. *J. Comput. Appl. Math.* **2009**, *231*, 657–663. [CrossRef]
9. Moaaz, O. New criteria for oscillation of nonlinear neutral differential equations. *Adv. Differ. Equ.* **2019**, *2019*, 484. [CrossRef]

10. Moaaz, O.; Elabbasy, E.M.; Qaraad, B. An improved approach for studying oscillation of generalized Emden–Fowler neutral differential equation. *J. Inequal. Appl.* **2020**, *2020*, 69. [CrossRef]
11. Philos, C. On the existence of nonoscillatory solutions tending to zero at ∞ for differential equations with positive delay. *Arch. Math. (Basel)* **1981**, *36*, 168–178. [CrossRef]
12. Sahiner, Y. On oscillation of second-order neutral type delay differential equations. *Appl. Math. Comput.* **2004**, *150*, 697–706.
13. Sun, Y.G.; Meng, F.W. Note on the paper of Dzurina and Stavroulakis. *Appl. Math. Comput.* **2006**, *174*, 1634–1641. [CrossRef]
14. Wang, P.G. Oscillation criteria for second-order neutral equations with distributed deviating arguments. *Comput. Math. Appl.* **2004**, *47*, 1935–1946. [CrossRef]
15. Xu, R.; Meng, F. Some new oscillation criteria for second order quasi-linear neutral delay differential equations. *Appl. Math. Comput.* **2006**, *182*, 797–803. [CrossRef]
16. Xu, R.; Meng, F. Oscillation criteria for second order quasi-linear neutral delay differential equations. *Appl. Math. Comput.* **2007**, *192*, 216–222. [CrossRef]
17. Xu, Z.T.; Weng, P.X. Oscillation of second-order neutral equations with distributed deviating arguments. *J. Comput. Appl. Math.* **2007**, *202*, 460–477. [CrossRef]
18. Zhang, B.G.; Zhou, Y. The distribution of zeros of solutions of differential equations with a variable delay. *J. Math. Anal. Appl.* **2001**, *256*, 216–228. [CrossRef]
19. Zhao, J.; Meng, F. Oscillation criteria for second-order neutral equations with distributed deviating argument. *Appl. Math. Comput.* **2008**, *206*, 485–493. [CrossRef]

Improved Approach for Studying Oscillatory Properties of Fourth-Order Advanced Differential Equations with p-Laplacian like Operator

Omar Bazighifan [1,2,*,†] and **Thabet Abdeljawad** [3,4,5,*,†]

1 Department of Mathematics, Faculty of Science, Hadhramout University, Hadhramout 50512, Yemen
2 Department of Mathematics, Faculty of Education, Seiyun University, Hadhramout 50512, Yemen
3 Department of Mathematics and General Sciences, Prince Sultan University, Riyadh 11586, Saudi Arabia
4 Department of Medical Research, China Medical University, Taichung 40402, Taiwan
5 Department of Computer Science and Information Engineering, Asia University, Taichung 40402, Taiwan
* Correspondence: o.bazighifan@gmail.com (O.B.); tabdeljawad@psu.edu.sa (T.A.)
† These authors contributed equally to this work.

Abstract: This paper aims to study the oscillatory properties of fourth-order advanced differential equations with p-Laplacian like operator. By using the technique of Riccati transformation and the theory of comparison with first-order delay equations, we will establish some new oscillation criteria for this equation. Some examples are considered to illustrate the main results.

Keywords: oscillation; advanced differential equations; p-Laplacian equations; comparison theorem

1. Introduction

In the last decades, many researchers from all fields of science, technology and engineering have devoted their attention to introducing more sophisticated analytical and numerical techniques to solve and analyze mathematical models arising in their fields.

Fourth-order advanced differential equations naturally appear in models concerning physical, biological, chemical phenomena applications in dynamical systems, mathematics of networks,and optimization. They also appear in the mathematical modeling of engineering problems to study electrical power systems, materials and energy, elasticity, deformation of structures, and soil settlement [1]. The p-Laplace equations have some applications in continuum mechanics, see for example [2–4].

An active and essential research area in the above investigations is to study the sufficient criterion for oscillation of delay differential equations. In fact, during this decade, Several works have been accomplished in the development of the oscillation theory of higher order delay and advanced equations by using the Riccati transformation and the theory of comparison between first and second-order delay equations, (see [5–12]). Further, the oscillation theory of fourth and second order delay equations has been studied and developed by using integral averaging technique and the Riccati transformation, (see [13–27]). The study of oscillation has been carried to fractional equations in the setting of fractional operators with singular and nonsingular kernels, as well (see [28,29] and the references therein).

We provide oscillation properties of the fourth order advanced differential equation with a p-Laplacian like operator

$$\left(b\left(v\right)\left|y'''\left(v\right)\right|^{p-2}y'''\left(v\right)\right)' + \sum_{i=1}^{j} q_i\left(v\right)g\left(y\left(\eta_i\left(v\right)\right)\right) = 0, \tag{1}$$

where $v \geq v_0$ and $j \geq 1$. Throughout this paper, we assume that:

(D_1) $p > 1$ is a real number,
(D_2) $q_i, \eta_i \in C([v_0, \infty), \mathbb{R}), q_i(v) \geq 0,$
(D_3) $\eta_i(v) \geq v,\ \lim_{v\to\infty} \eta_i(v) = \infty,\ i = 1, 2, .., j,$
(D_4) $g \in C(\mathbb{R}, \mathbb{R})$ such that

$$g(x) / |x|^{p-2} x \geq k > 0, \text{ for } x \neq 0. \tag{2}$$

(D_5) $b \in C^1([v_0, \infty), \mathbb{R}), b(v) > 0,\ b'(v) \geq 0$ and under the condition

$$\int_{v_0}^{\infty} \frac{1}{b^{1/(p-1)}(s)} \mathrm{d}s = \infty. \tag{3}$$

In fact, our aim in this paper is complete and improves the results in [5–7]. For the sake of completeness, we first recall and discuss these results. Li et al. [3] examined the oscillation of equation

$$\left(a(v) |z'''(v)|^{p-2} z'''(v)\right)' + \sum_{i=1}^{j} q_i(v) |w(\delta_i(v))|^{p-2} w(\delta_i(v)) = 0,$$

where $p > 1$ is a real number. The authors used the Riccati transformation and integral averaging technique. Park et al. [8] used Riccati technique to obtain necessary and sufficient conditions for oscillation of

$$\left(a(v) \left|w^{(\kappa-1)}(v)\right|^{p-2} w^{(\kappa-1)}(v)\right)' + q(v) g(w(\delta(v))) = 0,$$

where κ is even and under the condition

$$\int_{v_0}^{\infty} \frac{1}{a^{1/(p-1)}(s)} \mathrm{d}s = \infty.$$

Agarwal and Grace [5] considered the equation

$$\left(\left(y^{(\kappa-1)}(v)\right)^{\gamma}\right)' + q(v) y^{\gamma}(\eta(v)) = 0, \tag{4}$$

where κ is even and they proved it oscillatory if

$$\liminf_{v\to\infty} \int_{v}^{\eta(v)} (\eta(s) - s)^{\kappa-2} \left(\int_{s}^{\infty} q(v)\, dv\right)^{1/\gamma} ds > \frac{(\kappa-2)!}{e}. \tag{5}$$

Agarwal et al. in [6] studied Equation (4) and obtained the criterion of oscillation

$$\limsup_{v\to\infty} v^{\gamma(\kappa-1)} \int_{v}^{\infty} q(s)\, ds > ((\kappa-1)!)^{\gamma}. \tag{6}$$

Authors in [7] studied oscillatory behavior of (4) where $\gamma = 1$, κ is even and if there exists a function $\delta \in C^1([v_0, \infty), (0, \infty))$, also, they proved it oscillatory by using the Riccati transformation if

$$\int_{v_0}^{\infty} \left(\delta(s) q(s) - \frac{(\kappa-2)! (\delta'(s))^2}{2^{3-2\kappa} s^{\kappa-2} \delta(s)}\right) ds = \infty. \tag{7}$$

To compare the conditions, we apply the previous results to the equation

$$y^{(4)}(v) + \frac{q_0}{v^4} y(3v) = 0,\ v \geq 1, \tag{8}$$

1. By applying condition (5) on Equation (8), we get

$$q_0 > 13.6.$$

2. By applying condition (6) on Equation (8), we get

$$q_0 > 18.$$

3. By applying condition (7) on Equation (8), we get

$$q_0 > 576.$$

From the above we find the results in [6] improves results [7]. Moreover, the results in [5] improves results [6,7], we see this clearly in the Section 3. Thus, the motivation in studying this paper is complement and improve results [5–7].

We will need the following lemmas.

Lemma 1 ([18]). *If the function y satisfies $y^{(i)}(v) > 0$, $i = 0, 1, ..., n$, and $y^{(n+1)}(v) < 0$, then*

$$\frac{y(v)}{v^n/n!} \geq \frac{y'(v)}{v^{n-1}/(n-1)!}.$$

Lemma 2 ([10]). *Suppose that $y \in C^n([v_0, \infty), (0, \infty))$, $y^{(n)}$ is of a fixed sign on $[v_0, \infty)$, $y^{(n)}$ not identically zero and there exists a $v_1 \geq v_0$ such that*

$$y^{(n-1)}(v)\, y^{(n)}(v) \leq 0,$$

for all $v \geq v_1$. If we have $\lim_{v\to\infty} y(v) \neq 0$, then there exists $v_\lambda \geq v_1$ such that

$$y(v) \geq \frac{\lambda}{(n-1)!} v^{n-1} \left| y^{(n-1)}(v) \right|,$$

for every $\lambda \in (0, 1)$ and $v \geq v_\lambda$.

Lemma 3 ([21]). *Let γ be a ratio of two odd numbers, $V > 0$ and U are constants. Then*

$$Ux - Vx^{(\gamma+1)/\gamma} \leq \frac{\gamma^\gamma}{(\gamma+1)^{\gamma+1}} \frac{U^{\gamma+1}}{V^\gamma}, \quad V > 0.$$

Lemma 4 ([15]). *Assume that y is an eventually positive solution of (1). Then, there exist two possible cases:*

$$\begin{array}{ll} (\mathbf{S}_1) & y(v) > 0,\ y'(v) > 0,\ y''(v) > 0, y'''(v) > 0,\ y^{(4)}(v) \leq 0, \\ (\mathbf{S}_2) & y(v) > 0,\ y'(v) > 0,\ y''(v) < 0, y'''(v) > 0,\ y^{(4)}(v) \leq 0, \end{array}$$

for $v \geq v_1$, where $v_1 \geq v_0$ is sufficiently large.

2. Oscillation Criteria

In this section, we shall establish some oscillation criteria for equation (1).

Lemma 5. *Assume that y be an eventually positive solution of (1) and $(\mathbf{S}_1)$ holds. If*

$$\pi(v) := \delta(v) \left(\frac{b(v)\,(y'''(v))^{p-1}}{y^{p-1}(v)} \right), \tag{9}$$

where $\delta \in C^1([v_0,\infty),(0,\infty))$, *then*

$$\pi'(v) \le \frac{\delta'(v)}{\delta(v)}\pi(v) - k\delta(v)\sum_{i=1}^{j} q_i(v) - \frac{(p-1)\varepsilon v^2}{2(\delta(v)b(v))^{\frac{1}{(p-1)}}}\pi(v)^{\frac{p}{(p-1)}}, \tag{10}$$

for all $v > v_1$, *where* v_1 *large enough.*

Proof. Let y is an eventually positive solution of (1) and $(\mathbf{S}_1)$ holds. Thus, from Lemma 2, we get

$$y'(v) \ge \frac{\varepsilon}{2}v^2 y'''(v), \tag{11}$$

for every $\varepsilon \in (0,1)$ and for all large v. From (9), we see that $\pi(v) > 0$ for $v \ge v_1$, and

$$\begin{aligned}\pi'(v) &= \delta'(v)\frac{b(v)(y'''(v))^{p-1}}{y^{p-1}(v)} + \delta(v)\frac{\left(b(y''')^{p-1}\right)'(v)}{y^{p-1}(v)} \\ &\quad - (p-1)\delta(v)\frac{y^{p-2}(v)y'(v)b(v)(y'''(v))^{p-1}}{y^{2(p-1)}(v)}.\end{aligned}$$

Using (11) and (9), we obtain

$$\begin{aligned}\pi'(v) &\le \frac{\delta'(v)}{\delta(v)}\pi(v) + \delta(v)\frac{\left(b(v)(y'''(v))^{p-1}\right)'}{y^{p-1}(v)} \\ &\quad - (p-1)\delta(v)\frac{\varepsilon}{2}v^2\frac{b(v)(y'''(v))^{p}}{y^{p}(v)} \\ &\le \frac{\delta'(v)}{\delta(v)}\pi(v) + \delta(v)\frac{\left(b(v)(y'''(v))^{p-1}\right)'}{y^{p-1}(v)} \\ &\quad - \frac{(p-1)\varepsilon v^2}{2(\delta(v)b(v))^{\frac{1}{(p-1)}}}\pi(v)^{\frac{p}{(p-1)}}.\end{aligned} \tag{12}$$

From (1) and (12), we get

$$\pi'(v) \le \frac{\delta'(v)}{\delta(v)}\pi(v) - k\delta(v)\frac{\sum_{i=1}^{j} q_i(v)y^{p-1}(\eta_i(v))}{y^{p-1}(v)} - \frac{(p-1)\varepsilon v^2}{2(\delta(v)b(v))^{\frac{1}{(p-1)}}}\pi(v)^{\frac{p}{(p-1)}}.$$

Note that $y'(v) > 0$ and $\eta_i(v) \ge v$, thus, we find

$$\pi'(v) \le \frac{\delta'(v)}{\delta(v)}\pi(v) - k\delta(v)\sum_{i=1}^{j} q_i(v) - \frac{(p-1)\varepsilon v^2}{2(\delta(v)b(v))^{\frac{1}{(p-1)}}}\pi(v)^{\frac{p}{(p-1)}}.$$

The proof is complete. □

Lemma 6. *Assume that* y *be an eventually positive solution of (1) and* $(\mathbf{S}_2)$ *holds. If*

$$\xi(v) := \sigma(v)\frac{y'(v)}{y(v)}. \tag{13}$$

where $\sigma \in C^1([v_0,\infty),(0,\infty))$, *then*

$$\xi'(v) \le \frac{\sigma'(v)}{\sigma(v)}\xi(v) - \sigma(v)\int_v^\infty \left(\frac{k}{b(v)}\int_v^\infty \sum_{i=1}^{j} q_i(s)\,ds\right)^{1/(p-1)} dv - \frac{1}{\sigma(v)}\xi(v)^2, \tag{14}$$

for all $v > v_1$, *where* v_1 *large enough.*

Proof. Let y is an eventually positive solution of (1) and $(\mathbf{S}_2)$ holds. Integrating (1) from v to m and using $y'(v) > 0$, we obtain

$$b(m)(y'''(m))^{p-1} - b(v)(y'''(v))^{p-1} = -\int_v^m \sum_{i=1}^{j} q_i(s)\, g(y(\eta_i(s)))\,ds.$$

By virtue of $y'(v) > 0$ and $\eta_i(v) \ge v$, we get

$$b(m)(y'''(m))^{p-1} - b(v)(y'''(v))^{p-1} \le -ky^{p-1}(v)\int_v^u \sum_{i=1}^{j} q_i(s)\,ds.$$

Letting $m \to \infty$, we see that

$$b(v)(y'''(v))^{p-1} \ge ky^{p-1}(v)\int_v^\infty \sum_{i=1}^{j} q_i(s)\,ds$$

and so

$$y'''(v) \ge y(v)\left(\frac{k}{b(v)}\int_v^\infty \sum_{i=1}^{j} q_i(s)\,ds\right)^{1/(p-1)}.$$

Integrating again from v to ∞, we get

$$y''(v) + y(v)\int_v^\infty \left(\frac{k}{b(v)}\int_v^\infty \sum_{i=1}^{j} q_i(s)\,ds\right)^{1/(p-1)} dv \le 0. \tag{15}$$

From the definition of $\xi(v)$, we see that $\xi(v) > 0$ for $v \ge v_1$. By differentiating, we find

$$\xi'(v) = \frac{\sigma'(v)}{\sigma(v)}\xi(v) + \sigma(v)\frac{y''(v)}{y(v)} - \frac{1}{\sigma(v)}\xi(v)^2. \tag{16}$$

From (15) and (16), we obtain

$$\xi'(v) \le \frac{\sigma'(v)}{\sigma(v)}\xi(v) - \sigma(v)\int_v^\infty \left(\frac{k}{b(v)}\int_v^\infty \sum_{i=1}^{j} q_i(s)\,ds\right)^{1/(p-1)} dv - \frac{1}{\sigma(v)}\xi(v)^2.$$

The proof is complete. □

Theorem 1. *Assume that there exist positive functions* $\delta, \sigma \in C^1([v_0,\infty),(0,\infty))$ *such that*

$$\int_{v_0}^\infty \left(k\delta(s)\sum_{i=1}^{j} q_i(s) - \frac{2^{p-1}b(s)(\delta'(s))^p}{p^p(s^2\varepsilon\delta(s))^{p-1}}\right) ds = \infty, \tag{17}$$

for some $\varepsilon \in (0,1)$, *and either*

$$\int_{v_0}^\infty \sum_{i=1}^{j} q_i(s)\,ds = \infty \tag{18}$$

or

$$\int_{v_0}^{\infty}\left(\sigma(s)\int_{v}^{\infty}\left(\frac{k}{b(v)}\int_{v}^{\infty}\sum_{i=1}^{j}q_i(s)\,\mathrm{ds}\right)^{1/(p-1)}dv-\frac{1}{4\sigma(s)}\left(\sigma'(s)\right)^2\right)\mathrm{ds}=\infty. \tag{19}$$

Then every solution of (1) is oscillatory.

Proof. Assume that y is eventually positive solution of (1). Then, we can suppose that $y(v)$and $y(\eta_i(v))$ are positive for all $v \geq v_1$ sufficiently large. From Lemma 4, we have two possible cases $(\mathbf{S}_1)$ and $(\mathbf{S}_2)$.

Assume that case $(\mathbf{S}_1)$ holds. From Lemma 5, we get that (10) holds. Using Lemma 3 with

$$U=\delta'(v)/\delta(v),\ V=(p-1)\varepsilon v^2/\left(2(\delta(v)b(v))^{\frac{1}{(p-1)}}\right)\ \text{and}\ x=\pi(v),$$

we get

$$\frac{\delta'(v)}{\delta(v)}\pi(v)-\frac{(p-1)\varepsilon v^2}{2(\delta(v)b(v))^{\frac{1}{(p-1)}}}\pi(v)^{\frac{p}{(p-1)}}\leq-\frac{2^{p-1}b(v)(\delta'(v))^p}{p^p(v^2\varepsilon\delta(v))^{p-1}}. \tag{20}$$

From (10) and (20), we obtain

$$\pi'(v)\leq-k\delta(v)\sum_{i=1}^{j}q_i(v)+\frac{2^{p-1}b(v)(\delta'(v))^p}{p^p(v^2\varepsilon\delta(v))^{p-1}}.$$

Integrating from v_1 to v, we get

$$\int_{v_1}^{v}\left(k\delta(s)\sum_{i=1}^{j}q_i(s)-\frac{2^{p-1}b(s)(\delta'(s))^p}{p^p(s^2\varepsilon\delta(s))^{p-1}}\right)\mathrm{ds}\leq\pi(v_1),$$

for every $\varepsilon\in(0,1)$, which contradicts (17).

Let case $(\mathbf{S}_2)$ holds. Integrating (1) from m to v, we conclude that

$$-b(m)(y'''(m))^{p-1}=-\int_{m}^{v}\sum_{i=1}^{j}q_i(s)g(y(\eta_i(s)))ds.$$

By virtue of $y'(v)>0$ and $\eta_i(v)\geq v$, we get

$$\int_{m}^{v}\sum_{i=1}^{j}q_i(s)ds\leq\frac{b(m)(y'''(m))^{p-1}}{ky^{p-1}(m)},$$

which contradicts (18).

From Lemma 6, we get that (14) holds. Using Lemma 3 with

$$U=\sigma'(v)/\sigma(v),\ V=1/\sigma(v)\ \text{and}\ x=\xi(v),$$

we get

$$\frac{\sigma'(v)}{\sigma(v)}\xi(v)-\frac{1}{\sigma(v)}\xi^2(v)\leq-\frac{1}{4\sigma(v)}(\sigma'(v))^2. \tag{21}$$

From (14) and (21), we obtain

$$\xi'(v)\leq-\sigma(v)\int_{v}^{\infty}\left(\frac{k}{b(v)}\int_{v}^{\infty}\sum_{i=1}^{j}q_i(s)\,\mathrm{ds}\right)^{1/(p-1)}dv+\frac{1}{4\sigma(v)}(\sigma'(v))^2. \tag{22}$$

Integrating from v_1 to v, we get

$$\int_{v_1}^{v}\left(\sigma(s)\int_{v}^{\infty}\left(\frac{k}{b(v)}\int_{v}^{\infty}\sum_{i=1}^{j}q_i(s)\,\mathrm{d}s\right)^{1/\gamma}\mathrm{d}v-\frac{1}{4\sigma(s)}\left(\sigma'(s)\right)^2\right)\mathrm{d}s\leq\xi(v_1),$$

which contradicts (19). The proof is complete. □

When putting $\delta(v)=v^3$ and $\sigma(v)=v$ into Theorem 1, we get the following oscillation criteria:

Corollary 1. *Let (3) hold. Assume that*

$$\int_{v_0}^{\infty}\left(s^3\sum_{i=1}^{j}q_i(s)-\frac{2^{p-1}3^p s^{-3(p-1)+2}b(s)}{p^p\varepsilon^{p-1}}\right)\mathrm{d}s=\infty, \tag{23}$$

or some $\varepsilon\in(0,1)$. *If (18) holds and*

$$\int_{v_0}^{\infty}\left(s\int_{v}^{\infty}\left(\frac{k}{b(v)}\int_{v}^{\infty}\sum_{i=1}^{j}q_i(s)\,\mathrm{d}s\right)^{1/(p-1)}\mathrm{d}v-\frac{1}{4s}\right)\mathrm{d}s=\infty, \tag{24}$$

then every solution of (1) is oscillatory.

In the next theorem, we compare the oscillatory behavior of (1) with the first-order differential equations:

Theorem 2. *Assume that (3) holds. If the differential equations*

$$\theta'(v)+k\sum_{i=1}^{j}q_i(v)\left(\frac{\varepsilon v^2}{2b^{1/\gamma}(v)}\right)^{p-1}\theta(\eta(v))=0 \tag{25}$$

and

$$\phi'(v)+v\phi(v)\int_{v}^{\infty}\left(\frac{k}{b(v)}\int_{v}^{\infty}\sum_{i=1}^{j}q_i(s)\,\mathrm{d}s\right)^{1/(p-1)}\mathrm{d}v=0 \tag{26}$$

are oscillatory, then every solution of (1) is oscillatory.

Proof. Assume the contrary that y is a positive solution of (1). Then, we can suppose that $y(v)$and $y(\eta_i(v))$ are positive for all $v\geq v_1$ sufficiently large. From Lemma 4, we have two possible cases $(\mathbf{S}_1)$ and $(\mathbf{S}_2)$.

In the case where $(\mathbf{S}_1)$ holds, from Lemma 2, we get

$$y(v)\geq\frac{\varepsilon v^2}{2b^{1/(p-1)}(v)}\left(b^{1/(p-1)}(v)\,y'''(v)\right),$$

for every $\varepsilon\in(0,1)$ and for all large v. Thus, if we set

$$\theta(v)=b(v)\left(y'''(v)\right)^{p-1}>0,$$

then we see that ξ is a positive solution of the inequality

$$\theta'(v)+k\sum_{i=1}^{j}q_i(v)\left(\frac{\varepsilon v^2}{2b^{1/(p-1)}(v)}\right)^{p-1}\theta(\eta(v))\leq 0. \tag{27}$$

From ([27], Theorem 1), we conclude that the corresponding Equation (25) also has a positive solution, which is a contradiction.

In the case where $(\mathbf{S}_2)$ holds, from Lemma 1, we get

$$y(v) \geq v y'(v), \tag{28}$$

From (28) and (15), we get

$$y''(v) + v y'(v) \int_v^\infty \left(\frac{k}{b(v)} \int_v^\infty \sum_{i=1}^{j} q_i(s)\, \mathrm{d}s \right)^{1/(p-1)} \mathrm{d}v \leq 0.$$

Thus, if we set

$$\phi(v) = y'(v),$$

then we see that ξ is a positive solution of the inequality

$$\phi'(v) + v\phi(v) \int_v^\infty \left(\frac{k}{b(v)} \int_v^\infty \sum_{i=1}^{j} q_i(s)\, \mathrm{d}s \right)^{1/(p-1)} \mathrm{d}v \leq 0. \tag{29}$$

It is well known (see ([27], Theorem 1)) that the corresponding Equation (26) also has a positive solution, which is a contradiction. The proof is complete. □

Corollary 2. *Assume that (3) holds. If*

$$\liminf_{v\to\infty} \int_{\eta_i(v)}^{v} \sum_{i=1}^{j} q_i(s) \left(\frac{\varepsilon s^2}{2b^{1/(p-1)}(s)} \right)^{p-1} \mathrm{d}s > \frac{((n-1)!)^{p-1}}{\mathrm{e}} \tag{30}$$

and

$$\liminf_{v\to\infty} \int_{\eta_i(v)}^{v} v \int_v^\infty \left(\frac{k}{b(v)} \int_v^\infty \sum_{i=1}^{j} q_i(s)\, \mathrm{d}s \right)^{1/(p-1)} \mathrm{d}v\mathrm{d}s > \frac{1}{\mathrm{e}}, \tag{31}$$

then every solution of (1) is oscillatory.

3. Examples

For an application of Corollary 1, we give the following example:

Example 1. *Consider a differential equation*

$$y^{(4)}(v) + \frac{q_0}{v^4} y(2v) = 0,\ v \geq 1, \tag{32}$$

where $q_0 > 0$ is a constant. Note that $p = 2$, $b(v) = 1$, $q(v) = q_0/v^4$ and $\eta(v) = 2v$. If we set $k = 1$, then condition (23) becomes

$$\begin{aligned}
\int_{v_0}^\infty \left(s^3 \sum_{i=1}^{j} q_i(s) - \frac{2^{p-1} 3^p s^{-3(p-1)+2} b(s)}{p^p \varepsilon^{p-1}} \right) \mathrm{d}s &= \int_{v_0}^\infty \left(\frac{q_0}{s} - \frac{9}{2\varepsilon s} \right) ds \\
&= \left(q_0 - \frac{9}{2\varepsilon} \right) \int_{v_0}^\infty \frac{1}{s} ds \\
&= \infty \ \text{if} \ q_0 > 4.5
\end{aligned}$$

and condition (24) becomes

$$\int_{v_0}^{\infty}\left(s\int_{v}^{\infty}\left(\frac{k}{b(v)}\int_{v}^{\infty}\sum_{i=1}^{j}q_i(s)\,\mathrm{d}s\right)^{1/(p-1)}\mathrm{d}v-\frac{1}{4s}\right)\mathrm{d}s = \int_{v_0}^{\infty}\left(\frac{q_0}{6s}-\frac{1}{4s}\right)\mathrm{d}s$$
$$= \infty, \text{ if } q_0 > \frac{3}{2}.$$

Therefore, from Corollary 1, all solution Equation (32) is oscillatory if $q_0 > 4.5$.

Remark 1. *We compare our result with the known related criteria for oscillation of this equation are as follows:*

The condition	(5)	(6)	(7)	*our condition*
The criterion	$q_0 > 25.5$	$q_0 > 18$	$q_0 > 1728$	$q_0 > 4.5$

Therefore, it is clear that we see our result improves results [5–7].

For an application of Theorem 1, we give the following example.

Example 2. *Consider a differential equation*

$$(v\,(y'''(v)))' + \frac{a}{v^3}y(cv) = 0,\ v \geq 1, \tag{33}$$

where $c > 0$ *and* $a > 1$ *is a constant. Note that* $p = 2,\ b(v) = v,\ q(v) = a/v^3$.
If we set $k = 1,\ \delta(s) = \sigma(s) = s^2$, *then conditions (17) and (19) become*

$$\int_{v_0}^{\infty}\left(k\delta(s)\sum_{i=1}^{j}q_i(s)-\frac{2^{p-1}b(s)\,(\delta'(s))^p}{p^p\,(s^2\varepsilon\delta(s))^{p-1}}\right)\mathrm{d}s = \int_{v_0}^{\infty}\left(\frac{a}{s}-\frac{2}{s\varepsilon}\right)\mathrm{d}s$$
$$= \left(a-\frac{2}{\varepsilon}\right)\int_{v_0}^{\infty}\frac{1}{s}\mathrm{d}s$$
$$= \infty \text{ if } a > \frac{2}{\varepsilon}$$

and

$$\int_{v_0}^{\infty}\left(\sigma(s)\int_{v}^{\infty}\left(\frac{k}{b(v)}\int_{v}^{\infty}\sum_{i=1}^{j}q_i(s)\,\mathrm{d}s\right)^{1/(p-1)}\mathrm{d}v-\frac{1}{4\sigma(s)}\left(\sigma'(s)\right)^2\right)\mathrm{d}s = \int_{v_0}^{\infty}\left(\frac{a}{4}-\frac{1}{4}\right)\mathrm{d}s$$
$$= \infty, \text{ if } q_0 > 1.$$

for some constant $\varepsilon \in (0,\ 1)$. *Hence, by Theorem 1, every solution of Equation (33) is oscillatory if*

$$a > \frac{2}{\varepsilon}.$$

Remark 2. *By applying condition (23) in Equation (8), we find*

$$q_0 > 4.5,$$

while the conditions that we obtained in the introduction as follows:

The condition	(5)	(6)	(7)	*our condition*
The criterion	$q_0 > 13.6$	$q_0 > 18$	$q_0 > 576$	$q_0 > 4.5$

Therefore, our result improves results [5–7].

4. Conclusions

This paper is concerned with the oscillatory properties of the fourth-order differential equations with p-Laplacian like operators. New oscillation criteria are established, and they essentially improves the related contributions to the subject. In this paper the following methods were used:

1. Riccati transformations technique.
2. Method of comparison with first-order differential equations.

Further, in the future work we get some oscillation criteria of (1) under the condition $\int_{v_0}^{\infty} \frac{1}{b^{1/(p-1)}(s)} \mathrm{d}s < \infty$.

Author Contributions: O.B.: Writing original draft, writing review and editing. T.A.: Formal analysis, writing review and editing, funding and supervision. All authors have read and agreed to the published version of the manuscript.

Acknowledgments: The authors thank the reviewers for their useful comments, which led to the improvement of the content of the paper.

References

1. Hale, J.K. *Theory of Functional Differential Equations*; Springer-Verlag: New York, NY, USA, 1977.
2. Aronsson, G.; Janfalk, U. On Hele-Shaw flow of power-law fluids. *Eur. J. Appl. Math.* **1992**, *3*, 343–366.
3. Li, T.; Baculikova, B.; Dzurina, J.; Zhang, C. Oscillation of fourth order neutral differential equations with p-Laplacian like operators. Bound. *Value Probl.* **2014**, *56*, 41–58.
4. Zhang, C.; Agarwal, R.P.; Li, T. Oscillation and asymptotic behavior of higher-order delay differential equations with p -Laplacian like operators. *J. Math. Anal. Appl.* **2014**, *409*, 1093–1106.
5. Agarwal, R.; Grace, S.R. Oscillation theorems for certain functional differential equations of higher order. *Math. Comput. Model.* **2004**, *39*, 1185–1194.
6. Agarwal, R.; Grace, S.R.; O'Regan, D. Oscillation criteria for certain nth order differential equations with deviating arguments. *J. Math. Anal. Appl.* **2001**, *262*, 601–622.
7. Grace, S.R.; Lalli, B.S. Oscillation theorems for nth-order differential equations with deviating arguments. *Proc. Am. Math. Soc.* **1984**, *90*, 65–70.
8. Park, C.; Moaaz, O.; Bazighifan, O. Oscillation Results for Higher Order Differential Equations. *Axioms* **2020**, *9*, 14.
9. Baculikova, B.; Dzurina, J.; Graef, J.R. On the oscillation of higher-order delay differential equations. *Math. Slovaca* **2012**, *187*, 387–400.
10. Bazighifan, O.; Elabbasy, E.M.; Moaaz, O. Oscillation of higher-order differential equations with distributed delay. *J. Inequal. Appl.* **2019**, *55*, 1–9.
11. Moaaz, O.; Elabbasy, E.M.; Muhib, A. Oscillation criteria for even-order neutral differential equations with distributed deviating arguments. *Adv. Differ. Equ.* **2019**, *2019*, 297.
12. Zhang, C.; Agarwal, R.P.; Bohner, M.; Li, T. New results for oscillatory behavior of even-order half-linear delay differential equations. *Appl. Math. Lett.* **2013**, *26*, 179–183.
13. Bazighifan, O.; Ruggieri, M.; Scapellato, A. An Improved Criterion for the Oscillation of Fourth-Order Differential Equations. *Mathematics* **2020**, *8*, 610.
14. Agarwal, R.; Grace, S.; O'Regan, D. *Oscillation Theory for Difference and Functional Differential Equations*; Kluwer Acad. Publ.: Dordrecht, The Netherlands, 2000.
15. Bazighifan, O.; Cesarano, C. A Philos-Type Oscillation Criteria for Fourth-Order Neutral Differential Equations. *Symmetry* **2020**, *12*, 379.
16. Bazighifan, O. An Approach for Studying Asymptotic Properties of Solutions of Neutral Differential Equations. *Symmetry* **2020**, *12*, 555.
17. Bazighifan, O.; Cesarano, C. Some New Oscillation Criteria for Second-Order Neutral Differential Equations with Delayed Arguments. *Mathematics* **2019**, *7*, 619.
18. Bazighifan, O.; Postolache, M. An improved conditions for oscillation of functional nonlinear differential equations. *Mathematics* **2020**, *8*, 552.

19. Grace, S.R.; Bohner, M.; Liu, A. Oscillation criteria for fourth-order functional differential equations. *Math. Slovaca* **2013**, *63*, 1303–1320.
20. Cesarano, C.; Pinelas, S.; Al-Showaikh, F.; Bazighifan, O. Asymptotic Properties of Solutions of Fourth-Order Delay Differential Equations. *Symmetry* **2019**, *11*, 628.
21. Cesarano, C.; Bazighifan, O. Oscillation of fourth-order functional differential equations with distributed delay. *Axioms* **2019**, *7*, 61.
22. Cesarano, C.; Bazighifan, O. Qualitative behavior of solutions of second order differential equations. *Symmetry* **2019**, *11*, *777*.
23. Grace, S.; Dzurina, J.; Jadlovska, I.; Li, T. On the oscillation of fourth order delay differential equations. *Adv. Differ. Equ.* **2019**, *118*, 1–15.
24. Gyori, I.; Ladas, G. *Oscillation Theory of Delay Differential Equations with Applications*; Clarendon Press: Oxford, UK, 1991.
25. Moaaz, O.; Elabbasy, E.M.; Bazighifan, O. On the asymptotic behavior of fourth-order functional differential equations. *Adv. Differ. Equ.* **2017**, *261*, 1–13.
26. Moaaz, O.; Dassios, I.; Bazighifan, O.; Muhib, A. Oscillation Theorems for Nonlinear Differential Equations of Fourth-Order. *Mathematics* **2020**, *8*, 520.
27. Philos, C. On the existence of nonoscillatory solutions tending to zero at ∞ for differential equations with positive delay. *Arch. Math.* **1981**, *36*, 168–178.
28. Grace, S.R.; Agarwal, R.P.; Wong, P.J.Y.; Zafer, A. On the oscillation of fractional differential equations. *Frac. Calc. Appl. Anal.* **2012**, *15*, 222–231.
29. Abdalla, B.; Abdeljawad, T. On the oscillation of Caputo fractional differential equations with Mittag-Leffler nonsingular kernel. *Chaos Solitons Fractals* **2019**, *127*, 173–177.

14

Oscillatory Behavior of a Type of Generalized Proportional Fractional Differential Equations with Forcing and Damping Terms

Jehad Alzabut [1,*,†], James Viji [2,†], Velu Muthulakshmi [2,†] and Weerawat Sudsutad [3,†]

1 Department of Mathematics and General Sciences, Prince Sultan University, Riyadh 11586, Saudi Arabia
2 Department of Mathematics, Periyar University, Salem 636 011, Tamilnadu, India; vijimaths25@gmail.com (J.V.); vmuthumath@periyaruniversity.ac.in (V.M.)
3 Department of General Education, Navamindradhiraj University, Bangkok 10300, Thailand; weerawat@nmu.ac.th
* Correspondence: jalzabut@psu.edu.sa
† These authors contributed equally to this work.

Abstract: In this paper, we study the oscillatory behavior of solutions for a type of generalized proportional fractional differential equations with forcing and damping terms. Several oscillation criteria are established for the proposed equations in terms of Riemann-Liouville and Caputo settings. The results of this paper generalize some existing theorems in the literature. Indeed, it is shown that for particular choices of parameters, the obtained conditions in this paper reduce our theorems to some known results. Numerical examples are constructed to demonstrate the effectiveness of the our main theorems. Furthermore, we present and illustrate an example which does not satisfy the assumptions of our theorem and whose solution demonstrates nonoscillatory behavior.

Keywords: generalized proportional fractional operator; oscillation criteria; nonoscillatory behavior; damping and forcing terms

---

1. Introduction

Fractional calculus is a mathematical branch investigating the properties of derivatives and integrals of non-integer orders. The significance of this subject falls in the fact that the fractional derivative has the feature of nonlocal nature. This property makes these derivatives suitable to simulate more physical phenomena such as earthquake vibrations, polymers, and so forth; see, for example, References [1–10] and the references cited therein.

In recent years, there have appeared different types of fractional derivatives. However, it has been realized that most of these derivatives lose some of their basic properties that classical derivatives have such as the product rule and the chain rule. Fortunately, Khalil et al. [11] defined a new well-behaved fractional derivative, called the "conformable fractional derivative", which depends entirely on the classical limit definition of the derivative. Thereafter, researchers developed the conformable derivative and obtained different results exposing its features [12–14]. Recently, Jarad et al. [15] introduced the generalized proportional fractional (GPF) derivative of Caputo and Riemann-Liouville type involving exponential functions in their kernels. The GPF derivative not only preserves classical properties but also verifies semi group property and of nonlocal behavior. For recent results involving GPF derivative, one can refer to References [16–18].

In 2012, Grace et al. [19] initiated the study of oscillation theory for fractional differential equations. Thereafter, many researchers have investigated the oscillatory properties of fractional differential equations; see for instance References [20–25]. In 2019, Aphithana et al. [24] studied forced oscillatory properties of solutions to the conformable initial value problem of the form

$$\begin{cases} {}_aD^{1+\alpha,\rho}x(t) + p(t)_aD^{\alpha,\rho}x(t) + q(t)f(x(t)) = g(t), \quad t > a, \\ \lim\limits_{t \to a^+} {}_a\mathcal{J}^{j-\alpha,\rho}x(t) = b_j \quad (j = 1, 2, \dots, m), \end{cases}$$

where $m = \lceil \alpha \rceil$, $0 < \rho \leq 1$, $p, g \in \mathbb{C}(\mathbb{R}^+, \mathbb{R})$, $q \in \mathbb{C}(\mathbb{R}^+, \mathbb{R}^+)$, $f \in \mathbb{C}(\mathbb{R}, \mathbb{R})$ are continuous functions, ${}_aD^{\alpha,\rho}$ is the left conformable derivative of order $\alpha \in \mathbb{C}$ of x, $Re(\alpha) \geq 0$ in the Riemann-Liouville setting and ${}_a\mathcal{J}^{j-\alpha,\rho}$ is the left conformable integral operator of order $j - \alpha \in \mathbb{C}$, $b_j \in \mathbb{R}$, $j = 1, 2, \dots, m$.

They also studied the forced oscillation of conformable initial value problems in the Caputo setting of the form

$$\begin{cases} {}_a^CD^{1+\alpha,\rho}x(t) + p(t){}_a^CD^{\alpha,\rho}x(t) + q(t)f(x(t)) = g(t), \quad t > a, \\ {}_a^kD^{\rho}x(a) = b_k \quad (k = 0, 1, \dots, m-1), \end{cases}$$

where $m = \lceil \alpha \rceil$, $0 < \rho \leq 1$, and ${}_a^CD^{\alpha,\rho}$ is the left conformable derivative of order $\alpha \in \mathbb{C}$ of x, $Re(\alpha) \geq 0$ in the Caputo setting.

In 2020, Sudsutad et al. [26] established some oscillation criteria for the following generalized proportional fractional differential equation

$$\begin{cases} {}_aD^{\alpha,\rho}x(t) + \xi_1(t, x(t)) = \mu(t) + \xi_2(t, x(t)), \quad t > a \geq 0, \\ \lim\limits_{t \to a^+} {}_aI^{j-\alpha,\rho}x(t) = b_j \quad (j = 1, 2, \dots, n), \end{cases}$$

with $n = \lceil \alpha \rceil$, ${}_aD^{\alpha,\rho}$ is the generalized proportional fractional derivative operator of order $\alpha \in \mathbb{C}$, $Re(\alpha) \geq 0$, $0 < \rho \leq 1$ in the Riemann-Liouville setting and ${}_aI^{\alpha,\rho}$ is the generalized proportional fractional integral operator.

In this paper, motivated by the above papers, we establish some sufficient conditions for forced oscillation criteria of all solutions of the generalized proportional fractional (GPF) initial value problem with damping term in the Riemann-Liouville type of the form:

$$\begin{cases} {}_aD^{1+\alpha,\rho}y(l) + p(l)_aD^{\alpha,\rho}y(l) + q(l)f(y(l)) = g(l), \quad l > a \geq 0, \\ \lim\limits_{l \to a^+} {}_aI^{j-\alpha,\rho}y(l) = b_j \quad (j = 1, 2, \dots, m), \end{cases} \tag{1}$$

where $m = \lceil \alpha \rceil$, $0 < \rho \leq 1$, ${}_aD^{\alpha,\rho}$ is the left GPF derivative of order $\alpha \in \mathbb{C}$ of y, $Re(\alpha) \geq 0$ in the Riemann-Liouville setting and ${}_aI^{j-\alpha,\rho}$ is the left GPF integral of order $j - \alpha \in \mathbb{C}$, $Re(j - \alpha) > 0$, $b_j \in \mathbb{R}$, $j = 1, 2, \dots, m$ and $p, g \in \mathbb{C}(\mathbb{R}^+, \mathbb{R})$, $q \in \mathbb{C}(\mathbb{R}^+, \mathbb{R}^+)$, $f \in \mathbb{C}(\mathbb{R}, \mathbb{R})$.

Moreover, we study the forced oscillation criteria of all solutions of the GPF initial value problem with damping term in the Caputo type of the form

$$\begin{aligned} &{}_a^CD^{1+\alpha,\rho}y(l) + p(l){}_a^CD^{\alpha,\rho}y(l) + q(l)f(y(l)) = g(l), \quad l > a \geq 0, \\ &D^{k,\rho}y(a) = b_k \quad (k = 0, 1, \dots, n-1), \end{aligned} \tag{2}$$

where $n = \lceil \alpha \rceil$, $0 < \rho \leq 1$, ${}_a^CD^{\alpha,\rho}$ is the left GPF derivative of order $\alpha \in \mathbb{C}$ of y, $Re(\alpha) \geq 0$ in the Caputo setting and $D^{k,\rho} = \underbrace{D^{\rho}D^{\rho}\cdots D^{\rho}}_{\text{k times}}$, and D^{ρ} is the proportional derivative defined in Reference [13].

We claim that the results of this paper improve and generalize previously existing oscillation results in Reference [24].

Definition 1. *The solution y of problem* (1) *(respectively* (2)*) is called oscillatory if it has arbitrarily large zeros on* $(0,\infty)$*; otherwise, it is called nonoscillatory. An equation is called oscillatory if all its solutions are oscillatory.*

2. Preliminaries

In this section, we provide some basic definitions and results which will be used throughout this paper. For the justifications and proofs, the reader can consult References [13,15].

Definition 2. [15] *(Modified Conformable Derivatives).*
For $\rho \in [0,1]$*, let the functions* $k_0, k_1 : [0,1] \times \mathbb{R} \to [0,\infty)$ *be continuous such that for all* $l \in \mathbb{R}$ *we have*

$$\lim_{\rho\to 0^+} k_1(\rho,l) = 1, \quad \lim_{\rho\to 0^+} k_0(\rho,l) = 0, \quad \lim_{\rho\to 1^-} k_1(\rho,l) = 0, \quad \lim_{\rho\to 1^-} k_0(\rho,l) = 1, \tag{3}$$

and $k_1(\rho,l) \neq 0$, $\rho \in [0,1)$, $k_0(\rho,l) \neq 0$, $\rho \in (0,1]$.

Then, Anderson et al. [13] defined the modified conformable differential operator of order ρ by

$$D^\rho f(l) = k_1(\rho,l)f(l) + k_0(\rho,l)f'(l), \tag{4}$$

provided that the right-hand side exists at $l \in \mathbb{R}$ and $f'(l) = \frac{d}{dl}f$. The derivative given in (4) is called a proportional derivative. For more details about the control theory of the proportional derivatives and its component functions k_0 and k_1, we refer the reader to [27].

Of special interest, we shall restrict ourselves to the case when $k_1(\rho,l) = (1-\rho)$ and $k_0(\rho,l) = \rho$. Therefore, (4) becomes

$$D^\rho f(l) = (1-\rho)f(l) + \rho f'(l). \tag{5}$$

Notice that $\lim_{\rho\to 0^+} D^\rho f(l) = f(l)$ and $\lim_{\rho\to 1^-} D^\rho f(l) = f'(l)$. It is clear that the derivative (5) is somehow more general than the conformable derivative which does not tend to the original function as ρ tends to 0.

To find the associated integral to the proportional derivative in (5), we solve the following equation

$$D^\rho g(l) = (1-\rho)g(l) + \rho g'(l) = f(l), \quad l \geq a.$$

The above equation is a first order linear differential equation and its solution is given by

$$g(l) = \frac{1}{\rho}\int_a^l e^{\frac{\rho-1}{\rho}(l-s)} f(s)ds.$$

Define the proportional integral associated to D^ρ by

$${}_aI^{1,\rho} f(l) = \frac{1}{\rho}\int_a^l e^{\frac{\rho-1}{\rho}(l-s)} f(s)ds, \tag{6}$$

where we accept that ${}_aI^{0,\rho} f(l) = f(l)$.

Lemma 1. [15] *Let f be defined on* $[a,\infty)$ *and differentiable on* (a,∞) *and* $\rho \in (0,1]$*. Then, we have*

$${}_aI^{1,\rho} D^\rho f(l) = f(l) - e^{\frac{\rho-1}{\rho}(l-a)} f(a). \tag{7}$$

Definition 3. [15] *For* $\rho \in (0,1]$ *and* $\alpha \in \mathbb{C}$, $Re(\alpha) > 0$*, we define the left GPF integral of f by*

$$({}_aI^{\alpha,\rho} f)(l) = \frac{1}{\rho^\alpha \Gamma(\alpha)}\int_a^l e^{\frac{\rho-1}{\rho}(l-s)} (l-s)^{\alpha-1} f(s)ds = \rho^{-\alpha}\, e^{\frac{\rho-1}{\rho}l}\, {}_aI^\alpha\left(e^{\frac{1-\rho}{\rho}l} f(l)\right), \tag{8}$$

where ${}_aI^\alpha$ *is the left Riemann-Liouville fractional integral of order* α.

The right GPF integral ending at b, however, can be defined by

$$(I_b^{\alpha,\rho} f)(l) = \frac{1}{\rho^\alpha \Gamma(\alpha)} \int_l^b e^{\frac{\rho-1}{\rho}(s-l)} (s-l)^{\alpha-1} f(s) ds = \rho^{-\alpha} e^{\frac{\rho-1}{\rho} l} I_b^\alpha \left(e^{\frac{1-\rho}{\rho} l} f(l) \right), \tag{9}$$

where I_b^α *is the right Riemann-Liouville fractional integral of order* α.

Definition 4. *[15] For* $\rho \in (0,1]$ *and* $\alpha \in \mathbb{C}$, $Re(\alpha) \geq 0$, *we define the left GPF derivative of* f *by*

$$\begin{aligned} ({}_aD^{\alpha,\rho} f)(l) &= D^{n,\rho}{}_a I^{n-\alpha,\rho} f(l) \\ &= \frac{D_l^{n,\rho}}{\rho^{n-\alpha}\Gamma(n-\alpha)} \int_a^l e^{\frac{\rho-1}{\rho}(l-s)} (l-s)^{n-\alpha-1} f(s) ds. \end{aligned} \tag{10}$$

The right GPF derivative ending at b is defined by

$$\begin{aligned} (D_b^{\alpha,\rho} f)(l) &= {}_\ominus D^{n,\rho} I_b^{n-\alpha,\rho} f(l) \\ &= \frac{{}_\ominus D_l^{n,\rho}}{\rho^{n-\alpha}\Gamma(n-\alpha)} \int_l^b e^{\frac{\rho-1}{\rho}(s-l)} (s-l)^{n-\alpha-1} f(s) ds, \end{aligned} \tag{11}$$

where $n = [Re(\alpha)] + 1$.

If we let $\rho = 1$ in Definition 4 , then one can obtain the left and right Riemann-Liouville fractional derivatives as in [6]. Moreover, it is clear that

$$\lim_{\alpha \to 0} D^{\alpha,\rho} f(l) = f(l) \text{ and } \lim_{\alpha \to 1} D^{\alpha,\rho} f(l) = D^\rho f(l).$$

Lemma 2. *[15] Let* $Re(\alpha) > 0$, $n = -[-Re(\alpha)]$, $f \in L_1(a,b)$ *and* $({}_aI^{\alpha,\rho} f)(l) \in AC^n[a,b]$. *Then,*

$$({}_aI^{\alpha,\rho}{}_aD^{\alpha,\rho} f)(l) = f(l) - e^{\frac{\rho-1}{\rho}(l-a)} \sum_{j=1}^{n} ({}_aI^{j-\alpha,\rho} f)(a^+) \frac{(l-a)^{\alpha-j}}{\rho^{\alpha-j}\Gamma(\alpha+1-j)}. \tag{12}$$

Definition 5. *[13] (Partial Conformable Derivatives). Let* $\rho \in [0,1]$, *and let the functions* $k_0, k_1 : [0,1] \times \mathbb{R} \to [0,\infty)$ *be continuous and satisfy* (3). *Given a function* $f : \mathbb{R}^2 \to \mathbb{R}$ *such that* $\frac{\partial}{\partial l} f(l,s)$ *exists for each fixed* $s \in \mathbb{R}$, *define the partial differential operator* D_l^ρ *via*

$$D_l^\rho f(l,s) = k_1(\rho,l) f(l,s) + k_0(\rho,l) \frac{\partial}{\partial l} f(l,s). \tag{13}$$

Definition 6. *[13] (Conformable Exponential Function). Let* $\rho \in (0,1]$, *the points* $s, l, \in \mathbb{R}$ *with* $s \leq l$, *and let the function* $p : [s,l] \to \mathbb{R}$ *be continuous. Let* $k_0, k_1 : [0,1] \times \mathbb{R} \to [0,\infty)$ *be continuous and satisfy* (3), *with* p/k_0 *and* k_1/k_0 *Riemann integrable on* $[s,l]$. *Then the exponential function with respect to* D^ρ *in* (4) *is defined to be*

$$e_p(l,s) := e^{\int_s^l \frac{p(\tau)-k_1(\rho,\tau)}{k_0(\rho,\tau)} d\tau}, \quad e_0(l,s) := e^{-\int_s^l \frac{k_1(\rho,\tau)}{k_0(\rho,\tau)} d\tau}. \tag{14}$$

Using (4) *and* (14), *we have the following basic results.*

Lemma 3. *[13] (Basic Derivatives). Let the conformable differential operator* D^ρ *be given as in* (4), *where* $\rho \in [0,1]$. *Let the function* $p : [s,l] \to \mathbb{R}$ *be continuous. Let* $k_0, k_1 : [0,1] \times \mathbb{R} \to [0,\infty)$ *be continuous and satisfy* (3), *with* p/k_0 *and* k_1/k_0 *Riemann integrable on* $[s,l]$. *Assume the functions* f *and* g *are differentiable as needed. Then*

(i) $D^\rho[af + bg] = aD^\rho[f] + bD^\rho[g]$ *for all* $a, b \in \mathbb{R}$;

(ii) $D^{\rho}c = ck_1(\rho,\cdot)$ *for all constants* $c \in \mathbb{R}$;
(iii) $D^{\rho}[fg] = fD^{\rho}[g] + gD^{\rho}[f] - fgk_1(\rho,\cdot)$;
(iv) $D^{\rho}[f/g] = \frac{gD^{\rho}[f]-fD^{\rho}[g]}{g^2} + \frac{f}{g}k_1(\rho,\cdot)$;
(v) *for* $\rho \in (0,1]$ *and fixed* $s \in \mathbb{R}$, *the exponential function satisfies*

$$D_l^{\rho}[e_p(l,s)] = p(l)e_p(l,s)$$

for $e_p(l,s)$ *given in* (14);
(vi) *for* $\rho \in (0,1]$ *and for the exponential function* e_0 *given in* (14), *we have*

$$D^{\rho}\left[\int_a^l \frac{f(s)e_0(l,s)}{k_0(\rho,s)}ds\right] = f(l).$$

Definition 7. *[15] For* $\rho \in (0,1]$ *and* $\alpha \in \mathbb{C}$ *with* $Re(\alpha) \geq 0$, *we define the left GPF derivative of Caputo type starting at a by*

$$\begin{aligned} ({}_a^C D^{\alpha,\rho}f)(l) &= {}_aI^{n-\alpha,\rho}(D^{n,\rho}f)(l) \\ &= \frac{1}{\rho^{n-\alpha}\Gamma(n-\alpha)}\int_a^l e^{\frac{\rho-1}{\rho}(l-s)}(l-s)^{n-\alpha-1}(D^{n,\rho}f)(s)ds. \end{aligned} \tag{15}$$

The right GPF derivative of Caputo ending at b is defined by

$$\begin{aligned} ({}^C D_b^{\alpha,\rho}f)(l) &= I_b^{n-\alpha,\rho}\,({}_{\ominus}D^{n,\rho}f)(l) \\ &= \frac{1}{\rho^{n-\alpha}\Gamma(n-\alpha)}\int_l^b e^{\frac{\rho-1}{\rho}(s-l)}(s-l)^{n-\alpha-1}({}_{\ominus}D^{n,\rho}f)(s)ds, \end{aligned} \tag{16}$$

where $n = [Re(\alpha)] + 1$.

Lemma 4. *[15] For* $\rho \in (0,1]$ *and* $n = [Re(\alpha)] + 1$, *we have*

$${}_aI^{\alpha,\rho}({}_a^C D^{\alpha,\rho}f)(l) = f(l) - \sum_{k=0}^{n-1}\frac{(D^{k,\rho}f)(a)}{\rho^k k!}(l-a)^k e^{\frac{\rho-1}{\rho}}(l-a). \tag{17}$$

Proposition 1. *[15] Let* $\alpha, \beta \in \mathbb{C}$ *be such that* $Re(\alpha) \geq 0$ *and* $Re(\beta) > 0$. *Then, for any* $0 < \rho \leq 1$ *and* $n = [Re(\alpha)] + 1$, *we have*

(i) $\left({}_aI^{\alpha,\rho}e^{\frac{\rho-1}{\rho}l}(l-a)^{\beta-1}\right)(y) = \frac{\Gamma(\beta)}{\Gamma(\beta+\alpha)\rho^{\alpha}}e^{\frac{\rho-1}{\rho}y}(y-a)^{\beta+\alpha-1}, \quad Re(\alpha) > 0.$

(ii) $\left({}_aD^{\alpha,\rho}e^{\frac{\rho-1}{\rho}l}(l-a)^{\beta-1}\right)(y) = \frac{\rho^{\alpha}\Gamma(\beta)}{\Gamma(\beta-\alpha)}e^{\frac{\rho-1}{\rho}y}(y-a)^{\beta-\alpha-1}, \quad Re(\alpha) \geq 0.$

(iii) $\left({}_a^C D^{\alpha,\rho}e^{\frac{\rho-1}{\rho}l}(l-a)^{\beta-1}\right)(y) = \frac{\rho^{\alpha}\Gamma(\beta)}{\Gamma(\beta-\alpha)}e^{\frac{\rho-1}{\rho}y}(y-a)^{\beta-\alpha-1}, \quad Re(\alpha) > n.$

3. Oscillation Results via Riemann-Liouville Operator

In this section, we establish the oscillation criteria for the GPF initial value problem (1). We prove our results under the following assumption:

(H) $p \in \mathbb{C}(\mathbb{R}^+,\mathbb{R}), q \in \mathbb{C}(\mathbb{R}^+,\mathbb{R}^+), g \in \mathbb{C}(\mathbb{R}^+,\mathbb{R}), f \in C(\mathbb{R},\mathbb{R})$ with $\frac{f(u)}{u} > 0$ for all $u \neq 0$.

For our convenience, we set the following notations:

$$\Phi(l) := \Gamma(\alpha)e^{\frac{\rho-1}{\rho}(l-a)}\sum_{j=1}^{m}\frac{\rho^j b_j(l-a)^{\alpha-j}}{\Gamma(\alpha+1-j)}, \tag{18}$$

$$\Lambda(l,L) := \int_a^L e^{\frac{\rho-1}{\rho}(l-s)}(l-s)^{\alpha-1}\left(\frac{e^{\frac{\rho-1}{\rho}(s-l_1)}M + {}_{l_1}I^{1,\rho}\left(\rho g(s)V(s)\right)}{V(s)}\right)ds, \tag{19}$$

$$V(l) := \exp\int_{l_1}^{l}\frac{\rho p(\tau)-(1-\rho)}{\rho}d\tau, \tag{20}$$

$$M := {}_{l_1}D^{\alpha,\rho}y(l_1)V(l_1), \quad M \text{ is an arbitrary constant.} \tag{21}$$

Theorem 1. *Assume that* (H) *holds. If*

$$\liminf_{l\to\infty} l^{1-\alpha}\int_L^l e^{\frac{\rho-1}{\rho}(l-s)}(l-s)^{\alpha-1}\left[\frac{e^{\frac{\rho-1}{\rho}(s-l_1)}M + {}_{l_1}I^{1,\rho}\left(\rho g(s)V(s)\right)}{V(s)}\right]ds = -\infty \tag{22}$$

and

$$\limsup_{l\to\infty} l^{1-\alpha}\int_L^l e^{\frac{\rho-1}{\rho}(l-s)}(l-s)^{\alpha-1}\left[\frac{e^{\frac{\rho-1}{\rho}(s-l_1)}M + {}_{l_1}I^{1,\rho}\left(\rho g(s)V(s)\right)}{V(s)}\right]ds = \infty, \tag{23}$$

for every sufficiently large L, where $V(l)$ *and* M *are defined as in* (20) *and* (21) *respectively, then every solution of problem* (1) *is oscillatory.*

Proof. Suppose that $y(l)$ is a nonoscillatory solution of problem (1). Without loss of generality, let $L > a$ be large enough and $l_1 \geq L$ such that $y(l) > 0$ for all $l \geq l_1$. Using Lemma 3 (iii), Equations (5) and (13), we have

$$\begin{aligned}
D^\rho\left[{}_aD^{\alpha,\rho}y(l)V(l)\right] &= {}_aD^{\alpha,\rho}y(l)D^\rho V(l) + V(l)D^\rho\left({}_aD^{\alpha,\rho}y(l)\right) - (1-\rho)\,{}_aD^{\alpha,\rho}y(l)V(l)\\
&= {}_aD^{\alpha,\rho}y(l)D^\rho V(l) + V(l)\left[(1-\rho)\,{}_aD^{\alpha,\rho}y(l) + \rho\frac{d}{dl}\left({}_aD^{\alpha,\rho}y(l)\right)\right]\\
&\quad -(1-\rho)\,{}_aD^{\alpha,\rho}y(l)V(l)\\
&= \rho\left[{}_aD^{1+\alpha,\rho}y(l) + p(l)\,{}_aD^{\alpha,\rho}y(l)\right]V(l)\\
&= \rho\left[-q(l)f(y(l)) + g(l)\right]V(l)\\
&< \rho g(l)V(l).
\end{aligned}$$

Taking the proportional integral operator ${}_{l_1}I^{1,\rho}$ on both sides to the above inequality, we obtain

$${}_{l_1}I^{1,\rho}\left(D^\rho\left[{}_aD^{\alpha,\rho}y(l)V(l)\right]\right) < {}_{l_1}I^{1,\rho}\left(\rho g(l)V(l)\right). \tag{24}$$

Using Lemma 1 on the L.H.S of (24), we have

$${}_aD^{\alpha,\rho}y(l) < \frac{e^{\frac{\rho-1}{\rho}(l-l_1)}M + {}_{l_1}I^{1,\rho}\left(\rho g(l)V(l)\right)}{V(l)}.$$

Taking the left GPF integral operator ${}_aI^{\alpha,\rho}$on both sides to the above inequality, we get

$${}_aI^{\alpha,\rho}\left({}_aD^{\alpha,\rho}y(l)\right) < {}_aI^{\alpha,\rho}\left[\frac{e^{\frac{\rho-1}{\rho}(l-l_1)}M + {}_{l_1}I^{1,\rho}\left(\rho g(l)V(l)\right)}{V(l)}\right]. \tag{25}$$

Using Lemma 2 on the L.H.S of (25), we have

$$y(l) - e^{\frac{\rho-1}{\rho}(l-a)} \sum_{j=1}^{m} \frac{b_j (l-a)^{\alpha-j}}{\rho^{\alpha-j}\Gamma(\alpha+1-j)} < {}_a I^{\alpha,\rho} \left[\frac{e^{\frac{\rho-1}{\rho}(l-l_1)} M + {}_{l_1}I^{1,\rho}\left(\rho g(l) V(l)\right)}{V(l)} \right]. \tag{26}$$

Applying the left GPF integral formula on the R.H.S of (26) , we have

$$\begin{aligned} y(l) &< e^{\frac{\rho-1}{\rho}(l-a)} \sum_{j=1}^{m} \frac{b_j (l-a)^{\alpha-j}}{\rho^{\alpha-j}\Gamma(\alpha+1-j)} \\ &\quad + \frac{1}{\rho^{\alpha}\Gamma(\alpha)} \int_a^l e^{\frac{\rho-1}{\rho}(l-s)} (l-s)^{\alpha-1} \left[\frac{e^{\frac{\rho-1}{\rho}(s-l_1)} M + {}_{l_1}I^{1,\rho}\left(\rho\, g(s) V(s)\right)}{V(s)} \right] ds, \end{aligned}$$

for every sufficiently large L. If we multiply the above inequality by $\rho^{\alpha}\Gamma(\alpha)$, we get

$$\begin{aligned} \rho^{\alpha}\Gamma(\alpha) y(l) &< \Gamma(\alpha) e^{\frac{\rho-1}{\rho}(l-a)} \sum_{j=1}^{m} \frac{\rho^j b_j (l-a)^{\alpha-j}}{\Gamma(\alpha+1-j)} \\ &\quad + \int_a^L e^{\frac{\rho-1}{\rho}(l-s)} (l-s)^{\alpha-1} \left[\frac{e^{\frac{\rho-1}{\rho}(s-l_1)} M + {}_{l_1}I^{1,\rho}\left(\rho g(s) V(s)\right)}{V(s)} \right] ds \\ &\quad + \int_L^l e^{\frac{\rho-1}{\rho}(l-s)} (l-s)^{\alpha-1} \left[\frac{e^{\frac{\rho-1}{\rho}(s-l_1)} M + {}_{l_1}I^{1,\rho}\left(\rho g(s) V(s)\right)}{V(s)} \right] ds \\ &= \Phi(l) + \Lambda(l,L) + \int_L^l e^{\frac{\rho-1}{\rho}(l-s)} (l-s)^{\alpha-1} \left[\frac{e^{\frac{\rho-1}{\rho}(s-l_1)} M + {}_{l_1}I^{1,\rho}\left(\rho g(s) V(s)\right)}{V(s)} \right] ds, \end{aligned} \tag{27}$$

where $\Phi(l)$ and $\Lambda(l,L)$ are defined in (18) and (19), respectively.

Multiplying (27) by $l^{1-\alpha}$, we get

$$\begin{aligned} 0 &< l^{1-\alpha}\rho^{\alpha}\Gamma(\alpha) y(l) \\ &< l^{1-\alpha}\Phi(l) + l^{1-\alpha}\Lambda(l,L) \\ &\quad + l^{1-\alpha} \int_L^l e^{\frac{\rho-1}{\rho}(l-s)} (l-s)^{\alpha-1} \left[\frac{e^{\frac{\rho-1}{\rho}(s-l_1)} M + {}_{l_1}I^{1,\rho}\left(\rho g(s) V(s)\right)}{V(s)} \right] ds. \end{aligned} \tag{28}$$

Let us consider the following two cases for $L_1 \geq L$.

Case(i): Let $0 < \alpha \leq 1$. Then $m = 1$. Since $\left| e^{\frac{\rho-1}{\rho}(l-a)} \right| \leq 1$ and the function $h_1(l) = \left(\frac{l-a}{l}\right)^{\alpha-1}$ is decreasing for $\rho > 0\,, 0 < \alpha < 1$, we get for $l \geq L_1$,

$$\left| l^{1-\alpha}\Phi(l) \right| = \left| e^{\frac{\rho-1}{\rho}(l-a)} \rho b_1 \left(\frac{l-a}{l}\right)^{\alpha-1} \right| \leq \rho |b_1| \left(\frac{L_1-a}{L_1}\right)^{\alpha-1} := C_1(L_1), \tag{29}$$

and

$$\begin{aligned} \left|l^{1-\alpha}\Lambda(l,L)\right| &= \left|l^{1-\alpha}\int_a^L e^{\frac{\rho-1}{\rho}(l-s)}(l-s)^{\alpha-1}\left(\frac{e^{\frac{\rho-1}{\rho}(s-l_1)}M+{}_{l_1}I^{1,\rho}\left(\rho g(s)V(s)\right)}{V(s)}\right)ds\right| \\ &\le \int_a^L \left|e^{\frac{\rho-1}{\rho}(l-s)}\right|\left(\frac{l-a}{l}\right)^{\alpha-1}\left|\frac{e^{\frac{\rho-1}{\rho}(s-l_1)}M+{}_{l_1}I^{1,\rho}\left(\rho g(s)V(s)\right)}{V(s)}\right|ds \\ &\le \int_a^L \left(\frac{L_1-a}{L_1}\right)^{\alpha-1}\left|\frac{e^{\frac{\rho-1}{\rho}(s-l_1)}M+{}_{l_1}I^{1,\rho}\left(\rho g(s)V(s)\right)}{V(s)}\right|ds \\ &:= C_2(L,L_1). \end{aligned} \tag{30}$$

From (28), (29) and (30), we get, for $l \ge L_1$,

$$l^{1-\alpha}\int_L^l e^{\frac{\rho-1}{\rho}(l-s)}(l-s)^{\alpha-1}\left[\frac{e^{\frac{\rho-1}{\rho}(s-l_1)}M+{}_{l_1}I^{1,\rho}\left(\rho g(s)V(s)\right)}{V(s)}\right]ds \ge -[C_1(L_1)+C_2(L,L_1)].$$

Since the R.H.S of the above inequality is a negative constant, it follows that

$$\liminf_{l\to\infty} l^{1-\alpha}\int_L^l e^{\frac{\rho-1}{\rho}(l-s)}(l-s)^{\alpha-1}\left[\frac{e^{\frac{\rho-1}{\rho}(s-l_1)}M+{}_{l_1}I^{1,\rho}\left(\rho g(s)V(s)\right)}{V(s)}\right]ds > -\infty,$$

which leads to a contradiction with (22).

Case(ii): Let $\alpha > 1$. Then $m \ge 2$ and $\left(\frac{l-a}{l}\right)^{\alpha-1} < 1$ for $\alpha > 1$ and $\rho > 0$. Since $\left|e^{\frac{\rho-1}{\rho}(l-a)}\right| \le 1$ and the function $h_2(l) = (l-a)^{1-j}$ is decreasing for $j > 1$ and $\rho > 0$, for $l \ge L_1$, we have

$$\begin{aligned} \left|l^{1-\alpha}\Phi(l)\right| &= \left|l^{1-\alpha}\Gamma(\alpha)e^{\frac{\rho-1}{\rho}(l-a)}\sum_{j=1}^{m}\frac{\rho^j b_j(l-a)^{\alpha-j}}{\Gamma(\alpha+1-j)}\right| \\ &\le \Gamma(\alpha)\left(\frac{l-a}{l}\right)^{\alpha-1}\sum_{j=1}^{m}\frac{\rho^j|b_j|(l-a)^{1-j}}{\Gamma(\alpha+1-j)} \\ &\le \Gamma(\alpha)\sum_{j=1}^{m}\frac{\rho^j|b_j|(L_1-a)^{1-j}}{\Gamma(\alpha+1-j)} \\ &:= C_3(L_1), \end{aligned} \tag{31}$$

and

$$\begin{aligned} \left|l^{1-\alpha}\Lambda(l,L)\right| &= \left|\int_a^L e^{\frac{\rho-1}{\rho}(l-s)}\left(\frac{l-s}{l}\right)^{\alpha-1}\left(\frac{e^{\frac{\rho-1}{\rho}(s-l_1)}M+{}_{l_1}I^{1,\rho}\left(\rho g(s)V(s)\right)}{V(s)}\right)ds\right| \\ &\le \int_a^L \left|\frac{e^{\frac{\rho-1}{\rho}(s-l_1)}M+{}_{l_1}I^{1,\rho}\left(\rho g(s)V(s)\right)}{V(s)}\right|ds \\ &:= C_4(L). \end{aligned} \tag{32}$$

From (28), (31) and (32), we conclude that for $l \ge L_1$,

$$l^{1-\alpha}\int_L^l e^{\frac{\rho-1}{\rho}(l-s)}(l-s)^{\alpha-1}\left[\frac{e^{\frac{\rho-1}{\rho}(s-l_1)}M+{}_{l_1}I^{1,\rho}\left(\rho g(s)V(s)\right)}{V(s)}\right]ds \ge -[C_3(L_1)+C_4(L)].$$

Since, the R.H.S of the above inequality is a negative constant, it follows that

$$\liminf_{l\to\infty} l^{1-\alpha}\int_L^l e^{\frac{\rho-1}{\rho}(l-s)}(l-s)^{\alpha-1}\left[\frac{e^{\frac{\rho-1}{\rho}(s-l_1)}M + {}_{l_1}I^{1,\rho}\left(\rho g(s)V(s)\right)}{V(s)}\right]ds > -\infty,$$

which is a contradiction to (22).

Therefore, $y(l)$ is oscillatory. If $y(l)$ is eventually negative, by a similar argument, we get a contradiction with condition (23). Hence the theorem. □

4. Oscillation Results via Caputo Operator

In this section, we establish the oscillation criteria for the GPF initial value problem (2) under the assumption (H):

We set

$$\Psi(l) := \Gamma(\alpha)\sum_{k=0}^{m-1}\frac{b_k}{\rho^{k-\alpha}k!}(l-a)^k e^{\frac{\rho-1}{\rho}(l-a)}, \tag{33}$$

$$\Omega(l,L) := \int_a^L e^{\frac{\rho-1}{\rho}(l-s)}(l-s)^{\alpha-1}\left(\frac{e^{\frac{\rho-1}{\rho}(s-l_1)}M^* + {}_{l_1}I^{1,\rho}\left(\rho g(s)V(s)\right)}{V(s)}\right)ds, \tag{34}$$

$$M^* := {}_a^C D^{\alpha,\rho}y(a)V(a), \quad M^* \text{ is an arbitrary constant.} \tag{35}$$

Theorem 2. *Assume that* (H) *holds. If*

$$\liminf_{l\to\infty} l^{1-n}\int_L^l e^{\frac{\rho-1}{\rho}(l-s)}(l-s)^{\alpha-1}\left[\frac{e^{\frac{\rho-1}{\rho}(s-l_1)}M^* + {}_{l_1}I^{1,\rho}\left(\rho g(s)V(s)\right)}{V(s)}\right]ds = -\infty \tag{36}$$

and

$$\limsup_{l\to\infty} l^{1-n}\int_L^l e^{\frac{\rho-1}{\rho}(l-s)}(l-s)^{\alpha-1}\left[\frac{e^{\frac{\rho-1}{\rho}(s-l_1)}M^* + {}_{l_1}I^{1,\rho}\left(\rho g(s)V(s)\right)}{V(s)}\right]ds = \infty, \tag{37}$$

for every sufficiently large L, where $V(l)$ *and* M^* *are defined as in* (20) *and* (35)*, respectively, then every solution of problem* (2) *is oscillatory.*

Proof. Suppose that $y(l)$ is a nonoscillatory solution of problem (2). Without loss of generality, let $L > a$ be large enough and $l_1 \geq L$ such that $y(l) > 0$ for $l \geq l_1$. Using Lemma 3 (iii), Equations (5) and (13), we have

$$\begin{aligned}
D^\rho[{}_a^C D^{\alpha,\rho}y(l)V(l)] &= {}_a^C D^{\alpha,\rho}y(l)D^\rho V(l) + V(l)D^\rho\left({}_a^C D^{\alpha,\rho}y(l)\right) - (1-\rho){}_a^C D^{\alpha,\rho}y(l)V(l) \\
&= {}_a^C D^{\alpha,\rho}y(l)D^\rho V(l) + V(l)\left[(1-\rho){}_a^C D^{\alpha,\rho}y(l) + \rho\frac{d}{dl}\left({}_a^C D^{\alpha,\rho}y(l)\right)\right] \\
&\quad -(1-\rho){}_a^C D^{\alpha,\rho}y(l)V(l) \\
&= \rho\left[{}_a^C D^{1+\alpha,\rho}y(l) + p(l){}_a^C D^{\alpha,\rho}y(l)\right]V(l) \\
&= \rho\left[-q(l)f(y(l)) + g(l)\right]V(l) \\
&< \rho g(l)V(l).
\end{aligned}$$

Taking the proportional integral operator ${}_{l_1}I^{1,\rho}$ on both sides to the above inequality, we obtain

$$ {}_{l_1}I^{1,\rho}\left(D^{\rho}\left[{}_{a}^{C}D^{\alpha,\rho}y(l)V(l)\right]\right) < {}_{l_1}I^{1,\rho}\left(\rho g(l)V(l)\right). \tag{38}$$

Using Lemma (1) on the L.H.S of (38) , we have

$$ {}_{a}^{C}D^{\alpha,\rho}y(l) < \frac{e^{\frac{\rho-1}{\rho}(l-l_1)}M^{*} + {}_{l_1}I^{1,\rho}\left(\rho g(l)V(l)\right)}{V(l)}.$$

Applying the left GPF integral operator ${}_{a}I^{\alpha,\rho}$on both sides to the above inequality, we get

$$ {}_{a}I^{\alpha,\rho}\left({}_{a}^{C}D^{\alpha,\rho}y(l)\right) < {}_{a}I^{\alpha,\rho}\left[\frac{e^{\frac{\rho-1}{\rho}(l-l_1)}M^{*} + {}_{l_1}I^{1,\rho}\left(\rho g(l)V(l)\right)}{V(l)}\right]. \tag{39}$$

Using Lemma 4 on the L.H.S of (39), we have

$$ y(l) - \sum_{k=0}^{n-1}\frac{(D^{k,\rho}y)(a)}{\rho^{k}k!}(l-a)^{k}e^{\frac{\rho-1}{\rho}(l-a)} < {}_{a}I^{\alpha,\rho}\left[\frac{e^{\frac{\rho-1}{\rho}(l-l_1)}M^{*} + {}_{l_1}I^{1,\rho}\left(\rho g(l)V(l)\right)}{V(l)}\right]. \tag{40}$$

Applying the left GPF integral formula on the R.H.S of (40), we have

$$\begin{aligned} y(l) &< \sum_{k=0}^{n-1}\frac{b_k}{\rho^{k}k!}(l-a)^{k}e^{\frac{\rho-1}{\rho}(l-a)} \\ &\quad + \frac{1}{\rho^{\alpha}\Gamma(\alpha)}\int_{a}^{l} e^{\frac{\rho-1}{\rho}(l-s)}(l-s)^{\alpha-1}\left[\frac{e^{\frac{\rho-1}{\rho}(s-l_1)}M^{*} + {}_{l_1}I^{1,\rho}\left(\rho g(s)V(s)\right)}{V(s)}\right]ds, \end{aligned}$$

for every sufficiently large L. If we multiply the above inequality by $\rho^{\alpha}\Gamma(\alpha)$, we get

$$\begin{aligned} \rho^{\alpha}\Gamma(\alpha)y(l) &< \rho^{\alpha}\Gamma(\alpha)\sum_{k=0}^{n-1}\frac{b_k}{\rho^{k}k!}(l-a)^{k}\,e^{\frac{\rho-1}{\rho}(l-a)} \\ &\quad + \int_{a}^{L} e^{\frac{\rho-1}{\rho}(l-s)}(l-s)^{\alpha-1}\left[\frac{e^{\frac{\rho-1}{\rho}(s-l_1)}M^{*} + {}_{l_1}I^{1,\rho}\left(\rho g(s)V(s)\right)}{V(s)}\right]ds \\ &\quad + \int_{L}^{l} e^{\frac{\rho-1}{\rho}(l-s)}(l-s)^{\alpha-1}\left[\frac{e^{\frac{\rho-1}{\rho}(s-l_1)}M^{*} + {}_{l_1}I^{1,\rho}\left(\rho g(s)V(s)\right)}{V(s)}\right]ds \\ &= \Psi(l) + \Omega(l,L) \\ &\quad + \int_{L}^{l} e^{\frac{\rho-1}{\rho}(l-s)}(l-s)^{\alpha-1}\left[\frac{e^{\frac{\rho-1}{\rho}(s-l_1)}M^{*} + {}_{l_1}I^{1,\rho}\left(\rho g(s)V(s)\right)}{V(s)}\right]ds, \end{aligned} \tag{41}$$

where $\Psi(l)$ and $\Omega(l,L)$ are defined in (33) and (34), respectively.

Multiplying (41) by l^{1-n}, we get

$$\begin{aligned} 0 &< l^{1-n}\Gamma(\alpha)y(l) \\ &< l^{1-n}\Psi(l) + l^{1-n}\Omega(l,L) \\ &\quad + l^{1-n}\int_{L}^{l} e^{\frac{\rho-1}{\rho}(l-s)}(l-s)^{\alpha-1}\left[\frac{e^{\frac{\rho-1}{\rho}(s-l_1)}M^{*} + {}_{l_1}I^{1,\rho}\left(\rho g(s)V(s)\right)}{V(s)}\right]ds. \end{aligned} \tag{42}$$

Let us consider the following two cases for $L_1 \geq L$.

Case(i): Let $0 < \alpha \leq 1$. Then $n = 1$. Since $\left|e^{\frac{\rho-1}{\rho}(l-a)}\right| \leq 1$ and the function $h_3(l) = (l-s)^{\alpha-1}$ is decreasing for $0 < \alpha < 1$, we get for $l \geq L_1$,

$$\left|l^{1-n}\Psi(l)\right| = \left|l^{1-n}\Gamma(\alpha)\sum_{k=0}^{n-1}\frac{b_k}{\rho^{k-\alpha}k!}(l-a)^k e^{\frac{\rho-1}{\rho}(l-a)}\right| \leq \rho^\alpha\Gamma(\alpha)|b_0| := C_5(L), \tag{43}$$

and

$$\begin{aligned}
\left|l^{1-n}\Omega(l,L)\right| &= \left|l^{1-n}\int_a^L e^{\frac{\rho-1}{\rho}(l-s)}(l-s)^{\alpha-1}\left(\frac{e^{\frac{\rho-1}{\rho}(s-l_1)}M^* + {}_{l_1}I^{1,\rho}\left(\rho g(s)V(s)\right)}{V(s)}\right)ds\right| \\
&\leq \int_a^L (l-s)^{\alpha-1}\left|\frac{e^{\frac{\rho-1}{\rho}(s-l_1)}M^* + {}_{l_1}I^{1,\rho}\left(\rho g(s)V(s)\right)}{V(s)}\right| ds \\
&\leq \int_a^L (L_1-s)^{\alpha-1}\left|\frac{e^{\frac{\rho-1}{\rho}(s-l_1)}M^* + {}_{l_1}I^{1,\rho}\left(\rho g(s)V(s)\right)}{V(s)}\right| ds \\
&:= C_6(L,L_1).
\end{aligned} \tag{44}$$

Then, from (42) and $l \geq L_1$, we get

$$l^{1-n}\int_L^l e^{\frac{\rho-1}{\rho}(l-s)}(l-s)^{\alpha-1}\left[\frac{e^{\frac{\rho-1}{\rho}(s-l_1)}M^* + {}_{l_1}I^{1,\rho}\left(\rho g(s)V(s)\right)}{V(s)}\right] ds \geq -[C_5(L) + C_6(L,L_1)].$$

Since, the R.H.S of the above inequality is a negative constant, it follows that

$$\liminf_{l\to\infty} l^{1-n}\int_L^l e^{\frac{\rho-1}{\rho}(l-s)}(l-s)^{\alpha-1}\left[\frac{e^{\frac{\rho-1}{\rho}(s-l_1)}M^* + {}_{l_1}I^{1,\rho}\left(\rho g(s)V(s)\right)}{V(s)}\right] ds > -\infty,$$

which leads to a contradiction with the condition (36).

Case(ii): Let $\alpha > 1$. Then $n \geq 2$ and $\left(\frac{l-a}{l}\right)^{n-1} < 1$ for $n \geq 2$ and $\alpha > 1$. Since $\left|e^{\frac{\rho-1}{\rho}(l-a)}\right| \leq 1$ and the function $h_4(l) = (l-a)^{k-n+1}$ is decreasing for $k > n-1$ and for $l \geq L_1$, we have

$$\begin{aligned}
\left|l^{1-n}\Psi(l)\right| &= \left|l^{1-n}\,\Gamma(\alpha)\sum_{k=0}^{n-1}\frac{b_k}{\rho^{k-\alpha}k!}(l-a)^k e^{\frac{\rho-1}{\rho}(l-a)}\right| \\
&= \left|\rho^\alpha\Gamma(\alpha)\left(\frac{l-a}{l}\right)^{n-1}\sum_{k=0}^{n-1}\frac{b_k}{\rho^k k!}(l-a)^{k-n+1}e^{\frac{\rho-1}{\rho}(l-a)}\right| \\
&\leq \rho^\alpha\Gamma(\alpha)\sum_{k=0}^{n-1}\frac{|b_k|}{\rho^k k!}(l-a)^{k-n+1} \\
&\leq \rho^\alpha\Gamma(\alpha)\sum_{k=0}^{n-1}\frac{|b_k|(L_1-a)^{k-n+1}}{\rho^k k!} \\
&:= C_7(L_1),
\end{aligned} \tag{45}$$

and

$$\begin{aligned}
\left|l^{1-n}\Omega(l,L)\right| &= \left| l^{1-n}\int_a^L e^{\frac{\rho-1}{\rho}(l-s)}(l-s)^{\alpha-1}\left(\frac{e^{\frac{\rho-1}{\rho}(s-l_1)}M^* + {}_{l_1}I^{1,\rho}\left(\rho g(s)V(s)\right)}{V(s)}\right)ds\right| \\
&= \left| l^{\alpha-n}\int_a^L e^{\frac{\rho-1}{\rho}(l-s)}\left(\frac{l-s}{l}\right)^{\alpha-1}\left(\frac{e^{\frac{\rho-1}{\rho}(s-l_1)}M^* + {}_{l_1}I^{1,\rho}\left(\rho g(s)V(s)\right)}{V(s)}\right)ds\right| \\
&\le \int_a^L \left|\frac{e^{\frac{\rho-1}{\rho}(s-l_1)}M^* + {}_{l_1}I^{1,\rho}\left(\rho g(s)V(s)\right)}{V(s)}\right| ds \\
&:= C_8(L).
\end{aligned} \tag{46}$$

From Equations (42), (45) and (46), we conclude that for $l \ge L_1$,

$$l^{1-n}\int_L^l e^{\frac{\rho-1}{\rho}(l-s)}(l-s)^{\alpha-1}\left[\frac{e^{\frac{\rho-1}{\rho}(s-l_1)}M^* + {}_{l_1}I^{1,\rho}\left(\rho g(s)V(s)\right)}{V(s)}\right]ds \ge -[C_7(L_1)+C_8(L)].$$

Since, the R.H.S of the above inequality is a negative constant, it follows that

$$\lim_{l\to\infty}\inf l^{1-n}\int_L^l e^{\frac{\rho-1}{\rho}(l-s)}(l-s)^{\alpha-1}\left[\frac{e^{\frac{\rho-1}{\rho}(s-l_1)}M^* + {}_{l_1}I^{1,\rho}\left(\rho g(s)V(s)\right)}{V(s)}\right]ds > -\infty,$$

which contradicts the (36).

Therefore, $y(l)$ is oscillatory. If $y(l)$ is eventually negative, by a similar argument, we get a contradiction with condition (37). Hence the theorem. □

Remark 1. *If we put $\rho = 1$ in Theorem* (1) *and Theorem* (2)*, then they reduced to Theorem 3.1 and Theorem 4.1, respectively, of [24].*

5. Examples

This section include some examples for the illustration of our main results.

Example 1. *Consider the following GPF initial value problem*

$$\begin{cases} {}_0D^{\frac{3}{2},1}y(l) - {}_0D^{\frac{1}{2},1}y(l) + (l+7)^2(y+3)e^{\cos 2y} = e^{2l}\sin l, \quad l>0, \\ \lim_{l\to 0^+} {}_0I^{\frac{1}{2},1}y(l) = b_1. \end{cases} \tag{47}$$

Setting $\alpha = \frac{1}{2}$, $\rho = 1$, $a = 0$, $p(l) = -1$, $q(l) = (l+7)^2$, $f(y) = (y+3)e^{\cos 2y}$, $g(l) = e^{2l}\sin l$ and $V(l) = e^{l_1 - l}$. The assumption (H) is satisfied if $y(l) > 0$. Then,

$$\begin{aligned}
{}_{l_1}I^{1,\rho}\left(\rho g(s)V(s)\right) &= \frac{1}{\rho}\int_{l_1}^s e^{\frac{\rho-1}{\rho}(s-\tau)}\rho g(\tau)V(\tau)d\tau \\
&= \int_{l_1}^s e^{l_1+\tau}\sin\tau d\tau \\
&= \frac{e^{l_1+s}}{2}(\sin s - \cos s) - \frac{e^{2l_1}}{2}(\sin l_1 - \cos l_1) \\
&= \frac{\sqrt{2}e^{l_1+s}}{2}\sin\left(s-\frac{\pi}{4}\right) - \frac{e^{2l_1}}{2}(\sin l_1 - \cos l_1).
\end{aligned}$$

Set a point $l_1 = \frac{\pi}{2}$. Hence, we compute that

$$l^{1-\alpha} \int_0^l e^{\frac{\rho-1}{\rho}(l-s)} (l-s)^{\alpha-1} \left[\frac{e^{\frac{\rho-1}{\rho}(s-l_1)} M + {}_{l_1}I^{1,\rho}\left(\rho g(s) V(s)\right)}{V(s)} \right] ds$$

$$= l^{\frac{1}{2}} \int_0^l (l-s)^{-\frac{1}{2}} e^{s-\frac{\pi}{2}} \left[\left(M - \frac{e^{\pi}}{2} \right) + \frac{\sqrt{2}}{2} e^{\frac{\pi}{2}+s} \sin\left(s - \frac{\pi}{4}\right) \right] ds.$$

By setting $l - s = \tau^2$, we can get the above integral as

$$l^{\frac{1}{2}} \int_0^l (l-s)^{-\frac{1}{2}} e^{s-\frac{\pi}{2}} \left[\left(M - \frac{e^{\pi}}{2} \right) + \frac{\sqrt{2}}{2} e^{\frac{\pi}{2}+s} \sin\left(s - \frac{\pi}{4}\right) \right] ds$$

$$= l^{\frac{1}{2}} \int_{\sqrt{l}}^0 \frac{1}{\tau} e^{l-\tau^2-\frac{\pi}{2}} \left[\left(\frac{2M - e^{\pi}}{2} \right) + \frac{\sqrt{2}}{2} e^{\frac{\pi}{2}+l-\tau^2} \sin\left(l - \tau^2 - \frac{\pi}{4}\right) \right] (-2\tau) d\tau$$

$$= (2M - e^{\pi}) l^{\frac{1}{2}} e^{l-\frac{\pi}{2}} \int_0^{\sqrt{l}} e^{-\tau^2} d\tau + \sqrt{2} l^{\frac{1}{2}} e^{2l} \int_0^{\sqrt{l}} e^{-2\tau^2} \sin\left(l - \tau^2 - \frac{\pi}{4}\right) d\tau$$

$$= (2M - e^{\pi}) l^{\frac{1}{2}} e^{l-\frac{\pi}{2}} \int_0^{\sqrt{l}} e^{-\tau^2} d\tau + \sqrt{2} l^{\frac{1}{2}} e^{2l} \sin\left(l - \frac{\pi}{4}\right) \int_0^{\sqrt{l}} e^{-2\tau^2} \cos\tau^2 d\tau$$

$$-\sqrt{2} l^{\frac{1}{2}} e^{2l} \cos\left(l - \frac{\pi}{4}\right) \int_0^{\sqrt{l}} e^{-2\tau^2} \sin\tau^2 d\tau.$$

Let $l \to +\infty$ as the result of $|e^{-2\tau^2}\cos\tau^2| \leq e^{-2\tau^2}$, $|e^{-2\tau^2}\sin\tau^2| \leq e^{-2\tau^2}$ and $\lim_{l\to+\infty} \int_0^{\sqrt{l}} e^{-2\tau^2} d\tau = \frac{\sqrt{2\pi}}{4}$. Thus, we know that $\lim_{l\to+\infty} \int_0^{\sqrt{l}} e^{-2\tau^2}\cos\tau^2 d\tau$ and $\lim_{l\to+\infty} \int_0^{\sqrt{l}} e^{-2\tau^2}\sin\tau^2 d\tau$ are convergent.

Hence, we set $\lim_{l\to+\infty} \int_0^{\sqrt{l}} e^{-2\tau^2}\cos\tau^2 d\tau = A$ and $\lim_{l\to+\infty} \int_0^{\sqrt{l}} e^{-2\tau^2}\sin\tau^2 d\tau = B$. Select the sequence $\{l_k\} = \left\{ \frac{3\pi}{2} + \frac{\pi}{4} + 2k\pi - \arctan\left(-\frac{B}{A}\right) \right\}$, $\lim_{l\to+\infty} l_k = \infty$, then

$$\lim_{k\to+\infty} \left\{ l_k^{\frac{1}{2}} e^{l_k} \left[(2M - e^{\pi}) e^{-\frac{\pi}{2}} \int_0^{\sqrt{l_k}} e^{-\tau^2} d\tau + \sqrt{2} e^{l_k} \left(\sin\left(l_k - \frac{\pi}{4}\right) \int_0^{\sqrt{l_k}} e^{-2\tau^2} \cos\tau^2 d\tau \right.\right.\right.$$
$$\left.\left.\left. - \cos\left(l_k - \frac{\pi}{4}\right) \int_0^{\sqrt{l_k}} e^{-2\tau^2} \sin\tau^2 d\tau \right) \right] \right\}. \tag{48}$$

Firstly, we consider the following limit:

$$\lim_{k\to+\infty} \left\{ \sin\left(l_k - \frac{\pi}{4}\right) \int_0^{\sqrt{l_k}} e^{-2\tau^2} \cos\tau^2 d\tau - \cos\left(l_k - \frac{\pi}{4}\right) \int_0^{\sqrt{l_k}} e^{-2\tau^2} \sin\tau^2 d\tau \right\}$$

$$= A \cdot \lim_{k\to+\infty} \sin\left(\frac{3\pi}{2} + 2k\pi - \arctan\left(-\frac{B}{A}\right) \right) - B \cdot \lim_{k\to+\infty} \cos\left(\frac{3\pi}{2} + 2k\pi - \arctan\left(-\frac{B}{A}\right) \right)$$

$$= A \cdot \sin\left(\frac{3\pi}{2} - \arctan\left(-\frac{B}{A}\right) \right) - B \cdot \cos\left(\frac{3\pi}{2} - \arctan\left(-\frac{B}{A}\right) \right)$$

$$= -\sqrt{A^2 + B^2}.$$

Secondly, we know that $\lim_{k\to+\infty} e^{l_k} = +\infty$ and $\lim_{k\to+\infty} 2Me^{l_k}e^{-\frac{\pi}{2}}\int_0^{\sqrt{l_k}} e^{-\tau^2}d\tau = 2Me^{-\frac{\pi}{2}}\frac{\sqrt{\pi}}{2} = \sqrt{\pi}Me^{-\frac{\pi}{2}}$. Hence, for (48), we have

$$\begin{aligned}
&\lim_{k\to+\infty}\left\{l_k^{\frac{1}{2}}e^{l_k}\left[(2M-e^{\pi})e^{-\frac{\pi}{2}}\int_0^{\sqrt{l_k}}e^{-\tau^2}d\tau+\sqrt{2}e^{l_k}\left(\sin\left(l_k-\frac{\pi}{4}\right)\int_0^{\sqrt{l_k}}e^{-2\tau^2}\cos\tau^2d\tau\right.\right.\right.\\
&\left.\left.\left.-\cos\left(l_k-\frac{\pi}{4}\right)\int_0^{\sqrt{l_k}}e^{-2\tau^2}\sin\tau^2d\tau\right)\right]\right\}\\
=&\left[\sqrt{\pi}Me^{-\frac{\pi}{2}}+(+\infty)\left(-\sqrt{A^2+B^2}\right)\right]\\
=&-\infty.
\end{aligned}$$

Then, we obtain

$$\liminf_{l\to\infty} l^{1-\alpha}\int_0^l e^{\frac{\rho-1}{\rho}(l-s)}(l-s)^{\alpha-1}\left[\frac{e^{\frac{\rho-1}{\rho}(s-l_1)}M+{}_{l_1}I^{1,\rho}\left(\rho g(s)V(s)\right)}{V(s)}\right]ds=-\infty<0.$$

Similarly, selecting the sequence $\{l_r\}=\left\{\frac{3\pi}{2}+\frac{\pi}{4}+2r\pi-\arctan\left(-\frac{B}{A}\right)\right\}$, we can obtain

$$\limsup_{l\to\infty} l^{1-\alpha}\int_0^l e^{\frac{\rho-1}{\rho}(l-s)}(l-s)^{\alpha-1}\left[\frac{e^{\frac{\rho-1}{\rho}(s-l_1)}M+{}_{l_1}I^{1,\rho}\left(\rho g(s)V(s)\right)}{V(s)}\right]ds=+\infty>0.$$

Therefore, by Theorem 1 all solutions of the problem (47) are oscillatory.

Example 2. *Consider the following GPF Caputo initial value problem*

$$\begin{cases} {}_0^CD^{\frac{3}{2},1}y(l)-{}_0^CD^{\frac{1}{2},1}y(l)+e^{(l+1)^2}\ln(y^2+e)=e^{2l}\cos l, & l>0,\\ y(0)=b_0. \end{cases}\tag{49}$$

Setting $\alpha=\frac{1}{2}$, $\rho=1$, $a=0$, $p(l)=-1$, $q(l)=e^{(l+1)^2}$, $f(y)=\ln(y^2+e)$, $g(l)=e^{2l}\cos l$ and $V(l)=e^{l_1-l}$. The assumption (H) is satisfied if $y(l)>0$. Then, we get

$$\begin{aligned}
{}_{l_1}I^{1,\rho}\left(\rho g(s)V(s)\right) &= \frac{1}{\rho}\int_{l_1}^s e^{\frac{\rho-1}{\rho}(s-\tau)}\rho g(\tau)V(\tau)d\tau\\
&= \int_{l_1}^s e^{l_1+\tau}\cos\tau d\tau\\
&= \frac{e^{l_1+s}}{2}(\sin s+\cos s)-\frac{e^{2l_1}}{2}(\sin l_1+\cos l_1)\\
&= \frac{\sqrt{2}e^{l_1+s}}{2}\sin\left(s+\frac{\pi}{4}\right)-\frac{e^{2l_1}}{2}(\sin l_1+\cos l_1).
\end{aligned}$$

Set $l_1=\frac{\pi}{2}$ with $n=1$. Hence, we can compute that

$$\begin{aligned}
&l^{1-n}\int_0^l e^{\frac{\rho-1}{\rho}(l-s)}(l-s)^{\alpha-1}\left[\frac{e^{\frac{\rho-1}{\rho}(s-l_1)}M^*+{}_{l_1}I^{1,\rho}\left(\rho g(s)V(s)\right)}{V(s)}\right]ds\\
=&\int_0^l(l-s)^{-\frac{1}{2}}e^{s-\frac{\pi}{2}}\left[\left(M^*-\frac{e^{\pi}}{2}\right)+\frac{\sqrt{2}}{2}e^{\frac{\pi}{2}+s}\sin\left(s+\frac{\pi}{4}\right)\right]ds.
\end{aligned}$$

By setting $l - s = \tau^2$, we can get the above integral as

$$\begin{aligned}
&\int_0^l (l-s)^{-\frac{1}{2}} e^{s-\frac{\pi}{2}} \left[\left(M^* - \frac{e^\pi}{2} \right) + \frac{\sqrt{2}}{2} e^{\frac{\pi}{2}+s} \sin\left(s + \frac{\pi}{4} \right) \right] ds \\
= & \int_{\sqrt{l}}^0 \frac{1}{\tau} e^{l-\tau^2-\frac{\pi}{2}} \left[\left(\frac{2M^* - e^\pi}{2} \right) + \frac{\sqrt{2}}{2} e^{\frac{\pi}{2}+l-\tau^2} \sin\left(l - \tau^2 + \frac{\pi}{4} \right) \right] (-2\tau) d\tau \\
= & \left(2M^* - e^\pi\right) e^{l-\frac{\pi}{2}} \int_0^{\sqrt{l}} e^{-\tau^2} d\tau + \sqrt{2} e^{2l} \int_0^{\sqrt{l}} e^{-2\tau^2} \sin\left(l - \tau^2 + \frac{\pi}{4} \right) d\tau \\
= & \left(2M^* - e^\pi\right) e^{l-\frac{\pi}{2}} \int_0^{\sqrt{l}} e^{-\tau^2} d\tau + \sqrt{2} e^{2l} \sin\left(l + \frac{\pi}{4} \right) \int_0^{\sqrt{l}} e^{-2\tau^2} \cos \tau^2 d\tau \\
& - \sqrt{2} e^{2l} \cos\left(l + \frac{\pi}{4} \right) \int_0^{\sqrt{l}} e^{-2\tau^2} \sin \tau^2 d\tau.
\end{aligned}$$

Let $l \to +\infty$ as the result of $|e^{-2\tau^2} \cos \tau^2| \leq e^{-2\tau^2}$, $|e^{-2\tau^2} \sin \tau^2| \leq e^{-2\tau^2}$ and $\lim_{l\to+\infty} \int_0^{\sqrt{l}} e^{-2\tau^2} d\tau = \frac{\sqrt{2\pi}}{4}$. Thus, we know that $\lim_{l\to+\infty} \int_0^{\sqrt{l}} e^{-2\tau^2} \cos \tau^2 d\tau$ and $\lim_{l\to+\infty} \int_0^{\sqrt{l}} e^{-2\tau^2} \sin \tau^2 d\tau$ are convergent.

Hence, we can set $\lim_{l\to+\infty} \int_0^{\sqrt{l}} e^{-2\tau^2} \cos \tau^2 d\tau = A$ and $\lim_{l\to+\infty} \int_0^{\sqrt{l}} e^{-2\tau^2} \sin \tau^2 d\tau = B$. Select the sequence $\{l_k\} = \left\{ \frac{7\pi}{2} - \frac{\pi}{4} + 2k\pi - \arctan\left(-\frac{B}{A}\right) \right\}$, $\lim_{l\to+\infty} l_k = \infty$, then we compute the following term:

$$\begin{aligned}
&\lim_{k\to+\infty} \left\{ e^{l_k} \left[\left(2M^* - e^\pi\right) e^{-\frac{\pi}{2}} \int_0^{\sqrt{l_k}} e^{-\tau^2} d\tau + \sqrt{2} e^{l_k} \left(\sin\left(l_k + \frac{\pi}{4} \right) \int_0^{\sqrt{l_k}} e^{-2\tau^2} \cos \tau^2 d\tau \right.\right.\right. \\
&\left.\left.\left. - \cos\left(l_k + \frac{\pi}{4} \right) \int_0^{\sqrt{l_k}} e^{-2\tau^2} \sin \tau^2 d\tau \right) \right] \right\}. \qquad (50)
\end{aligned}$$

Firstly, we consider the following limit:

$$\begin{aligned}
&\lim_{k\to+\infty} \left\{ \sin\left(l_k + \frac{\pi}{4} \right) \int_0^{\sqrt{l_k}} e^{-2\tau^2} \cos \tau^2 d\tau - \cos\left(l_k + \frac{\pi}{4} \right) \int_0^{\sqrt{l_k}} e^{-2\tau^2} \sin \tau^2 d\tau \right\} \\
= & A \cdot \lim_{k\to+\infty} \sin\left(\frac{7\pi}{2} + 2k\pi - \arctan\left(-\frac{B}{A}\right) \right) - B \cdot \lim_{k\to+\infty} \cos\left(\frac{7\pi}{2} + 2k\pi - \arctan\left(-\frac{B}{A}\right) \right) \\
= & A \cdot \sin\left(\frac{7\pi}{2} - \arctan\left(-\frac{B}{A}\right) \right) - B \cdot \cos\left(\frac{7\pi}{2} - \arctan\left(-\frac{B}{A}\right) \right) \\
= & -\sqrt{A^2 + B^2}.
\end{aligned}$$

Secondly, we know that $\lim_{k\to+\infty} e^{l_k} = +\infty$ and $\lim_{k\to+\infty} 2M^* e^{l_k} e^{-\frac{\pi}{2}} \int_0^{\sqrt{l_k}} e^{-\tau^2} d\tau = 2M^* e^{-\frac{\pi}{2}} \frac{\sqrt{\pi}}{2} = \sqrt{\pi} M^* e^{-\frac{\pi}{2}}$. Hence, for (50), we have

$$\begin{aligned}
&\lim_{k\to+\infty} \left\{ e^{l_k} \left[\left(2M^* - e^\pi\right) e^{-\frac{\pi}{2}} \int_0^{\sqrt{l_k}} e^{-\tau^2} d\tau + \sqrt{2} e^{l_k} \left(\sin\left(l_k + \frac{\pi}{4} \right) \int_0^{\sqrt{l_k}} e^{-2\tau^2} \cos \tau^2 d\tau \right.\right.\right. \\
&\left.\left.\left. - \cos\left(l_k + \frac{\pi}{4} \right) \int_0^{\sqrt{l_k}} e^{-2\tau^2} \sin \tau^2 d\tau \right) \right] \right\} \\
= & \left[\sqrt{\pi} M^* e^{-\frac{\pi}{2}} + (+\infty) \left(-\sqrt{A^2 + B^2} \right) \right] \\
= & -\infty.
\end{aligned}$$

Then, we obtain

$$\liminf_{l\to\infty} l^{1-n}\int_0^l e^{\frac{\rho-1}{\rho}(l-s)}(l-s)^{\alpha-1}\left[\frac{e^{\frac{\rho-1}{\rho}(s-l_1)}M^* + {}_{l_1}I^{1,\rho}\left(\rho g(s)V(s)\right)}{V(s)}\right]ds = -\infty < 0.$$

Similarly, selecting the sequence $\{l_r\} = \left\{\frac{\pi}{2} - \frac{\pi}{4} + 2r\pi - \arctan\left(-\frac{B}{A}\right)\right\}$, we can obtain

$$\limsup_{l\to\infty} l^{1-n}\int_0^l e^{\frac{\rho-1}{\rho}(l-s)}(l-s)^{\alpha-1}\left[\frac{e^{\frac{\rho-1}{\rho}(s-l_1)}M^* + {}_{l_1}I^{1,\rho}\left(\rho g(s)V(s)\right)}{V(s)}\right]ds = +\infty > 0.$$

Therefore, by Theorem 2 all solutions of the problem (49) are oscillatory.

Example 3. *Consider the following GPF Riemann-Liouville initial value problem*

$$\begin{cases} {}_0D^{\frac{3}{2},1}y(l) + \sqrt{l}\left(\frac{4}{\sqrt{\pi}} + \frac{e^{3\sqrt{-y}}}{(-y)^{\frac{1}{4}}}\right) = e^{3l}, \quad l > 0, \\ \lim_{l\to 0^+} {}_0I^{\frac{1}{2},1}y(l) = 0. \end{cases} \tag{51}$$

Setting $\alpha = \frac{1}{2}$, $\rho = 1$, $a = p(l) = 0$, $q(l) = \sqrt{l}$, $f(y) = \frac{4}{\sqrt{\pi}} + \frac{e^{3\sqrt{-y}}}{(-y)^{\frac{1}{4}}}$, $g(l) = e^{3l}$ and $V(l) = 1$. The assumption (H) is satisfied if $y(l) > 0$. Then,

$${}_{l_1}I^{1,\rho}\left(\rho g(s)V(s)\right) = \frac{1}{\rho}\int_{l_1}^s e^{\frac{\rho-1}{\rho}(s-\tau)}\rho g(\tau)V(\tau)d\tau = \int_{l_1}^s e^{3\tau}d\tau = \frac{e^{3s}}{3} - \frac{e^{3l_1}}{3} = \frac{1}{3}\left(e^{3s} - e^{3l_1}\right).$$

By setting $l_1 = \frac{1}{3}$ and $l - s = \tau^2$, it follows that

$$\begin{aligned} & l^{1-\alpha}\int_0^l e^{\frac{\rho-1}{\rho}(l-s)}(l-s)^{\alpha-1}\left[\frac{e^{\frac{\rho-1}{\rho}(s-l_1)}M + {}_{l_1}I^{1,\rho}\left(\rho g(s)V(s)\right)}{V(s)}\right]ds \\ = \; & l^{\frac{1}{2}}\int_0^l (l-s)^{-\frac{1}{2}}\left[M + \frac{1}{3}\left(e^{3s} - e\right)\right]ds. \\ = \; & l^{\frac{1}{2}}\left[\left(M - \frac{e}{3}\right)\int_0^l (l-s)^{-\frac{1}{2}}ds + \frac{1}{3}\int_0^l (l-s)^{-\frac{1}{2}}e^{3s}ds\right] \\ = \; & 2l^{\frac{1}{2}}\left[\left(M - \frac{e}{3}\right)\sqrt{l} + \frac{1}{3}e^{3l}\int_0^{\sqrt{l}} e^{-3\tau}d\tau\right]. \end{aligned}$$

However, the condition (22) does not holds since

$$\begin{aligned} & \liminf_{l\to\infty} l^{1-\alpha}\int_0^l e^{\frac{\rho-1}{\rho}(l-s)}(l-s)^{\alpha-1}\left[\frac{e^{\frac{\rho-1}{\rho}(s-a)}M + {}_{l_1}I^{1,\rho}\left(\rho g(s)V(s)\right)}{V(s)}\right]ds \\ = \; & \liminf_{l\to\infty}\left\{2l^{\frac{1}{2}}\left[\left(M - \frac{e}{3}\right)\sqrt{l} + \frac{1}{3}e^{3l}\int_0^{\sqrt{l}} e^{-3\tau}d\tau\right]\right\} \\ = \; & \left[\left(M - \frac{e}{3}\right)(+\infty) + (+\infty)\frac{\sqrt{\pi}}{2}\right] = \infty. \end{aligned}$$

Using Proposition 1 (ii) with $\alpha = \frac{3}{2}$, $\beta = 3$ and $\rho = 1$, we get ${}_aD^{\alpha,\rho}y(l) = -\frac{4\sqrt{l}}{\sqrt{\pi}}$, it is easy to verify that $y(l) = -l^2$ is a nonoscillatory solution of (51). Figure 1 demonstrates the solution $y(l) = -l^2$.

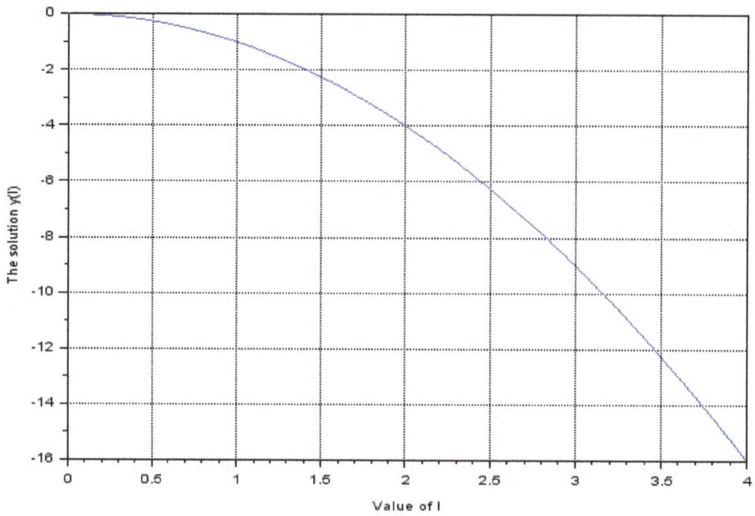

Figure 1. The nonoscillatory behavior of the solution $y(l) = -l^2$.

6. Conclusions

In this paper, the oscillatory behavior of solutions of generalized proportional fractional initial value problem is studied. Forced and damped oscillation results are obtained via GPF operators in the frame of Riemann-Liouville and Caputo settings. The main theorems of this paper improve and generalize some existing oscillation theorems reported in the literature. In particular, for the choice of $\rho = 1$, our contributions obtained using GPF operators cover the results discussed in Reference [24] which are obtained via conformable operators. At the end, we presented some numerical examples with particular values of parameters to illustrate the validity of the proposed results. Interestingly, we provided an example demonstrating that the failure of any condition forces the existence of a nonoscillatory solution. This justifies the advantage of our findings.

We believe that the results of this paper are of great importance for the audience of interested researchers. Several types of oscillation conditions could be generalized by considering respective equations within GPF derivatives.

Author Contributions: All authors contributed equally and significantly to this paper. All authors have read and approved the final version of the manuscript.

Acknowledgments: J. Alzabut would like to thank Prince Sultan University for funding this work through research group Nonlinear Analysis Methods in Applied Mathematics (NAMAM) group number RG-DES-2017-01-17. The second author was supported by DST-INSPIRE Scheme (No.DST/INSPIRE Fellowship/2018/IF180260) New Delhi, India. The third author was supported by DST-FIST Scheme (Grant No. SR/FST/MSI-115/2016), New Delhi, India. The fourth author was partially supported by Navamindradhiraj University Research Fund (NURS), Navamindradhiraj University, Thailand.

References

1. Miller, K.S.; Ross, B. *An Introduction to the Fractional Calculus and Fractional Differential Equations;* Wiley: New York, NY, USA, 1993.
2. Podlubny, I. *Fractional Differential Equations;* Academic Press: San Diego, CA, USA, 1999.
3. Diethelm, K. The Analysis of Fractional Differential Equations. In *Lecture Notes in Mathematics;* Springer: Berlin, Germany, 2010.
4. Lakshmikantham, V.; Leela, S.; Devi, J.V. *Theory of Fractional Dynamic Systems;* Cambridge Acadamic Publishers: Cambridge, UK, 2009.
5. Kilbas, A.A.; Srivastava, H.M.; Trujillo, J.J. *Theory and Application of Fractional Differential Equations;* North Holland Mathematics Studies; Elsever: Amsterdam, The Netherlands, 2006; Volume 204.

6. Samko, S.G.; Kilbas, A.A.; Marichev, O.L. *Fractional Integral and Derivatives: Theory and Applications*; Gordon and Breach: Yverdon, Switzerland, 1993.
7. Oldham, K.B.; Spanier, J. *Fractional Calculus: Theory and Applications, Differentiation and Integration to Arbitrary Order*; Academic Press: New York, NY, USA, 1974.
8. Hilfer, R. *Applications of Fractional Calculus in Physics*; Word Scientific: Singapore, 2000.
9. Martínez-García, M.; Zhang, Y.; Gordon, T. Memory pattern identification for feedback tracking control in human-machine systems. *Hum. Factors* **2019**. [CrossRef] [PubMed]
10. Martínez Garcia, M. Modelling Human-Driver Behaviour Using a Biofidelic Approach. Doctoral Dissertation, University of Lincoln, Lincoln, UK, 2018.
11. Khalil, R.; Al Horani, M.; Yousef, A.; Sababheh, M. A new definition of fractional derivative. *J. Comput. Appl. Math.* **2014**, *264*, 65–70. [CrossRef]
12. Abdeljawad, T. On conformable Calculus. *J. Comput. Appl. Math.* **2015**, *279*, 57–66. [CrossRef]
13. Anderson, D.R.; Ulness, D.J. Newly defined conformable derivatives. *Adv. Dyn. Syst. Appl.* **2015**, *10*, 109–137.
14. Caputo, M.; Fabrizio, M. A new definition of fractional derivative without singular kernal. *Prog. Frac. Differ. Appl.* **2015**, *1*, 73–85.
15. Jarad, F.; Abdeljawad, T.; Alzabut, J. Generalized fractional derivatives generated by a class of local proportional derivatives. *Eur. Phys. J. Spec. Top.* **2017**, *226*, 3457–3471. [CrossRef]
16. Alzabut, J.; Abdeljawad, T.; Jarad, F.; Sudsutad, W. A Gronwall inequality via the generalized proportional fractional derivative with applications. *J. Inequal. Appl.* **2019**, *2019*, 101. [CrossRef]
17. Alzabut, J.; Sudsutad, W.; Kayar, Z.; Baghani, H. A new Gronwall-Bellman inequality in a frame of generalized proportional fractional derivative. *Mathematics* **2019** , *7*, 747. [CrossRef]
18. Shammakh, W.; Alzumi, H.Z. Existence results for nonlinear fractional boundary value problem involving generalized proportional derivative. *Adv. Differ. Equ.* **2019**, *2019*, 94. [CrossRef]
19. Grace, S.; Agarwal, R.; Wong, P.; Zafer, A. On the oscillation of fractional differential equations. *Fract. Calc. Appl. Anal.* **2012**, *15*, 222–231. [CrossRef]
20. Chen, D.X.; Qu, P.X.; Lan, Y.H. Forced oscillation of certain fractional differential equations. *Adv. Differ. Equ.* **2013**, *2013*, 125. [CrossRef]
21. Feng, Q.H.; Meng, F.W. Oscillation of solutions to nonlinear forced fractional differential equations. *Electron. J. Differ. Equ.* **2013**, *2013*, 169.
22. Pavithra, S.; Muthulaksmi, V. Oscillatory behavior for a class of fractional differential equations. *Int. J. Pure Appl. Math.* **2017**, *115*, 93–107.
23. Abdalla, B. Oscillation of differential equations in the frame of nonlocal fractional derivatives generated by conformable derivatives. *Adv. Differ. Equ.* **2018**, *2018*, 107. [CrossRef]
24. Aphithana, A.; Ntouyas, S.K.; Tariboon, J. Forced oscillation of fractional differential equations via conformable erivatives with damping term. *Bound. Value Probl.* **2019**, *2019*, 47. [CrossRef]
25. Alzabut, J.; Manikandan, S.; Muthulakshmi, V.; Harikrishnan, S. Oscillation criteria for a class of nonlinear conformable fractional damped dynamic equations on time scales. *J. Nonlinear Funct. Anal.* **2020**, *2020*, 10.
26. Sudsutad, W.; Alzabut, J.; Tearnbucha, C.; Thaiprayoon, C. On the oscillation of differential equations in frame of generalized proportional fractional derivatives. *AIMS Math.* **2020**, *5*, 856–871. [CrossRef]
27. Anderson, D.R. Second-order self-adjoint differential equations using a proportional-derivative controller. *Commun. Appl. Nonlinear Anal.* **2017**, *24*, 17–48.

15

A Class of Quantum Briot–Bouquet Differential Equations with Complex Coefficients

Rabha W. Ibrahim [1,2], Rafida M. Elobaid [3,*] and Suzan J. Obaiys [4]

1 Informetrics Research Group, Ton Duc Thang University, Ho Chi Minh City 758307, Vietnam; rabhaibrahim@tdtu.edu.vn
2 Faculty of Mathematics & Statistics, Ton Duc Thang University, Ho Chi Minh City 758307, Vietnam
3 Department of General Sciences, Prince Sultan University, Riyadh 12345, Saudi Arabia
4 School of Mathematical and Computer Sciences, Heriot-Watt University Malaysia, Putrajaya 62200, Malaysia; s.obaiys@hw.ac.uk
* Correspondence: robaid@psu.edu.sa

Abstract: Quantum inequalities (QI) are local restraints on the magnitude and range of formulas. Quantum inequalities have been established to have a different range of applications. In this paper, we aim to introduce a study of QI in a complex domain. The idea basically, comes from employing the notion of subordination. We shall formulate a new q-differential operator (generalized of Dunkl operator of the first type) and employ it to define the classes of QI. Moreover, we employ the q-Dunkl operator to extend the class of Briot–Bouquet differential equations. We investigate the upper solution and exam the oscillation solution under some analytic functions.

Keywords: differential operator; unit disk; univalent function; analytic function; subordination; q-calculus; fractional calculus; fractional differential equation; q-differential equation

MSC: 30C45

---

1. Introduction

Quantum calculus exchanges the traditional derivative by a difference operator, which permits dealing with sets of non-differentiable curves and admits several formulas. The most common formula of quantum calculus is constructed by the q-operator (q-indicates for the quantum), which is created by the Jackson q-difference operator [1] as follows: let δ_q be the q-calculus which is formulated by

$$\delta_q(\Upsilon(\xi)) = \Upsilon(q\xi) - \Upsilon(\xi),$$

then the derivatives of functions are presented as fractions by the q-derivative

$$D_q(\Upsilon(\xi)) = \frac{\delta_q(\Upsilon(\xi))}{\delta_q(\xi)} = \frac{\Upsilon(q\xi) - \Upsilon(\xi)}{(q-1)\xi}, \quad \xi \neq 0.$$

For example, the q-derivative of the function ξ^n (for some positive integer n) is

$$D_q(\xi^n) = \frac{q^n - 1}{q-1}\xi^{n-1} = [n]_q\, \xi^{n-1}, \quad [n]_q = \frac{q^n - 1}{q-1}.$$

Recently, quantum inequalities (differential and integral) have extensive applications not only in mathematical physics but also in other sciences. In variation problems, Cruz et al. [2] presented a new variational calculus created by the general quantum difference operator of D_q. Rouze and Datta used the quantum functional inequalities to describe the transportation cost functional inequality [3]. Giacomo and Trevisan proved the conditional Entropy Power Inequality for Gaussian quantum systems [4]. Bharti et al. provided a novel algorithm to self-test local quantum systems employing non-contextuality quantum inequalities [5]. The class of quantum energy inequalities is studied by Fewster and Kontou [6]. In the control system, Ibrahim et al. established different classes of quantum differential inequalities [7]. In quantum information processing, Mao et al formulated a new quantum key distribution based on quantum inequalities [8].

In this investigation, we formulate a novel q-differential operator of complex coefficients and discuss its behavior in view of the theory of geometric functions. The suggested q-differential operator indicates a generalization of well-known differential operators in the open unit disk, such as the Dunkl operator and the Sàlàgean operator. It will be considered in some subclasses of starlike functions. Quantum inequalities involve the q-differential operator and some special functions are studied. Sharpness of QI is studied in the sequel. As an application, we employ the q-differential operator to define the q-Briot–Bouquet differential equations (q-BBE). Special cases are discussed and compared with recent works. Moreover, we illustrate a set of examples of q-BBE (for q = 1/2) and exam its oscillated solutions.

2. Related Works

The quantum calculus receives the attention of many investigators. This calculus, for the first time appeared in complex analysis by Ismail et al. [9]. They defined a class of complex analytic functions dealing with the inequality condition $|\Upsilon(q\xi)| < |\Upsilon(\xi)|$ on the open unit disk. Grinshpan [10] presented some interesting outcomes filled with geometric observations are of very significant in the univalent function theory. Newly, q-calculus becomes very attractive in the field of special functions. Srivastava and Bansal [11] presented a generalization of the well-known Mittag–Leffler functions and they studied the sufficient conditions under which it is close-to-convex in the open unit disk. Srivastava et al. [12] established a new subclass of normalized smooth and starlike functions in $\cup$. Mahmood et al. [13] introduced a family of q-starlike functions which are based on the Ruscheweyh-type q-derivative operator. Shi et al. [14] examined some recent problems concerning the concept of q-starlike functions. Ibrahim and Darus [15] employed the notion of quantum calculus and the Hadamard product to amend an extended Sàlàgean q-differential operator. Srivastava [16] developed many functions and classes of smooth functions based on the q-calculus. The q-Subordination inequality presented by Ul-Haq et al. [17]. Govindaraj and Sivasubramanian [18] as well as Ibrahim et al. [7] used the quantum calculus and the Hadamard product to deliver some subclasses of analytic functions involving the modified Sàlàgean q-differential operator and the generalized symmetric Sàlàgean q-differential operator respectively.

3. q-Differential Operator

Assume that $\wedge$ is the set of the smooth functions formulating by the followed power series

$$\Upsilon(\xi) = \xi + \sum_{n=2}^{\infty} \Upsilon_n \xi^n, \quad \xi \in \cup = \{\xi : |\xi| < 1\}.$$

For a function $\Upsilon \in \wedge$, the Sàlàgean operator expansion is formulated by the expansion

$$\varsigma^m \Upsilon(\xi) = \xi + \sum_{n=2}^{\infty} n^m \Upsilon_n \xi^n.$$

For $\Upsilon \in \wedge$, we get

$$D_q \Upsilon(\xi) = \sum_{n=1}^{\infty} \Upsilon_n [n]_q \xi^{n-1}, \quad \xi \in \cup, \Upsilon_1 = 1.$$

Now, let $\Upsilon \in \Lambda$, the Sàlàgean q-differential operator [18] is formulated by

$$\varsigma_q^0 \Upsilon(\xi) = \Upsilon(\xi),\ \varsigma_q^1 \Upsilon(\xi) = \xi D_q \Upsilon(\xi), \ldots,\ \varsigma_q^m \Upsilon(\xi) = \xi D_q \left(\varsigma_q^{m-1} \Upsilon(\xi)\right),$$

where m is a positive integer. A calculation associated by the formula of D_q, yields $\varsigma_q^m \Upsilon(\xi) = \Upsilon(\xi) * \Theta_q^m(\xi)$, where $*$ is the convolution product,

$$\Theta_q^m(\xi) = \xi + \sum_{n=2}^{\infty} [n]_q^m \xi^n$$

and

$$\varsigma_q^m \Upsilon(\xi) = \xi + \sum_{n=2}^{\infty} [n]_q^m \Upsilon_n \xi^n.$$

Next, we present the q-differential operator as follows:

$$\begin{aligned}
{}_q\Lambda_\lambda^0 \Upsilon(\xi) &= \Upsilon(\xi) \\
{}_q\Lambda_\lambda^1 \Upsilon(\xi) &= \xi D_q \Upsilon(\xi) + \Big((\lambda \Upsilon(\xi) - \xi) - \lambda(\Upsilon(-\xi) + \xi) \Big), \\
&\vdots \\
{}_q\Lambda_\lambda^m \Upsilon(\xi) &= {}_q\Lambda_\lambda({}_q\Lambda_\lambda^{m-1} \Upsilon(\xi)) \\
&= \xi + \sum_{n=2}^{\infty} \left([n]_q + ((-1)^{n+1} + 1)\lambda\right)^m \Upsilon_n \xi^n,
\end{aligned} \tag{1}$$

where $\lambda \in \mathbb{C}$. For $\lambda = 0, q \to 1^-$, the operator subjects to the Sàlàgean operator [19]. In addition, the operator ${}_q\Lambda_\lambda^m$ represents to the q-Dunkl operator of first rank [20], such that the value of λ is the Dunkl parameter. The term $([n]_q + \lambda(1 + (-1)^{n+1}))^m$ indicates a major law in oscillation study (see [21]). Furthermore, the term $e^{2i\pi}$ is denoting the quantum number q when $\hbar = 1$. That is there is a connection between the definition of the ${}_q\Lambda_\lambda^m$ and its coefficients.

Two functions Υ and $\curlywedge$ in Λ are subordinated ($\Upsilon \prec \curlywedge$), if there occurs a Schwarz function $\zeta \in \cup$ with $\zeta(0) = 0$ and $|\zeta(\xi)| < 1$, whenever $\Upsilon(\xi) = \curlywedge(\zeta(\xi))$ for all $\xi \in \cup$ (see [22]). Literally, the subordination inequality is indicated the equality at the origin and inclusion regarding $\cup$.

Definition 1. *Assume that $\lambda \in \mathbb{C}$, $m \in \mathbb{N}$ and $\Upsilon \in \Lambda$. Then it is in the set ${}_q\mathfrak{S}_m^*(\lambda, \varsigma)$ if and only if*

$$\frac{\xi({}_q\Lambda_\lambda^m \Upsilon(\xi))'}{{}_q\Lambda_\lambda^m \Upsilon(\xi)} \prec \varsigma(\xi), \quad \xi \in \cup,$$

where ς is univalent function of a positive real part in $\cup$ realizing $\varsigma(0) = 1$, $\Re(\varsigma'(\xi)) > 0$.

The set of functions ${}_q\mathfrak{S}_m^*(\alpha, \lambda, \varsigma)$ is an extension of types related to the Ma and Minda classes (see [23–27]). Next result indicates the upper and lower bound of a convex formula involving special functions, which will be useful in the next section.

Theorem 1. *Assume that $0 \le \beta \le 1$. Then for $\varsigma \in \cup$ as follows:*

1. $\varsigma(\xi) = (1 - \beta)\sqrt{1 + \xi} + \beta,$
2. $\varsigma(\xi) = (1 - \beta)e^{\xi} + \beta,$
3. $\varsigma(\xi) = (1 - \beta)(1 + \sin(\xi)) + \beta,$
4. $\varsigma(\xi) = (1 - \beta)e^{e^{\xi} - 1} + \beta,$

satisfying

$$\min_{|\xi|=r} \Re(\varsigma(\xi)) = \varsigma(-r) = \min_{|\xi|=r} |\varsigma(\xi)|, \quad r < 1$$

and

$$\max_{|\xi|=r} \Re(\varsigma(\xi)) = \varsigma(r) = \max_{|\xi|=r} |\varsigma(\xi)|, \quad r < 1.$$

Proof. For $\vartheta \in (0, 2\pi)$, $\Re(\varsigma(re^{i\vartheta})) = (1-\beta)e^{r\cos(\vartheta)}\cos(r\sin(\vartheta)) + \beta$, the first and second formula can be found in [25]. In the same manner, we show the third formal. When $\beta = 0$ this implies that $\varsigma(\xi) = 1 + \sin(\xi)$ (see [24]). Obviously,

$$\sin(re^{i\vartheta}) = \sin(r\cos(\vartheta))\cosh(r\sin(\vartheta)) + i\cos(r\cos(\vartheta))\sinh(r\sin(\vartheta))$$

thus, this yields

$$\Re(\varsigma(\xi)) = \sin(r\cos(\vartheta))\cosh(r\sin(\vartheta)) + 1.$$

Now, let $r \to 0$, this leads to

$$\min_{|\xi|=r} \Re(\varsigma(\xi)) = 1 - \sin(r) = \min_{|\xi|=r} |\varsigma(\xi)| = 1.$$

Consequently, we indicate that

$$|\sin(re^{i\vartheta})|^2 = \cos^2(r\cos\vartheta)\sinh 2(r\sin\vartheta) + \sin^2 2(r\cos\vartheta)\cosh 2(r\sin r) \le \sinh^2(r);$$

thus, this yields

$$\max_{|\xi|=r} \Re(\varsigma(\xi)) = 1 + \sin(r) = \max_{|\xi|=r} |\varsigma(\xi)| \le 1 + \sinh^2(r).$$

Extend the above outcome, for $\beta > 0$, we obtain

$$\min_{|\xi|=r} \Re(\varsigma(\xi)) = \beta + (1-\beta)(1 - \sin(r)) = \min_{|\xi|=r} |\varsigma(\xi)| = 1,$$

and

$$\max_{|\xi|=r} \Re(\varsigma(\xi)) = \beta + (1-\beta)(1 + \sin(r)) = \max_{|\xi|=r} |\varsigma(\xi)| \le \beta + (1-\beta)(1 + \sinh^2(r)).$$

Similarly, for the last assertion,where for $\beta = 0$, we have a result in [27]. □

The next result can be found in [22].

Lemma 1. *Suppose that $\tau > 0$ and $\varsigma \in \mathfrak{H}[1, n]$. Then for two constants $\wp > 0$ and $v > 0$ with $v = v(\wp, \tau, n)$ are achieving*

$$\varsigma(\xi) + \tau\xi\varsigma'(\xi) \prec \left[\frac{1+\xi}{1-\xi}\right]^{v} \Rightarrow \varsigma(\xi) \prec \left[\frac{1+\xi}{1-\xi}\right]^{\wp}.$$

Lemma 2. *Consider $\varphi(\xi)$ is a convex function in $\cup$ and $h(\xi) = \varphi(\xi) + nv(\xi\varphi'(\xi))$ for $v > 0$ and n is a positive integer. If $\varrho \in \mathfrak{H}[\varphi(0), n]$, and*

$$\varrho(\xi) + v\xi\varrho'(\xi) \prec h(\xi), \quad \xi \in \cup,$$

then $\varrho(\xi) \prec \varphi(\xi)$, and this outcome is sharp.

4. q-Subordination Relations

In this section, we deal with the set ${}_q\mathfrak{S}^*_m(\lambda, \varsigma)$ for some ς.

Theorem 2. *Assume that ${}_q\mathfrak{S}^*_m(\lambda, \varsigma)$ fulfills the next relation:*

$${}_q\mathfrak{S}^*_m(\lambda, \varsigma) \subset {}_q\mathfrak{S}^*_m(\lambda, \gamma) \subset {}_q\mathfrak{S}^*_m(\lambda),$$

where γ is non-negative real number (depending on β) and ς is one of the form in Theorem 1 and

$$ {}_q\mathfrak{S}^*_m(\lambda,\gamma) := \{\Upsilon \in \bigwedge : \Re\Big(\frac{\xi({}_q\Lambda^m_\lambda \Upsilon(\xi))'}{{}_q\Lambda^m_\lambda \Upsilon(\xi)}\Big) > \gamma, \gamma \geq 0\}; $$

$$ {}_q\mathfrak{S}^*_m(\lambda) := \{\Upsilon \in \bigwedge : \Re\Big(\frac{\xi({}_q\Lambda^m_\lambda \Upsilon(\xi))'}{{}_q\Lambda^m_\lambda \Upsilon(\xi)}\Big) > 0\}. $$

Proof. Suppose that $\Upsilon \in_q \mathfrak{S}^*_m(\lambda,\varsigma)$ and $\varsigma(\xi) = (1-\beta)\sqrt{1+\xi}+\beta$. This implies the inequality

$$ \frac{\xi({}_q\Lambda^m_\lambda \Upsilon(\xi))'}{{}_q\Lambda^m_\lambda \Upsilon(\xi)} \prec (1-\beta)\sqrt{1+\xi}+\beta, \quad \xi \in \cup. $$

According to Theorem 1, one can find

$$ \min_{|\xi|=1^-} \Re((1-\beta)(\xi+1)^{0.5}+\beta) < \Re\left(\frac{\xi({}_q\Lambda^m_\lambda \Upsilon(\xi))'}{{}_q\Lambda^m_\lambda \Upsilon(\xi)}\right) < \max_{|\xi|=1^+} \Re((1-\beta)(\xi+1)^{0.5}+\beta, $$

which indicates

$$ \beta < \Re\Big(\frac{\xi({}_q\Lambda^m_\lambda \Upsilon(\xi))'}{{}_q\Lambda^m_\lambda \Upsilon(\xi)}\Big) < (1-\beta)\sqrt{2}+\beta. $$

Hence, we have

$$ \Re\Big(\frac{\xi({}_q\Lambda^m_\lambda \Upsilon(\xi))'}{{}_q\Lambda^m_\lambda \Upsilon(\xi)}\Big) > \beta := \gamma \geq 0, $$

which leads to the requested result. Assume that $\varsigma(\xi) = (1-\beta)\,e^{\xi}+\beta$, then we conclude the next minimization and maximization inequality

$$ \min_{|\xi|=1} \Re((1-\beta)e^{\xi}+\beta) < \Re\Big(\frac{\xi({}_q\Lambda^m_\lambda \Upsilon(\xi))'}{{}_q\Lambda^m_\lambda \Upsilon(\xi)}\Big) < \max_{|\xi|=1} \Re((1-\beta)e^{\xi}+\beta), $$

which implies

$$ ((1-\beta)\frac{1}{e}+\beta) < \Re\Big(\frac{\xi({}_q\Lambda^m_\lambda \Upsilon(\xi))'}{{}_q\Lambda^m_\lambda \Upsilon(\xi)}\Big) < ((1-\beta)e+\beta), $$

that is

$$ \Re\Big(\frac{\xi({}_q\Lambda^m_\lambda \Upsilon(\xi))'}{{}_q\Lambda^m_\lambda \Upsilon(\xi)}\Big) > ((1-\beta)\frac{1}{e}+\beta) := \gamma \geq 0. $$

Now, suppose that $\varsigma(\xi) = (1-\beta)\,(1+\sin(\xi)+\beta)$, which implies that

$$ \min_{|\xi|=1} \Re((1-\beta)(1+\sin(\xi))+\beta) < \Re\Big(\frac{\xi({}_q\Lambda^m_\lambda \Upsilon(\xi))'}{{}_q\Lambda^m_\lambda \Upsilon(\xi)}\Big) < \max_{|\xi|=1} \Re((1-\beta)(1+\sin(\xi))+\beta). $$

A calculation yields

$$ (0.158(1-\beta)+\beta) < \Re\Big(\frac{\xi({}_q\Lambda^m_\lambda \Upsilon(\xi))'}{{}_q\Lambda^m_\lambda \Upsilon(\xi)}\Big) < (1.841(1-\beta)+\beta), $$

this yields

$$ \Re\Big(\frac{\xi({}_q\Lambda^m_\lambda \Upsilon(\xi))'}{{}_q\Lambda^m_\lambda \Upsilon(\xi)}\Big) > (0.158(1-\beta)+\beta) := \gamma \geq 0. $$

□

Remark 1. *In Theorem 2,*

- *When* $m = 0, \beta = 0, \varsigma(\xi) = 1 + \sin\xi \Longrightarrow {}_q\mathfrak{S}_0^*(\lambda, 1 + \sin\xi) \subset {}_q\mathfrak{S}_0^*(\lambda, \gamma) \subset {}_q\mathfrak{S}_0^*(\lambda)$ *for non-negative* γ *(see [24]);*
- *When* $m = 0 \Longrightarrow {}_q\mathfrak{S}_0^*(\lambda, \varsigma) \subset {}_q\mathfrak{S}_0^*(\lambda, \gamma) \subset {}_q\mathfrak{S}_0^*(\lambda)$ *for all* ς *in Theorem 2 and non-negative real number* γ *(see [25]);*
- *When* $m = 0, \beta = 0, \varsigma(\xi) = e^{\xi} \Longrightarrow {}_q\mathfrak{S}_0^*(\lambda, e^{\xi}) \subset {}_q\mathfrak{S}_0^*(\lambda, \gamma) \subset {}_q\mathfrak{S}_0^*(\lambda)$ *for all non-negative* γ *(see [28]);*
- *When* $m = 0, \beta = 0, \varsigma(\xi) = (\xi + 1)^{0.5} \Longrightarrow {}_q\mathfrak{S}_0^*(\lambda, (\xi + 1)^{0.5}) \subset {}_q\mathfrak{S}_0^*(\lambda, \gamma) \subset {}_q\mathfrak{S}_0^*(\lambda)$ *for all non-negative* γ *(see [28]).*

Next outcome shows the inclusion relation between the class ${}_q\mathfrak{S}_m^*(\lambda, \sigma)$ and other geometric class.

Theorem 3. *The set* ${}_q\mathfrak{S}_m^*(\lambda, \varsigma)$ *satisfies the inclusion:*

$${}_q\mathfrak{S}_m^*(\lambda, \varsigma) \subset {}_q\mathfrak{M}_m(\lambda, \alpha) := \{\Upsilon \in \bigwedge : \Re\Big(\frac{\xi({}_q\Lambda_\lambda^m \Upsilon(\xi))'}{{}_q\Lambda_\lambda^m \Upsilon(\xi)}\Big) < \alpha,\ \alpha > 1\},$$

where ς *is given in Theorem 1 .*

The class ${}_q\mathfrak{M}_m(\lambda, \alpha)$ is an extension of the Uralegaddi set (see [29])

$$\mathfrak{M}(\alpha) := \{\Upsilon \in \bigwedge : \Re\Big(\frac{\xi(\Upsilon(\xi))'}{\Upsilon(\xi)}\Big) < \alpha,\ \alpha > 1\}.$$

Proof. Suppose that $\Upsilon \in {}_q\mathfrak{S}_m^*(\lambda, \varsigma)$, where ς is termed in Theorem 1. Then we obtain

$$\Re\Big(\frac{\xi({}_q\Lambda_\lambda^m \Upsilon(\xi))'}{{}_q\Lambda_\lambda^m \Upsilon(\xi)}\Big) < (1-\beta)\sqrt{2} + \beta := \alpha,$$

$$\Re\Big(\frac{\xi({}_q\Lambda_\lambda^m \Upsilon(\xi))'}{{}_q\Lambda_\lambda^m \Upsilon(\xi)}\Big) < (1-\beta)e + \beta := \alpha$$

and

$$\Re\Big(\frac{\xi({}_q\Lambda_\lambda^m \Upsilon(\xi))'}{{}_q\Lambda_\lambda^m \Upsilon(\xi)}\Big) < 1.841(1-\beta) + \beta := \alpha,$$

Thus, $\Upsilon \in {}_q\mathfrak{M}_m(\Upsilon(\xi), \alpha)$. □

Remark 2. *We have the following special cases from Theorem 3,*

- $m = 0, \beta = 0, \varsigma(\xi) = 1 + \sin\xi \Longrightarrow_q \mathfrak{S}_0^*(\lambda, 1 + \sin\xi) \subset {}_q\mathfrak{M}_0(\lambda, \alpha)$ *for* $\alpha > 1$ *(see [24]);*
- $m = 0, \varsigma(\xi) = \beta + (1-\beta)e^{\xi} \Longrightarrow_q \mathfrak{S}_0^*(\lambda, (1-\beta)e^{\xi}) + \beta \subset {}_q\mathfrak{M}_0(\lambda, \alpha)$ *for* $\alpha > 1$ *(see [25], Theorem 2.5);*
- $m = 0, \varsigma(\xi) = (1-\beta)((\xi+1)^{0.5}) + \beta \Longrightarrow_q \mathfrak{S}_0^*(\lambda, \beta + (1-\beta)(\xi+1)^{0.5}) + \beta) \subset {}_q\mathfrak{M}_0(\lambda, \alpha)$ *for* $\alpha > 1$ *(see [25], Theorem 2.6).*
- $m = 0, \beta = 0, \varsigma(\xi) = ((\xi+1)^{0.5}) \Longrightarrow_q \mathfrak{S}_0^*(\lambda, (\xi+1)^{0.5}) \subset {}_q\mathfrak{M}_0(\lambda, \alpha)$ *where* $\alpha > 1$ *(see [25], Corollary 2.7).*

The following theorem confirms the belonging of a normalized function in the class ${}_q\mathfrak{S}_m^*(\lambda, \varsigma)$, where ς indicates the Janowski formula of order $\wp > 0$.

Theorem 4. *If* $\Upsilon \in \bigwedge$ *satisfies the subordination*

$$\Big(\frac{\xi({}_q\Lambda_\lambda^m \Upsilon(\xi))'}{{}_q\Lambda_\lambda^m \Upsilon(\xi)}\Big)\Big(2 + \frac{\xi({}_q\Lambda_\lambda^m \Upsilon(\xi))''}{({}_q\Lambda_\lambda^m \Upsilon(\xi))'} - \frac{\xi({}_q\Lambda_\lambda^m \Upsilon(\xi))'}{{}_q\Lambda_\lambda^m \Upsilon(\xi)}\Big) \prec \Big(\frac{\xi+1}{1-\xi}\Big)^{\tau}$$

then $\Upsilon \in_q \mathfrak{S}^*_m(\lambda, \varsigma)$, *where* $\varsigma(\xi) = \left(\frac{\xi+1}{1-\xi}\right)^{\wp}$ *for* $\wp > 0, \tau > 0$.

Proof. In virtue of Lemma 1, a computation yields

$$\begin{aligned}
&\Big(\frac{\xi({}_q\Lambda^m_\lambda \Upsilon(\xi))'}{{}_q\Lambda^m_\lambda \Upsilon(\xi)}\Big) + \xi\Big(\frac{\xi({}_q\Lambda^m_\lambda \Upsilon(\xi))'}{{}_q\Lambda^m_\lambda \Upsilon(\xi)}\Big)' \\
&= \Big(\frac{\xi({}_q\Lambda^m_\lambda \Upsilon(\xi))'}{{}_q\Lambda^m_\lambda \Upsilon(\xi)}\Big)\Big(\frac{\xi({}_q\Lambda^m_\lambda \Upsilon(\xi))''}{({}_q\Lambda^m_\lambda \Upsilon(\xi))'} - \frac{\xi({}_q\Lambda^m_\lambda \Upsilon(\xi))'}{{}_q\Lambda^m_\lambda \Upsilon(\xi)} + 2\Big) \\
&\prec \left(\frac{\xi+1}{1-\xi}\right)^{\tau}.
\end{aligned}$$

Now, according to Lemma 1, we attain

$$\Big(\frac{\xi({}_q\Lambda^m_\lambda \Upsilon(\xi))'}{{}_q\Lambda^m_\lambda \Upsilon(\xi)}\Big) \prec \left(\frac{\xi+1}{1-\xi}\right)^{\wp} := \varsigma(\xi),$$

which indicates that $\Upsilon \in_q \mathfrak{S}^*_m(\lambda, \varsigma)$. □

The next result shows the iteration inequality including the q-differential operator ${}_q\Lambda^m_\lambda$ and ${}_q\Lambda^{m+1}_\lambda$.

Theorem 5. *Suppose that φ is convex with $\varphi(0) = 0$ and g is defined as follows:*

$$g(\xi) = \varphi(\xi) + \left(\frac{1}{1-\ell}\right)(\xi\varphi'(\xi)), \quad \ell \in (0,1), \xi \in \mathbb{U}.$$

If $\Upsilon \in \Lambda$ fulfills the inequality

$$\Big(\frac{\xi}{{}_q\Lambda^{m+1}_\lambda \Upsilon(\xi)}\Big)^{\ell} \frac{{}_q\Lambda^m_\lambda \Upsilon(\xi)}{1-\ell}\Big(\frac{({}_q\Lambda^{m+1}_\lambda \Upsilon(\xi))'}{{}_q\Lambda^{m+1}_\lambda \Upsilon(\xi)} - \ell\frac{({}_q\Lambda^m_\lambda \Upsilon(\xi))'}{{}_q\Lambda^m_\lambda \Upsilon(\xi)}\Big) \prec g(\xi)$$

then

$$\Big(\frac{{}_q\Lambda^{m+1}_\lambda \Upsilon(\xi)}{\xi}\Big)\Big(\frac{\xi}{{}_q\Lambda^{m+1}_\lambda \Upsilon(\xi)}\Big)^{\ell} \prec \varphi(\xi).$$

Proof. For all $\xi \in \mathbb{U}$, we define

$$\varrho(\xi) = \Big(\frac{{}_q\Lambda^{m+1}_\lambda \Upsilon(\xi)}{\xi}\Big)\Big(\frac{\xi}{{}_q\Lambda^{m+1}_\lambda \Upsilon(\xi)}\Big)^{\ell}.$$

Please note that the term

$$\Big(\frac{\xi}{{}_q\Lambda^{m+1}_\lambda \Upsilon(\xi)}\Big)^{\ell} = \left(\frac{\xi}{\xi + \sum_{n=2}^{\infty}\left([n]_q + (1+(-1)^{n+1})\lambda\right)^{m+1}}\right)^{\ell} = 1 + \ldots;$$

therefore, $\Big(\frac{\xi}{{}_q\Lambda^{m+1}_\lambda \Upsilon(\xi)}\Big)^{\ell}\Big|_{\xi=0} = 1$. A differentiation implies that

$$\begin{aligned}
&\Big(\frac{\xi}{{}_q\Lambda^{m+1}_\lambda \Upsilon(\xi)}\Big)^{\ell}\left(\frac{{}_q\Lambda^m_\lambda \Upsilon(\xi)}{1-\ell}\right)\Big(\frac{({}_q\Lambda^{m+1}_\lambda \Upsilon(\xi))'}{{}_q\Lambda^{m+1}_\lambda \Upsilon(\xi)} - \ell\frac{({}_q\Lambda^m_\lambda \Upsilon(\xi))'}{{}_q\Lambda^m_\lambda \Upsilon(\xi)}\Big) \\
&= \left(\frac{1}{1-\ell}\right)(\xi\varrho'(\xi)) + \varrho(\xi)
\end{aligned}$$

Consequently, we indicate that

$$\left(\frac{1}{1-\ell}\right)(\xi\varrho'(\xi))+\varrho(\xi)\prec g(\xi)=\left(\frac{1}{1-\ell}\right)(\xi\,\varphi'(\xi))+\varphi(\xi).$$

In virtue of Lemma 2, we conclude that $\varrho(\xi)\prec g(\xi)$, which leads to

$$\left(\frac{{}_q\Lambda_\lambda^{m+1}\Upsilon(\xi)}{\xi}\right)\left(\frac{\xi}{{}_q\Lambda_\lambda^{m+1}\Upsilon(\xi)}\right)^{\ell}\prec\varphi(\xi).$$

□

5. Q-Differential Equations

This section deals with a class of differential equations type complex Briot–Bouquet (see [30,31] for recent works) and its analytic solutions. The main formula of BBE is $\frac{\xi(\Upsilon(\xi))'}{\Upsilon(\xi)}=\mathrm{Y}(\xi)$. The operator (1) can be used to extend q-BBE as follows:

$$\frac{\xi({}_q\Lambda_\lambda^m\Upsilon(\xi))'}{{}_q\Lambda_\lambda^m\Upsilon(\xi)}=\mathrm{Y}(\xi),\quad\xi\in\cup,\Upsilon\in\bigwedge,\tag{2}$$

where $\mathrm{Y}(\xi)\in\mathcal{C}$ (the set of univalent and convex in $\cup$). The aim is to discuss the maximum outcome of (2) by using $q-$inequalities.

Theorem 6. *Suppose that $\Upsilon\in\bigwedge$ and $\mathrm{Y}(\xi)\in\mathcal{C}$ satisfy*

$$\frac{\xi\,({}_q\Lambda_\lambda^m\Upsilon(\xi))'}{{}_q\Lambda_\lambda^m\Upsilon(\xi)}\prec\mathrm{Y}(\xi).\tag{3}$$

Then the maximum solution of (3) *is*

$${}_q\Lambda_\lambda^m\Upsilon(\xi)\prec\left(\exp\left(\int_0^\xi\frac{\mathrm{Y}(\Phi(\iota))-1}{\iota}d\iota\right)\right)\xi,$$

where $\Phi(\xi)$ is smooth in $\cup$, such that $\Phi(0)=0$, $|\Phi(\xi)|<1$ and it is the upper limit in the above integral. Also, for $|\xi|=\iota$, ${}_q\Lambda_\lambda^m\Upsilon(\xi)$ achieves the inequality

$$\exp\left(\int_0^1\frac{\mathrm{Y}(\Phi(-\iota))-1}{\iota}d\iota\right)\le\left|\frac{{}_q\Lambda_\lambda^m\Upsilon(\xi)}{\xi}\right|\le\exp\left(\int_0^1\frac{\mathrm{Y}(\Phi(\iota))-1}{\iota}d\iota\right).$$

Proof. By the definition of the subordination, inequality (3) achieves that there exists a Schwarz function Φ satisfying $|\Phi(\xi)|<1$, $\Phi(0)=0$ and

$$\frac{\xi({}_q\Lambda_\lambda^m\Upsilon(\xi))'}{{}_q\Lambda_\lambda^m\Upsilon(\xi)}=\mathrm{Y}(\Phi(\xi)),\quad\xi\in\cup.$$

This implies

$$\frac{({}_q\Lambda_\lambda^m\Upsilon(\xi))'}{{}_q\Lambda_\lambda^m\Upsilon(\xi)}-\frac{1}{\xi}=\frac{\mathrm{Y}(\Phi(\xi))-1}{\xi}.$$

Integrating both sides yields

$$\log_q\Lambda_\lambda^m\Upsilon(\xi)-\log\xi=\int_0^\xi\frac{\mathrm{Y}(\Phi(\iota))-1}{\iota}d\iota.$$

A calculation indicates

$$\log\left(\frac{{}_q\Lambda_\lambda^m \Upsilon(\xi)}{\xi}\right) = \int_0^{\xi} \frac{\Upsilon(\Phi(\iota)) - 1}{\iota} d\iota. \tag{4}$$

Then, we have

$${}_q\Lambda_\lambda^m \Upsilon(\xi) \prec \xi \exp\Big(\int_0^{\xi} \frac{\Upsilon(\Phi(\iota)) - 1}{\iota} d\iota\Big)$$

for some Schwarz function. In addition, by the behavior of the function Υ on the disk $0 < |\xi| < \iota < 1$ we have

$$\Upsilon(-\iota|\xi|) \le \Re(\Upsilon(\Phi(\iota\xi))) \le \Upsilon(\iota|\xi|), \quad \iota \in (0,1),$$

and

$$\Upsilon(-\iota) \le \Upsilon(-\iota\,|\xi|), \quad \Upsilon(\iota\,|\xi|) \le \Upsilon(\iota).$$

Thus, we conclude that

$$\int_0^1 \frac{\Upsilon(\Phi(-\iota|\xi|)) - 1}{\iota} d\iota \le \Re\Big(\int_0^1 \frac{\Upsilon(\Phi(\iota)) - 1}{\iota} d\iota\Big) \le \int_0^1 \frac{\Upsilon(\Phi(\iota|\xi|)) - 1}{\iota} d\iota,$$

which implies

$$\int_0^1 \frac{\Upsilon(\Phi(-\iota|\xi|)) - 1}{\iota} d\iota \le \log\left|\frac{{}_q\Lambda_\lambda^m \Upsilon(\xi)}{\xi}\right| \le \int_0^1 \frac{\Upsilon(\Phi(\iota|\xi|)) - 1}{\iota} d\iota,$$

and

$$\exp\Big(\int_0^1 \frac{\Upsilon(\Phi(-\iota|\xi|)) - 1}{\iota} d\iota\Big) \le \left|\frac{{}_q\Lambda_\lambda^m \Upsilon(\xi)}{\xi}\right| \le \exp\Big(\int_0^1 \frac{\Upsilon(\Phi(\iota|\xi|)) - 1}{\iota} d\iota\Big).$$

We conclude that

$$\exp\Big(\int_0^1 \frac{\Upsilon(\Phi(-\iota)) - 1}{\iota} d\iota\Big) \le \left|\frac{{}_q\Lambda_\lambda^m \Upsilon(\xi)}{\xi}\right| \le \exp\Big(\int_0^1 \frac{\Upsilon(\Phi(\iota)) - 1}{\iota} d\iota\Big).$$

□

Next result presents the condition on the coefficients of the normalized function Υ to satisfy the upper bound in Theorem 6.

Theorem 7. *Suppose that $\Upsilon \in \Lambda$ has non-negative coefficients. If $\Upsilon \in C$ in Equation (2) and $\Re(\lambda) > 0$ then there is a solution satisfying the maximum bound inequality*

$${}_q\Lambda_\lambda^m \Upsilon(\xi) \prec \xi \exp\Big(\int_0^{\xi} \frac{\Upsilon(\Phi(\iota)) - 1}{\iota} d\iota\Big), \tag{5}$$

where $\Phi(\xi)$ is smooth with $|\Phi(\xi)| < 1$ and $\Phi(0) = 0$.

Proof. By the condition of the theorem, we get the following assertions:

$$\begin{aligned}
&\Re\left(\frac{\xi({}_q\Lambda_\lambda^m \Upsilon(\xi))'}{{}_q\Lambda_\lambda^m \Upsilon(\xi)}\right) > 0\\
\Leftrightarrow\ &\Re\left(\frac{\xi + \sum_{n=2}^{\infty} n[[n]_q + \lambda(1+(-1)^{n+1})]^m\ \Upsilon_n\,\xi^n}{\xi + \sum_{n=2}^{\infty}[[n]_q + \lambda(1+(-1)^{n+1})]^m\ \Upsilon_n\,\xi^n}\right) > 0\\
\Leftrightarrow\ &\Re\left(\frac{1 + \sum_{n=2}^{\infty} n[[n]_q + \lambda(1+(-1)^{n+1})]^m\ \Upsilon_n\,\xi^{n-1}}{1 + \sum_{n=2}^{\infty}[[n]_q + \lambda(1+(-1)^{n+1})]^m\ \Upsilon_n\,\xi^{n-1}}\right) > 0\\
\Leftrightarrow\ &\left(\frac{1 + \sum_{n=2}^{\infty} n[[n]_q + \lambda(1+(-1)^{n+1})]^m\ \Upsilon_n}{1 + \sum_{n=2}^{\infty}[[n]_q + \lambda(1+(-1)^{n+1})]^m\ \Upsilon_n}\right) > 0\\
\Leftrightarrow\ &\left(1 + \sum_{n=2}^{\infty} n\ [[n]_q +\ (1+\ (-1)^{n+1})\ \lambda]^m\ \Upsilon_n\right) > 0.
\end{aligned}$$

Moreover, we have $({}_q\Lambda_\lambda^m \Upsilon)(0) = 0$, which leads to

$$\frac{\xi({}_q\Lambda_\lambda^m \Upsilon(\xi))'}{{}_q\Lambda_\lambda^m \Upsilon(\xi)} \in \mathcal{P}.$$

Hence the proof. □

We illustrate an example to find the upper and oscillation solution of q-BBE when $q = 1/2$, see Tables 1 and 2.

Table 1. The upper bound solution of q-BBE for different $\Upsilon(\xi)$.

$\Upsilon(\xi)$	Upper Solution	Graph	Polynomial
$\cos(\xi)$	$(\xi * \exp(\sin(\xi) - \xi))$	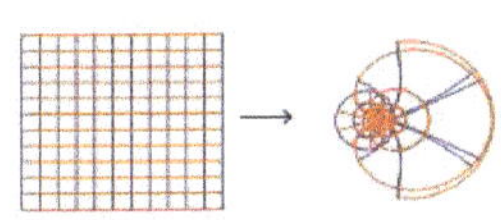	$\xi - \xi^4/6 + O(\xi^6)$
$1 - \sin(\xi)$	$(\xi * \exp(\cos(\xi) + \xi - 1))$	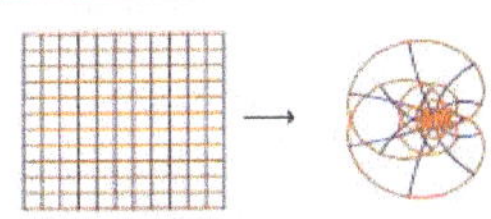	$\xi + \xi^2 - \frac{\xi^4}{3} - \frac{\xi^5}{24} + O(\xi^6)$
$1/(1-\xi)$	$(\xi * \exp(-\xi - \log(1-\xi)))$	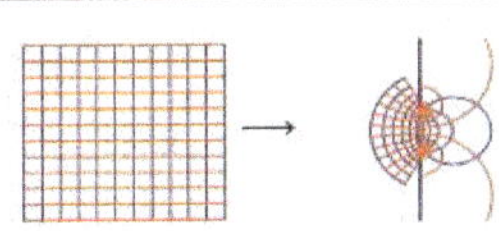	$\xi + \frac{\xi^3}{2} + \frac{\xi^4}{3} + \frac{3\xi^5}{8} + O(\xi^6)$
$1/(1-\xi)^2$	$(\xi * \exp(\frac{\xi^2}{1-\xi}))$	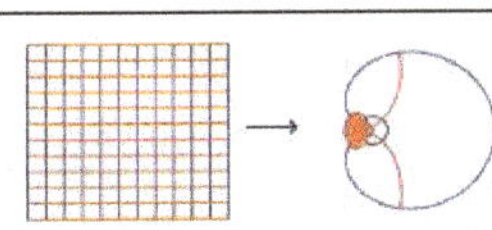	$\xi + \xi^3 + \xi^4 + \frac{3\xi^5}{2} + O(\xi^6)$
$1 - \xi$	$(\xi * \exp(-\xi^2/2))$	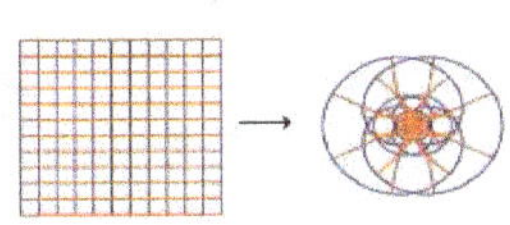	$\xi - \frac{\xi^3}{2} + \frac{\xi^5}{8} - \frac{\xi^7}{48} + O(\xi^9)$

Table 2. The oscillation solution for different $Y(\xi)$.

$\frac{1}{2}$-BBE (2)	Oscillation Solution	Graph
$\cos(\xi)$	$c_1 e^{Ci(\xi)}$	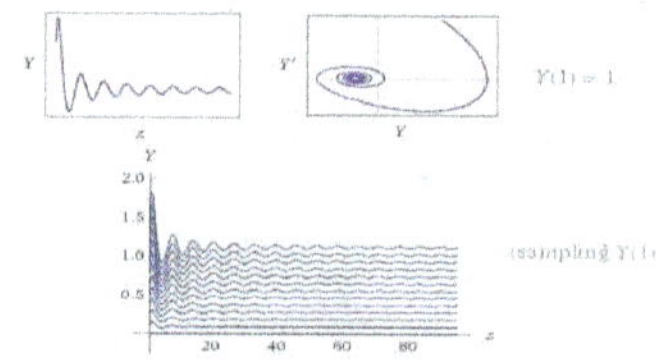
$1-\sin(\xi)$	$c_1 \xi e^{(-Si(\xi))}$	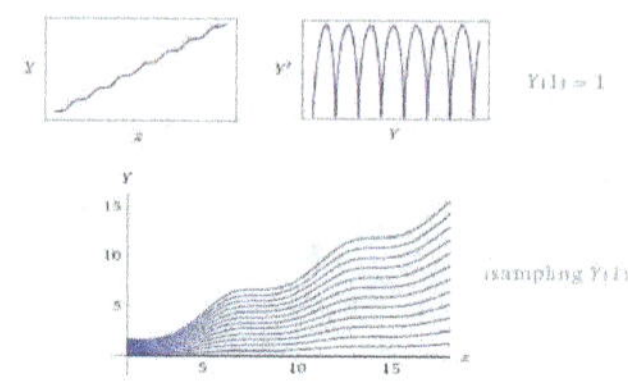
$1/(1-\xi)$	$c_1 e^{(1/(1-\xi))}\xi$	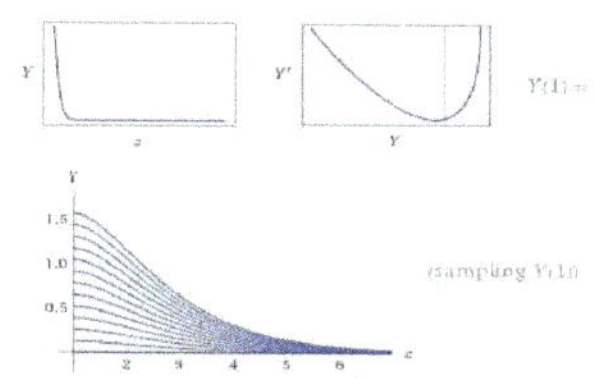
$1-\xi$	$c_1 e^{(-\xi)}\xi$	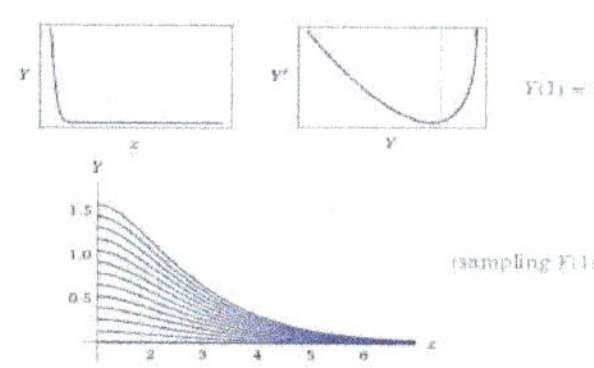

Where Ci is the cos integral function and Si is the sin integral function. The first example of $\frac{1}{2}$-BBE of (2) is $Y(\xi) = \cos(\xi)$ which has an oscillation solution with one branch point at the origin and has a local maximum at $\xi = \frac{\pi}{2} + 2n\pi$ and local minimum at $\xi = \frac{\pi}{2} - 2n\pi$. While, for $Y(\xi) = 1 - \sin(\xi)$, the oscillation solution has no branch point in the disk. Moreover, for $Y(\xi) = 1/(1-\xi)$ the oscillation solution has no branch point. Finally, when $Y(\xi) = 1 - \xi$, the oscillation solution has a global maximum equal to $1/e$ at $\xi = 1$.

6. Conclusions

From above, we conclude that in view of the quantum calculus, some generalized differential operators in the open unit disk can have connections (coefficients) convergence of quantum numbers. These numbers might change the behavior of the operator and its classes of analytic functions. We investigated the oscillation solution and asymptotic solutions of different differential equations of the Briot–Bouquet type. For future work, one can employ the q-operator (1) in different classes of analytic functions such as the meromorphic and multivalent functions (see [32–34]).

Author Contributions: Investigation, R.M.E.; Methodology, R.W.I.; Writing original draft, R.W.I.; Writing review and editing, R.M.E. and S.J.O. All authors have read and agreed to the published version of the manuscript.

Acknowledgments: The authors would like to thank both anonymous reviewers and editor for their helpful advice.

References

1. Kac, V. *Ch. Pokman*; Quantum Calculus; Springer: New York, NY, USA, 2002.
2. Da Cruz, A.M.C.B.; Natalia, M. General quantum variational calculus. *Stat. Optim. Inf. Comput.* **2018**, *6*, 22–41. [CrossRef]
3. Cambyse, R.; Datta, N. Concentration of quantum states from quantum functional and transportation cost inequalities. *J. Math. Phys.* **2019**, *60*, 012202.
4. Giacomo, D.P.; Trevisan, D. The conditional entropy power inequality for bosonic quantum systems. *Commun. Math. Phys.* **2018**, *360*, 639–662.
5. Bharti, K.; Ray, M.; Varvitsiotis, A.; Warsi, N.A.; Cabello, A.; Kwek, L.C. Robust self-testing of quantum systems via noncontextuality inequalities. *Phys. Rev. Lett.* **2019**, *122*, 250403. [CrossRef]
6. Fewster, C.J.; Eleni-Alexandra, K. Quantum strong energy inequalities. *Phys. Rev. D* **2019**, *99*, 045001. [CrossRef]
7. Ibrahim, R.W.; Hadid, S.B.; Momani, S. Generalized Briot-Bouquet differential equation by a quantum difference operator in a complex domain. *Int. J. Dyn. Control* **2020**, 1–10. [CrossRef]
8. Mao, Q.P.; Wang, L.; Zhao, S.M. Decoy-state round-robin differential-phase-shift quantum key distribution with source errors. *Quantum Inf. Process.* **2020**, *19*, 56. [CrossRef]
9. Ismail, M.E.H.; Merkes, E.; Styer, D. A generalization of starlike functions. *Complex Var. Theory Appl.* **1990**, *14*, 77–84. [CrossRef]
10. Grinshpan, A.Z. Logarithmic geometry, exponentiation, and coefficient bounds in the theory of univalent functions and nonoverlapping domains. In *Handbook of Complex Analysis: Geometric Function Theory*; Kühnau, R., Ed.; Elsevier: Amsterdam, The Netherlands, 2002; pp. 273–332.
11. Srivastava, H.M.; Bansal, D.E. Close-to-convexity of a certain family of q-Mittag-Leffler functions. *J. Nonlinear Var. Anal.* **2017**, *1*, 61–69.
12. Srivastava, H.M.; Ahmad, Q.Z.; Khan, N.; Khan, N.; Khan, B. Hankel and Toeplitz determinants for a subclass of q-starlike functions associated with a general conic domain. *Mathematics* **2019**, *7*, 181. [CrossRef]
13. Mahmood, S.; Srivastava, H.M.; Khan, N.; Ahmad, Q.Z.; Khan, B.; Ali, I. Upper bound of the third Hankel determinant for a subclass of q-starlike functions. *Symmetry* **2019**, *11*, 347. [CrossRef]
14. Shi, L.; Khan, Q.; Srivastava, G.; Liu, J.L.; Arif, M. A study of multivalent q-starlike functions connected with circular domain. *Mathematics* **2019**, *7*, 670. [CrossRef]
15. Ibrahim, R.W.; Darus, M. On a class of analytic functions associated to a complex domain concerning q-differential-difference operator. *Adv. Differ. Equ.* **2019**, *2019*, 515. [CrossRef]
16. Srivastava, H.M. Operators of basic (or q-) calculus and fractional q-calculus and their applications in geometric function theory of complex analysis. *Iran. J. Sci. Technol. Trans. A Sci.* **2020**, *44*, 327–344. [CrossRef]
17. Ul-Haq, M.; Raza, M.; Arif, M.; Khan, Q.; Tang, H. q-Analogue of Differential Subordinations. *Mathematics* **2019**, *7*, 724. [CrossRef]
18. Govindaraj, M.; Sivasubramanian, S. On a class of analytic functions related to conic domains involving q-calculus. *Anal. Math.* **2017**, *43*, 475–487. [CrossRef]
19. Sàlàgean, G.S. Subclasses of univalent functions. In *Complex Analysis-Fifth Romanian-Finnish Seminar*; Springer: Berlin/Heidelberg, Germany, 1983; pp. 362–372.
20. Dunkl, C.F. Differential-difference operators associated to reflection groups. *Trans. Am. Math. Soc.* **1989**, *311*, 167–183. [CrossRef]
21. Genest, V.X.; Lapointe, A.; Vinet, L. The Dunkl-Coulomb problem in the plane. *Phys. Lett. A* **2015**, *379*, 923–927. [CrossRef]
22. Miller, S.S.; Mocanu, P.T. *Differential Subordinations: Theory and Applications*; CRC Press: Boca Raton, FL, USA, 2000.
23. Ma, W.C.; Minda, D. A unified treatment of some special classes of univalent functions. In Proceedings of the Conference on Complex Analysis, Tianjin, China, 19–23 June 1992.
24. Cho, N.E.; Kumar, V.; Kumar, S.S.; Ravichandran, V. Radius problems for starlike functions associated with the sine function. *Bull. Iran. Math. Soc.* **2019**, *45*, 213–232. [CrossRef]
25. Khatter, K.; Ravichandran, V.; Kumar, S.S. Starlike functions associated with exponential function and the lemniscate of Bernoulli. Revista de la Real Academia de Ciencias Exactas, Fisicas y Naturales. Serie A. *Matematicas* **2019**, *113*, 233–253.

26. Mendiratta, R.; Nagpal, S.; Ravichandran, V. On a subclass of strongly starlike functions associated with exponential function. *Bull. Malays. Math. Sci. Soc.* **2015**, *38*, 365–386. [CrossRef]
27. Kumar, V.; Cho, N.E.; Ravichandran, V.; Srivastava, H.M. Sharp coefficient bounds for starlike functions associated with the Bell numbers. *Math. Slovaca* **2019**, *69*, 1053–1064. [CrossRef]
28. Kanas, S.; Wisniowska, A. Conic domains and starlike functions. *Revue Roumaine de Mathematiques Pures et Appliquees* **2000**, *45*, 647–658.
29. Uralegaddi, B.A.; Ganigi, M.D.; Sarangi, S.M. Univalent functions with positive coefficients. *Tamkang J. Math* **1994**, *25*, 225–230.
30. Ibrahim, R.W.; Jahangiri, J.M. Conformable differential operator generalizes the Briot-Bouquet differential equation in a complex domain. *AIMS Math.* **2019**, *4*, 1582–1595. [CrossRef]
31. Ibrahim, R.W.; Elobaid, R.M.; Obaiys, S.J. Symmetric Conformable Fractional Derivative of Complex Variables. *Mathematics* **2020**, *8*, 363. [CrossRef]
32. Noor, K.I.; Badar, R.S. On a class of quantum alpha-convex functions. *J. Appl. Math. Inform.* **2018**, *36*, 541–548.
33. Ahuja, O.; Anand, S.; Jain, N.K. Bohr Radius Problems for Some Classes of Analytic Functions Using Quantum Calculus Approach. *Mathematics* **2020**, *8*, 623. [CrossRef]
34. Arif, M.; Barkub, O.; Srivastava, H.M.; Abdullah, S.; Khan, S.A. Some Janowski Type Harmonic q-Starlike Functions Associated with Symmetrical Points. *Mathematics* **2020**, *8*, 629. [CrossRef]

16

On the Stability of la Cierva's Autogiro

M. Fernández-Martínez [1,‡] and Juan L.G. Guirao [2,3,*,‡]

1 MDE-UPCT, University Centre of Defence at the Spanish Air Force Academy, 30720 Santiago de la Ribera, Región de Murcia, Spain; manuel.fernandez-martinez@cud.upct.es
2 Departamento de Matematica Aplicada y Estadística, Universidad Politécnica de Cartagena, 30203 Cartagena, Spain
3 Department of Mathematics, Faculty of Science, King Abdulaziz University, 80203, Jeddah 21589, Saudi Arabia
* Correspondence: juan.garcia@upct.es.
‡ These authors contributed equally to this work.

Abstract: In this paper, we rediscover in detail a series of unknown attempts that some Spanish mathematicians carried out in the 1930s to address a challenge posed by Mr. la Cierva in 1934, which consisted of mathematically justifying the stability of la Cierva's autogiro, the first practical use of the direct-lift rotary wing and one of the first helicopter type aircraft.

Keywords: la Cierva's autogiro; la Cierva's equation; stability; differential equation with periodic coefficients

---

1. Introduction

The autogiro was the first practical use of the direct-lift rotary wing, where a windmilling rotor replaces the wing of the airplane, and the propulsive force is generated by a propeller. Interestingly, the autogiro allows a very slow flight and also behaves like an airplane in cruise. This kind of aircraft was developed by the Spanish aeronautical engineer Mr. Juan de la Cierva y Codorníu (Murcia (Spain), 1895–Croydon (UK), 1936), who also coined the term "autogiro". The origins of the autogiro come back to 1919, when an airplane that had been designed by Mr. la Cierva crashed due to stall near the ground. This fact encouraged him to design an aircraft with both a low landing speed and take-off.

Mr. la Cierva evolved the autogiro over the years. Firstly, the C-3 autogiro, which included a five-bladed rigid rotor, was built in 1922. The use of articulated rotor blades on the autogiro was suggested later, and Mr. la Cierva was the first to successfully apply a flap hinge in a rotary-wing aircraft. The C-4 autogiro (1923), which equipped a four-bladed rotor with flap hinges on the blades, was proved to fly with success. Thereafter, in 1924, it was built the C-6 autogiro with a rotor consisting of four flapping blades. This type, which is considered to be the first successful model of la Cierva's autogiro, took part in a demonstration at the Royal Aircraft Establishment the next year (c.f. [1]).

The Cierva Autogiro Company was founded in 1925 in UK by Mr. la Cierva, and about 500 autogiros were built in the next decade, many of them under license of the Cierva Company. In this regard, and for illustration purposes, Figure 1 depicts an autogiro constructed under license by Pitcairn in the United States (c.f. [2]). In those times, the autogiro was described as an easy to handle and fast aircraft, ahead of its time, which could land almost without rolling and take off in less than 30 m, and being able to stop off in the air, just to name some of its features. Certainly, the autogiro developments had an effect on the subsequent helicopter developments. Presently, however, the aircraft design seems to have evolved differently from the times of la Cierva's autogiro. In fact, novel settings consisting of combinations of four or more electric motors driving blades of carbon fiber will allow for less pollution and noise, and also lead to higher efficient aircraft. From a mathematical viewpoint, several problems related to modern aviation have been addressed by means of Fractional Calculus (c.f. [3]), path planning algorithms (c.f. [4]), or non-linear hyperbolic partial differential equations (c.f. [5]), to name some groundbreaking techniques.

Figure 1. The picture above (public domain) shows a PCA-2 autogiro built in the United States by Pitcairn under license of the Cierva Company. This unit was used by the National Advisory Committee for Aeronautics (NACA) for research purposes on rotor systems (c.f. [2]).

One of the first versions of la Cierva's aircraft, the C-3 autogiro, exhibited a certain tendency to fall over side-ways [1]. This issue made him to pay special attention to several aspects related to the stability of the autogiro. In this regard, in 1934, he attended a lecture at the Escuela Superior Aerotécnica (Madrid - Spain), and posed the following linear differential equation with periodic coefficients [6]:

$$m\,\frac{\mathrm{d}^2\,\Theta}{\mathrm{d}\,\varphi^2}+\left(\frac{3}{4}+\lambda\,\sin\varphi\right)\frac{\mathrm{d}\,\Theta}{\mathrm{d}\,\varphi}+\left(m+\lambda\,\cos\varphi+\frac{3}{4}\,\lambda^2\,\sin(2\varphi)\right)\,\Theta=0, \tag{1}$$

where φ is the azimuthal angle of the autogiro's blade, Θ is a function of φ that measures the angle of deviation of the blade with respect to its position of dynamic equilibrium when rotating, λ is a parameter that provides a relationship between the forward speed of the aircraft and the peripheral speed, and m is the ratio of the mass of the air volume (assumed to be contained in a rectangular parallelepiped with sides equal to the radius of the rotor and the width of the blade, twice) to the mass of the blade. The periodic nature of the coefficients of that equation is clear due to the autogiro's blade movement.

Following [7], we shall refer to Equation (1) as la Cierva's equation hereafter. It is worth mentioning that Mr. la Cierva appeared interested in mathematically determine whether the expression that bears his name admits convergent solutions since it could imply positive consequences concerning the stability of the autogiro. However, that expression resisted the attempts by Spanish and British mathematicians to that date, and in fact, some articles requiring the attention of mathematicians to address that equation can be found in the press of the time (c.f., e.g., [6]).

Next, let us provide some further comments regarding the parameters λ and m that are involved in Equation (1). Firstly, notice that λ increases as the speed does. In this way, Mr. la Cierva posed $\lambda=1$ as an appropriate limit value, thus taking into account future evolutions of the autogiro, the so-called ultrarrapid autogiro. On the other hand, Mr. la Cierva suggested the parameter m to vary in the range $[0.15, 1]$, depending on the aircraft. However, for a given autogiro, that parameter remains constant except in the case of large variations concerning the air density. As such, $m = 0.5$ was then considered to be an acceptable average value.

As stated in [7], Mr. la Cierva was especially interested in mathematically justifying the stability of the movement of the blades of the autogiro rather than quantitatively integrating Equation (1) for certain initial conditions. It is worth mentioning that such a stability had been fully verified in all the autogiros that had been assembled until then, and was also expected for higher speeds of values of the parameter λ. As such, the problem regarding the stability of la Cierva's autogiro could be mathematically stated in the following terms: does Θ go to zero as φ is increased regardless of the initial conditions? Regarding the latter, the reader may think of possible gusts of wind that could affect the movement of the blasts of the aircraft.

The main goal of this paper is to unveil the unknown attempts that some Spanish mathematicians carried out in the 1930s to solve the problem of the stability of la Cierva's autogiro. As such, the structure of this paper is as follows. Section 2 contains some preliminaries regarding differential equations with periodic coefficients. In this way, the concepts of characteristic exponent, characteristic number, and characteristic equation will be introduced. Section 3 describes in detail the first attempt of Prof. Orts y Aracil to analytically integrate Equation (1). Section 4 develops the calculations made by Prof. Orts y Aracil leading to sufficient conditions to guarantee that Equation (1) possesses convergent solutions. Shortly thereafter, the renowned Spanish engineer and mathematician Pedro Puig Adam (Barcelona (Spain), 1900–Madrid (Spain), 1960), Ph.D. in mathematics in 1921, published a qualitative approach regarding the stability of la Cierva's autogiro. Their calculations, which we have described in detail, have been included in Section 5 together with numerical calculations we have carried out in Mathematica. On the other hand, Section 6 contains some results that Puig-Adam obtained in regard to the reduced la Cierva's equation. Finally, Section 7 presents some additional remarks to complete the present study.

2. Preliminaries

In this section, we recall the basics on differential equations with periodic coefficients, thus paying special attention to the key concepts of characteristic exponent, characteristic number, and characteristic equation associated with a differential equation with periodic coefficients.

Firstly, it is clear that the so-called la Cierva's equation (c.f. Equation (1)) stands as a particular case of the following expression:

$$\frac{d^2 y(x)}{d x^2} + p_1(x) \frac{d y(x)}{d x} + p_2(x)\, y(x) = 0, \tag{2}$$

where $p_1(x)$ and $p_2(x)$ are continuous and $\omega-$periodic functions (with $\omega = 2\pi$ in the case of la Cierva's equation). Furthermore, if $y(x)$ is a solution of Equation (2), then $y(x+\omega)$ also is.

Let $y_1(x)$ and $y_2(x)$ be two linearly independent solutions of Equation (2). Hence, $y_1(x+\omega)$ and $y_2(x+\omega)$ also are. Thus, we can write

$$\begin{aligned} y_1(x+\omega) &= a_{11}\, y_1(x) + a_{12}\, y_2(x) \\ y_2(x+\omega) &= a_{21}\, y_1(x) + a_{22}\, y_2(x). \end{aligned} \tag{3}$$

Moreover, the coefficients $a_{ij} : i, j = 1, 2$ in Equation (3) could be calculated just by assigning particular values to the independent variable x.

Let $a \in \mathbb{R}$ and $\varphi(x)$ be a $\omega-$periodic function. Then the logarithmic derivative of the function $\eta(x) := e^{ax}\, \varphi(x)$ (i.e., $\frac{\eta'(x)}{\eta(x)}$) is also $\omega-$periodic, though $\eta(x)$ is not. In fact, it holds that

$$\eta(x+\omega) = e^{a(x+\omega)}\, \varphi(x+\omega) = e^{a\omega}\, e^{ax}\, \varphi(x) = e^{a\omega}\, \eta(x) \tag{4}$$

for all $x \in \text{dom}(\eta)$. Thus, if the variable x is increased in ω units, then the image of $x+\omega$ by η coincides with $\eta(x)$ multiplied by a factor equal to $s := e^{a\omega}$. In this context, a is named the characteristic exponent, whereas the factor s is known as the characteristic number. Notice that either the characteristic number or the characteristic exponent provides information about whether $\eta(x)$ goes to zero as $x \to \infty$. In particular, if $|s| < 1$, then $\mu(x) \to 0$ as $x \to \infty$, which means that the oscillations of the movement of the autogiro blade would get dampened. In fact, the amplitude of the oscillations of that blade would be multiplied by a factor less than the unit each new rotation. As such, we are interested in the calculation of those characteristic numbers, s.

Let $\varphi(x)$ be a ω−periodic solution of Equation (2). Then we can write $\eta(x)$ as a linear combination of both $y_1(x)$ and $y_2(x)$, namely

$$\eta(x) = C_1\, y_1(x) + C_2\, y_2(x). \tag{5}$$

Hence, we have that

$$\begin{aligned}
\eta(x+\omega) &= C_1\, y_1(x+\omega) + C_2\, y_2(x+\omega) \\
&= C_1\,(a_{11}\, y_1(x) + a_{12}\, y_2(x)) + C_2\,(a_{21}\, y_1(x) + a_{22}\, y_2(x)) \\
&= (C_1\, a_{11} + C_2\, a_{21})\, y_1(x) + (C_1\, a_{12} + C_2\, a_{22})\, y_2(x) \\
&= s\,\eta(x) = s\, C_1\, y_1(x) + s\, C_2\, y_2(x),
\end{aligned} \tag{6}$$

where the identity at Equation (5) has been used in the first equality, Equation (3) has been applied in the second identity, the fourth one is a consequence of $\eta(x)$ assumed to be ω−periodic and Equation (4), and the last identity is due to $\eta(x)$ being a particular solution of Equation (2) (c.f. Equation (5)). By identifying coefficients between the expressions at both the third and fifth equalities of Equation (6), it holds that

$$\begin{aligned}
C_1\,(a_{11} - s) + C_2\, a_{21} &= 0 \\
C_1\, a_{12} + C_2\,(a_{22} - s) &= 0.
\end{aligned} \tag{7}$$

Therefore, the so-called characteristic equation stands from the following expression:

$$\begin{vmatrix} a_{11} - s & a_{21} \\ a_{12} & a_{22} - s \end{vmatrix} = 0, \tag{8}$$

which is equivalent to

$$s^2 - (a_{11} {+} _{22})\, s + [a_{11}\, a_{22} - a_{12}\, a_{21}] = 0. \tag{9}$$

Assume that the polynomial in Equation (9) possesses two distinct roots, s_1 and s_2. If both of them are introduced in Equation (7), then a pair of specific values for each constant C_1 and C_2 will be obtained, thus leading to a pair of functions, $\eta_1(x)$ and $\eta_2(x)$ (c.f. Equation (5)) satisfying the condition at Equation (4). Accordingly, each solution of Equation (2) could be written as a linear combination of the functions $\eta_i(x) : i = 1, 2$. Following the above, the next result holds.

Theorem 1. *If the polynomial in Equation (8) has two distinct roots being less than the unit in absolute value, then $\eta_i(x) : i = 1, 2$ go to zero as x goes to infinity. More generally, any solution of Equation (2) would go to zero as x goes to infinity.*

A consequence of Theorem 1 is that the movement of the blade of the autogiro will be in equilibrium regardless the initial conditions.

We conclude this section by providing the statement of a known result concerning harmonic combinations of periodic functions. In fact,

Theorem 2. *Let $\alpha, \beta \in \mathbb{R}$ with $\alpha \neq 0$. Then*

$$\alpha \sin x + \beta \cos x = A \sin(x + \gamma),$$

where $\gamma = \arctan\left(\frac{\beta}{\alpha}\right)$ *and* $A = \operatorname{sgn}(\alpha)\sqrt{\alpha^2 + \beta^2}$.

The proof of that result becomes straightforward by using that $\sin(x+\gamma) = \sin x \cos\gamma + \cos x \sin\gamma$, and identiying coefficients with those from $\alpha \sin x + \beta \cos x$. This result will be applied in forthcoming Section 6.

3. Towards a Particular Solution of la Cierva's Equation

In this section, we revisit in detail a first approach that Prof. José Mª Orts y Aracil (Paterna, Valencia (Spain), 1891–Barcelona (Spain), 1968) contributed in [8] to mathematically determine a particular solution to Equation (1). First, it is clear that la Cierva's equation can be rewritten as follows:

$$\frac{d^2\Theta}{d\varphi^2} + \frac{1}{m}\left(\frac{3}{4} + \lambda \sin\varphi\right)\frac{d\Theta}{d\varphi} + \frac{1}{m}\left(m + \lambda\cos\varphi + \frac{3}{4}\lambda^2 \sin(2\varphi)\right)\Theta = 0. \tag{10}$$

Let $\Theta = u\, e^v$, where both u and v are functions of φ. Then it is clear that

$$\Theta' = e^v\left(u' + v'u\right) \qquad \Theta'' = e^v\left(u'' + 2\,v'u' + (v'^2 + v'')u\right). \tag{11}$$

If we replace the expressions at Equation (11) in Equation (10), then we have

$$\begin{aligned} e^v\left(u'' + 2\,v'u' + (v'^2 + v'')\,u\right) &+ \frac{1}{m}\left(\frac{3}{4} + \lambda\sin\varphi\right)\left(u' + v'u\right)e^v \\ &+ \frac{1}{m}\left(m + \lambda\cos\varphi + \frac{3}{4}\lambda^2\sin(2\varphi)\right)u\,e^v = 0, \end{aligned}$$

which is equivalent to

$$\begin{aligned} u'' &+ \left(2v' + \frac{1}{m}\left(\frac{3}{4} + \lambda\sin\varphi\right)\right)u' \\ &+ \left(v'' + v'^2 + \frac{1}{m}\left(\frac{3}{4} + \lambda\sin\varphi\right)v' + 1 + \frac{\lambda}{m}\cos\varphi + \frac{3\lambda^2}{4m}\sin(2\varphi)\right)u = 0. \end{aligned} \tag{12}$$

Next, we cancel the coefficient of u' in Equation (12). In fact,

$$2v' + \frac{1}{m}\left(\frac{3}{4} + \lambda\sin\varphi\right) = 0 \Leftrightarrow v' = -\frac{1}{2m}\left(\frac{3}{4} + \lambda\sin\varphi\right). \tag{13}$$

Hence, it is clear that

$$v = \frac{1}{2m}\left(\lambda\cos\varphi - \frac{3}{4}\varphi\right) \text{ and } v'' = -\frac{\lambda}{2m}\cos\varphi. \tag{14}$$

As such, Equation (12) can be reduced as follows:

$$u'' + p(\varphi)\,u = 0,$$

where

$$p(\varphi) = v'' + v'^2 + \frac{1}{m}\left(\frac{3}{4} + \lambda\sin\varphi\right)v' + 1 + \frac{\lambda}{m}\cos\varphi + \frac{3\lambda^2}{4m}\sin(2\varphi). \tag{15}$$

If we replace the expressions in both Equations (13) and (14) into Equation (15), then

$$p(\varphi) = -\frac{1}{4m^2}\left(\frac{9}{16} + \lambda^2 \sin^2\varphi + \frac{3\lambda}{2}\sin\varphi\right) + \frac{\lambda}{2m}\cos\varphi + 1 + \frac{3\lambda^2}{4m}\sin(2\varphi). \tag{16}$$

Moreover, if we replace $\sin^2\varphi$ by $\frac{1}{2}(1-\cos(2\varphi))$ in Equation (16), then we have

$$\begin{aligned} p(\varphi) = 1 &- \frac{9}{64m^2} - \frac{\lambda^2}{8m^2} + \frac{\lambda}{2m}\cos\varphi + \frac{\lambda^2}{8m^2}\cos(2\varphi) \\ &- \frac{3\lambda}{8m^2}\sin\varphi + \frac{3\lambda^2}{4m}\sin(2\varphi). \end{aligned}$$

As such, Equation (12) can be expressed in the following terms:

$$\frac{\mathrm{d}^2 u}{\mathrm{d}\varphi^2} = (a_0 + a_1\cos\varphi + a_2\cos(2\varphi) + b_1\sin\varphi + b_2\sin(2\varphi))\,u, \tag{17}$$

where

$$\begin{aligned} a_0 &= \frac{9+8\lambda^2}{64\,m^2} - 1, \quad a_1 = -\frac{\lambda}{2\,m}, \quad a_2 = -\frac{\lambda^2}{8\,m^2} \\ b_1 &= \frac{3}{8}\frac{\lambda}{m^2}, \quad b_2 = -\frac{3}{4}\frac{\lambda^2}{m}. \end{aligned} \tag{18}$$

Additionally, by writing $u = e^{\int z\,\mathrm{d}\varphi}$, the expression in Equation (17) can be rewritten as follows:

$$\frac{\mathrm{d}z}{\mathrm{d}\varphi} + z^2 = a_0 + a_1\cos\varphi + a_2\cos(2\varphi) + b_1\sin\varphi + b_2\sin(2\varphi), \tag{19}$$

which leads to a Ricatti type equation. The next expression was suggested by Prof. Orts y Aracil as a potential solution of Equation (19):

$$z_1 = \alpha + \beta\sin\varphi + \gamma\cos\varphi, \tag{20}$$

where α, β, and γ are three constants that can be determined by introducing Equation (20) in the former Equation (19) and identifying coefficients in both sides of that expression. As such, we obtain that

$$\begin{aligned} a_0 &= \alpha^2 + \tfrac{1}{2}(\beta^2+\gamma^2), \quad a_1 = \beta + 2\,\alpha\,\gamma, \quad a_2 = \tfrac{1}{2}(\gamma^2-\beta^2) \\ b_1 &= 2\,\alpha\,\beta - \gamma, \quad b_2 = \beta\,\gamma. \end{aligned} \tag{21}$$

Next, we observe that

$$a_2^2 + b_2^2 = \left(\frac{1}{2}(\gamma^2+\beta^2)\right)^2 \geq 0,$$

so $\alpha^2 + \sqrt{a_2^2+b_2^2} = \alpha^2 + \frac{1}{2}(\gamma^2+\beta^2) = a_0$. Therefore,

$$a_0 \geq \frac{1}{2}(\gamma^2+\beta^2) = \sqrt{a_2^2+b_2^2}.$$

On the other hand, it is clear that

$$\sqrt{a_2^2+b_2^2} - a_2 = \frac{1}{2}(\gamma^2+\beta^2) - \frac{1}{2}(\gamma^2-\beta^2) = \beta^2 \geq 0.$$

Furthermore, it holds that $a_2 + \sqrt{a_2^2 + b_2^2} = \frac{1}{2}(\gamma^2 - \beta^2) + \frac{1}{2}(\gamma^2 + \beta^2) = \gamma^2 \geq 0$. All the calculations above lead to the following values of the parameters α, β, and γ of the particular solution at Equation (19):

$$\alpha = \sqrt{a_0 - \sqrt{a_2^2 + b_2^2}}, \quad \beta = \sqrt{\sqrt{a_2^2 + b_2^2} - a_2}, \quad \gamma = \sqrt{a_2 + \sqrt{a_2^2 + b_2^2}}.$$

Going beyond, it is possible to reduce the parameters α, β, and γ in Equation (21), thus leading to a pair of relationships among the coefficients a_i and b_j for $i = 0, 1, 2$ and $j = 1, 2$. Recall that a_i and b_j can be expressed, in turn, in terms of λ and m (c.f. Equation (18)). In fact, the following expressions hold.

$$\begin{aligned} 1327104\, m^8 + 359424\, m^6 + 35712\, m^4 - 324\, m^2 &= 0 \\ \lambda^2 &= \frac{9 + 16\, m^2 - (9 - 48\, m^2)\sqrt{1 + 36\, m^2}}{8\sqrt{1 + 36\, m^2}\,(1 - \sqrt{1 + 36\, m^2})}. \end{aligned} \tag{22}$$

If the eight order polynomial in m at Equation (22) is divided by $12\, m^2$ (under the assumption that $m \neq 0$), and the change of variable $t = 48\, m^2$ is considered, then the following third order polynomial stands:

$$t^3 + 13\, t^2 + 62\, t - 27 = 0. \tag{23}$$

By Bolzano's Theorem, it is clear that the polynomial in Equation (23) possesses a root, say t_1, in the subinterval $[0.4007, 0.4008]$. That root could be approximated by some numerical method, though in [8], t_1 was considered merely as the middle point of that subinterval, i.e., $t_1 = 0.40075$. Since $t_1 = 48\, m_1^2$, then we have $m_1 \simeq 0.0914$. Hence, the second expression in Equation (22) leads to $\lambda_1 \simeq 0.7249$. With the values of both parameters m and λ estimated, the coefficients α, β, and γ of the particular solution of Equation (19) given by Equation (20) can be calculated by Equation (18). In fact, that particular solution remains as follows:

$$z_1 = 3.8391 + 4.1036\, \sin\varphi + 1.0511\, \cos\varphi. \tag{24}$$

Also, we have $u_1 = \exp(3.8391\,\varphi + 1.0511\, \sin\varphi - 4.1036\, \cos\varphi)$, and hence,

$$\begin{aligned} \Theta_1 = \exp(&3.8391\,\varphi + 1.0511\, \sin\varphi \\ &+ 0.6898\left(0.0914\, \cos\varphi - \frac{3}{4}\,\varphi\right) - 4.1036\, \cos\varphi), \end{aligned}$$

stands as a particular solution of Equation (10), the differential equation which models the equilibrium of the blade of la Cierva's autogiro.

As Prof. Orts y Aracil commented, the approach contributed in this section threw a value of $\lambda_1 = 0.7249$ lying within the range suggested by Mr. Herrera in [6], i.e., the subinterval $[0, 1]$, though the value of $m_1 = 0.091$ appears out of its corresponding range, the subinterval $[0.15, 1]$. In this regard, it was argued that the problem of the equilibrium of la Cierva's autogiro had been addressed from a mathematical (and not an Saee) viewpoint.

4. Sufficient Conditions on the Existence of Convergent Solutions

In this section, sufficient conditions are provided to guarantee the existence of convergent solutions for la Cierva's equation, an issue that was further addressed by Prof. Orts y Aracil in [9]. With this aim, we start by sketching an alternative approach to that one described in Section 3 with the aim to integrate the expression at Equation (10).

First of all, let us denote by $p(\varphi)$ the continuous and periodic function that appears at the right term of Equation (19), i.e.,

$$p(\varphi) = a_0 + a_1 \cos\varphi + a_2 \cos(2\varphi) + b_1 \sin\varphi + b_2 \sin(2\varphi), \tag{25}$$

which allows rewriting Equation (17) as follows:

$$\frac{d^2 u}{d\varphi^2} = p(\varphi)\, u. \tag{26}$$

Such a kind of differential equations can be integrated by means of a characteristic equation of the form

$$s^2 - As + 1 = 0, \text{ where} \tag{27}$$

$$\begin{aligned} A &= 2 + \sum_{n=1}^{+\infty} \left[F_n(2\pi) + f_n'(2\pi)\right], \quad F_n(\varphi) = \int_0^\varphi d\varphi \int_0^\varphi p(\varphi)\, F_{n-1}(\varphi)\, d\varphi, \\ f_n(\varphi) &= \int_0^\varphi d\varphi \int_0^\varphi p(\varphi)\, f_{n-1}(\varphi)\, d\varphi, \quad F_0(\varphi) = 1, \text{ and } f_0(\varphi) = \varphi \end{aligned} \tag{28}$$

(c.f. ([10] [item 49, p. 402]) and ([11] [Chapter 3, Section 55])). Moreover, if $p(\varphi) \geq 0$ for all $\varphi > 0$, then it holds that $F_n(\varphi), f_n(\varphi), f_n'(\varphi) > 0$ for all $\varphi > 0$ and all $n \in \mathbb{N}$. Hence, $A > 2$ and the expression at Equation (27) possesses two positive roots, say s_1 and s_2, with one of them being greater (resp., smaller) than the unit and the other being smaller (resp., greater) than the unit.

A fundamental system of solutions for Equation (26) is provided by the functions

$$u_1 = e^{\frac{\varphi}{2\pi} l s_1} \cdot \alpha(\varphi), \qquad u_2 = e^{\frac{\varphi}{2\pi} l s_2} \cdot \beta(\varphi), \tag{29}$$

where $\alpha(\varphi)$ and $\beta(\varphi)$ are $2\pi-$periodic continuous functions.

Hence, one of the integrals at Equation (29), say u_1, goes to zero as $\varphi \to \infty$, and so does Θ. In this way, the so-called Liapounov's condition can be stated as follows (c.f. [10]).

Theorem 3 (Liapounov's condition). *The second order differential equation in Equation (26) admits a convergent integral as $\varphi \to \infty$, if and only if, $p(\varphi) \geq 0$ for all $\varphi > 0$.*

Following the above, our next goal is to verify that sufficient condition. To deal with, let us apply the change of variable $x = \tan(\frac{\varphi}{2})$ to the periodic function at Equation (25). As such, we have

$$\begin{aligned} p(\varphi) &= a_0 + a_1 \cos\varphi + a_2 \frac{1 - \tan^2\varphi}{1 + \tan^2\varphi} + b_1 \sin\varphi + b_2 \frac{2\tan\varphi}{1 + \tan^2\varphi} \\ &= a_0 + a_1 \frac{1 - x^2}{1 + x^2} + a_2 \frac{1 - 6x^2 + x^4}{(1 + x^2)^2} + b_1 \frac{2x}{1 + x^2} + b_2 \frac{4x(1 - x^2)}{(1 + x^2)^2}, \end{aligned} \tag{30}$$

and hence, we can write $p(\varphi) = \frac{1}{(1+x^2)^2} q(x)$, where

$$\begin{aligned} q(x) &= a_0 (1 + x^2)^2 + a_1 (1 - x^2)(1 + x^2) + a_2 (1 - 6x^2 + x^4) \\ &\quad + 2 b_1 x (1 + x^2) + 4 b_2 x (1 - x^2) \\ &= (a_0 + a_1 + a_2) + 2(b_1 + 2b_2)\, x + 2(a_0 - 3a_2)\, x^2 \\ &\quad + 2(b_1 - 2b_2)\, x^3 + (a_0 - a_1 + a_2)\, x^4. \end{aligned} \tag{31}$$

Notice that the first equality at Equation (30) has been applied that

$$\sin(2\varphi) = \frac{2\,\tan\varphi}{1+\tan^2\varphi}, \quad \cos(2\varphi) = \frac{1-\tan^2\varphi}{1+\tan^2\varphi},$$

whereas the second identity at that expression holds since that change of variable implies that

$$\cos\varphi = \frac{1-x^2}{1+x^2}, \quad \sin\varphi = \frac{2x}{1+x^2}, \quad \tan\varphi = \frac{2\,\tan(\frac{\varphi}{2})}{1-\tan^2(\frac{\varphi}{2})} = \frac{2x}{1-x^2}.$$

Moreover, by writing

$$q(x) = c_4 + c_3\,x + c_2\,x^2 + c_1\,x^3 + c_0\,x^4, \tag{32}$$

we can identify coefficients with those ones at the right side of Equation (31). In fact,

$$\begin{aligned} c_0 &= a_0 - a_1 + a_2 = \frac{9}{64\,m^2} - 1 + \frac{\lambda}{2\,m} \\ c_1 &= 2\,(b_1 - 2\,b_2) = \frac{3}{4}\frac{\lambda}{m^2} + 3\frac{\lambda^2}{m} \\ c_2 &= 2\,(a_0 - 3\,a_2) = \frac{9}{32\,m^2} - 2 + \frac{\lambda^2}{m^2} \\ c_3 &= 2\,(b_1 + 2\,b_2) = \frac{3}{4}\frac{\lambda}{m^2} - 3\frac{\lambda^2}{m} \\ c_4 &= a_0 + a_1 + a_2 = \frac{9}{64\,m^2} - \frac{\lambda}{2\,m} - 1, \end{aligned} \tag{33}$$

where Equation (18) allows writing the c_i's in terms of the parameters λ and m.

On the other hand, a necessary condition to get $p(\varphi) \geq 0$ for all $\varphi > 0$ consists of both coefficients c_0 and c_4 of the polynomial at Equation (32) being positive. In this way, Equation (33) implies that

$$\begin{aligned} &c_4 > 0 \Leftrightarrow 32m\,(\lambda + 2m) < 9 \Leftrightarrow X^2 - Y^2 < 9. \\ &c_0 > 0 \Leftrightarrow 32m\,(2m - \lambda) < 9 \Leftrightarrow m < \frac{3}{8} \Leftrightarrow X - Y < 3, \end{aligned} \tag{34}$$

where $X := 8m + 2\lambda$ and $Y := 2\lambda$. Observe that $X, Y > 0$ since both parameters m and λ are positive. In fact, regarding the second equivalence at the first line of Equation (34), just observe that we can write

$$\begin{aligned} 9 > 32\,m\,(\lambda + 2m) &= 64\,m^2 + 32\,m\lambda \\ &= 8^2 m^2 + 2 \times 16\,m\lambda + (2\lambda)^2 - (2\lambda)^2 \\ &= (8m + 2\lambda)^2 - (2\lambda)^2 = X^2 - Y^2. \end{aligned}$$

Thus, the condition $c_4 > 0$ is equivalent to a point at the first quadrant, (X, Y), located above the hyperbola $X^2 - Y^2 = 9$.

5. Puig-Adam's Qualitative Approach

In this section, we revisit in detail the approach contributed by Puig-Adam in [7] to approach the solutions of la Cierva's equation from a qualitative viewpoint.

According to the contents of Section 2, we are interested in obtaining two particular solutions of la Cierva's equation (c.f. Equation (10)), say $y_1(x)$ and $y_2(x)$. Let them be given by the following initial conditions:

$$\begin{aligned} y_1(0) &= 1, \quad & y_1'(0) &= 0 \\ y_2(0) &= 0, \quad & y_2'(0) &= 1. \end{aligned} \tag{35}$$

It is clear that $y_1(x)$ and $y_2(x)$ would be independent since their Wronskian at 0 is distinct from zero, $W(y_1, y_2)(0) = 1$. Moreover, from Equation (3), we have that

$$\begin{aligned} y_1'(x+\omega) &= a_{11}\, y_1'(x) + a_{12}\, y_2'(x) \\ y_2'(x+\omega) &= a_{21}\, y_1'(x) + a_{22}\, y_2'(x) \end{aligned} \tag{36}$$

for all x. If we particularize both Equations (3) and (36) in $x = 0$, then

$$\begin{aligned} y_1(\omega) &= a_{11}, & y_1'(\omega) &= a_{12} \\ y_2(\omega) &= a_{21}, & y_2'(\omega) &= a_{22}. \end{aligned}$$

Moreover, from Equation (9), the characteristic equation holds from the following expression:

$$s^2 - (y_1(\omega) + y_2'(\omega))\, s + [y_1(\omega)\, y_2'(\omega) - y_1'(\omega)\, y_2(\omega)] = 0. \tag{37}$$

As such, Equation (37) would be fully determined once $y_i(\omega)$ and their derivatives, $y_i'(\omega)$ for $i = 1, 2$, have been calculated. It is also worth mentioning that the coefficients of the characteristic polynomial are independent from the initial conditions that were selected, i.e., such coefficients only depend on the coefficients of the given differential equation. In particular, notice that the independent term of Equation (37), which coincides with $W(y_1, y_2)(\omega)$, can be calculated in terms of $p_1(x)$ (recall Equation (2)), by means of the following expression (c.f., e.g., [11]):

$$W(y_1, y_2)(x) = W(y_1, y_2)(x_0)\, \exp\left[-\int_{x_0}^{x} p_1(x)\, \mathrm{d}x \right]. \tag{38}$$

In fact, observe that the former expression can be justified just by identifying the differential equation in Equation (2) with the next one:

$$W(y, y_1, y_2)(x) = 0. \tag{39}$$

In fact, Equation (39) is equivalent to

$$\begin{aligned} (y_1(x)\, y_2'(x) - y_1'(x)\, y_2(x))\, y''(x) + (y_1''(x)\, y_2(x) - y_1(x)\, y_2''(x))\, y'(x) \\ + (y_1'(x)\, y_2''(x) - y_1''(x)\, y_2'(x))\, y(x) = 0, \end{aligned}$$

which leads to

$$y''(x) + \frac{y_1''(x)\, y_2(x) - y_1(x)\, y_2''(x)}{y_1(x)\, y_2'(x) - y_1'(x)\, y_2(x)}\, y'(x) + \frac{y_1'(x)\, y_2''(x) - y_1''(x)\, y_2'(x)}{y_1(x)\, y_2'(x) - y_1'(x)\, y_2(x)}\, y(x) = 0 \tag{40}$$

since $y_1(x)$ and $y_2(x)$ have been assumed to be independent solutions (and hence, $W(y_1, y_2)(x) \neq 0$ for all x). Thus, if the expressions in both Equations (2) and (40) coincide term by term, then it holds that

$$p_1(x) = \frac{y_1''(x)\, y_2(x) - y_1(x)\, y_2''(x)}{y_1(x)\, y_2'(x) - y_1'(x)\, y_2(x)} = -\frac{W'(y_1, y_2)(x)}{W(y_1, y_2)(x)}.$$

Following the above, it holds that the independent term of Equation (37) can be obtained just by applying Equation (38) in the open interval $(x_0, x) = (0, \omega)$. Since $W(y_1, y_2)(0) = 1$, then we have that

$$\begin{aligned} W(y_1, y_2)(2\pi) &= \exp\left[-\int_0^{2\pi} p_1(x)\, \mathrm{d}x \right] \\ &= \exp\left[-\int_0^{2\pi} \frac{1}{m} \left(\frac{3}{4} + \lambda \sin x \right) \mathrm{d}x \right] = e^{-\frac{3}{2m}\pi}, \end{aligned} \tag{41}$$

where it has been used that $p_1(x) = \frac{1}{m}\left(\frac{3}{4} + \lambda \sin x\right)$ and $\omega = 2\pi$ in the case of la Cierva's equation (c.f. Equation (10)). Hence, the characteristic polynomial associated to la Cierva's equation remains as follows:

$$s^2 - (y_1(2\pi) + y_2'(2\pi))\, s + e^{-\frac{3}{2m}\pi} = 0. \quad (42)$$

Interestingly, it holds that the independent term, $e^{-\frac{3}{2m}\pi}$, does not depend on the forward speed.

However, to fully determine the characteristic polynomial at Equation (42), it becomes necessary to know the values of both functions $y_1(x)$ and $y_2'(x)$ at $x = 2\pi$. With this aim, Puig-Adam, instead of carrying out a power series expansion in regard to the periodic coefficients of the starting equation, for instance, preferred to apply a (second order) Runge-Kutta numerical approach to each particular solution, $y_1(x)$ and $y_2(x)$, of la Cierva's equation in the closed bounded interval $[0, 2\pi]$ with parameters $m = 0.5$ and $\lambda = 1$, that according to Puig-Adam, had been suggested by Mr. la Cierva. In [7], it was stated that the trapezoidal method had been applied. In this paper, though, we shall apply a explicit midpoint method (also known as modified Euler method), which appears implemented in Mathematica. In any case, both of them are second-order approaches.

In this way, and similarly to [7], Figures 1 and 2 depicts our approximations to each particular solution of Mr. la Cierva's equation, $y_1(x)$ with initial conditions $y_1(0) = 1, y_1'(0) = 0$, and $y_2(x)$ with initial conditions $y_2(0) = 0, y_2'(0) = 1$ (c.f. Equation (35)), as provided by the second-order (Runge-Kutta explicit) midpoint approach on the interval $[0, 2\pi]$, which corresponds to a turn of the blade of the autogiro.

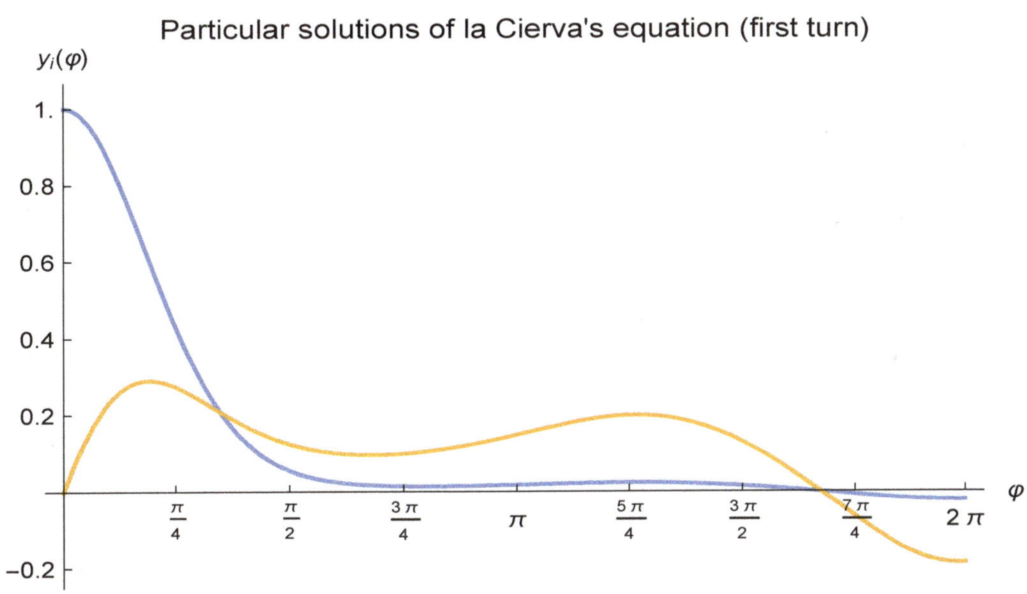

Figure 2. Second order Runge-Kutta approximations (obtained by the explicit midpoint method) to each particular solution of la Cierva's equation according to the procedure described in Section 5, i.e., $y_1(\varphi)$ (blue line) and $y_2(\varphi)$, where φ varies in the range $[0, 2\pi]$, which means a turn of the blade of the autogiro, and the choice of parameters was as suggested by Mr. la Cierva, i.e., $m = 0.5$ and $\lambda = 1$.

According to our numerical calculations, it holds that

$$y_1(2\pi) = -0.0222528, \qquad y_2'(2\pi) = 0.0230689,$$

and hence, the characteristic polynomial at Equation (42) remains as follows:

$$s^2 - 0.000816093\, s + e^{-\frac{3}{2m}\pi} = 0. \quad (43)$$

As such, it holds that the polynomial at Equation (43) possesses two complex (conjugated) roots, namely $s_1 = 0.000408046 - 0.00897402\, i$ and $s_2 = 0.000408046 + 0.00897402\, i$. Since $|s_i| = 0.00898329 \ll 1$ for

$i = 1, 2$, it can be guaranteed that the blade movement of la Cierva's autogiro behaves quite stably for that choice of parameters.

Remark 1. *It is worth mentioning that, in the original study carried out by Puig-Adam, the following values were obtained by the numerical approach carried out therein:* $y_1(2\pi) = -0.013$ *and* $y_2'(2\pi) = 0.04197$, *which led to the next characteristic equation:*

$$s^2 - 0.02897\, s + e^{-\frac{3}{2m}\pi} = 0,$$

whose real roots are $t_1 = 0.00312209$ *and* $t_2 = 0.0258479$. *Such results mainly differ from ours in the nature of the roots of the characteristic polynomial. That issue was mainly caused by the approximation errors made due to the limitations of the calculation systems available in the 1930s. It is also true that we have approximated the coefficients* $y_1(2\pi)$ *and* $y_2'(2\pi)$ *by the midpoint method instead of the trapezoidal approach used by Puig-Adam. However, both are second-order approaches, so they should lead to close results.*

Recall also that $W(y_1, y_2)(2\pi) = e^{-3\pi} \simeq 8.0699518 \times 10^{-5}$ (c.f. Equation (41)). Alternatively, if we calculate an approximation to that Wronskian by means of the expression appeared at Equation (37) and the values of the coefficients provided by the numerical approach used by Puig-Adam, then we have

$$\begin{aligned} W(y_1, y_2)(2\pi) &= y_1(2\pi)\, y_2'(2\pi) - y_1'(2\pi)\, y_2(2\pi) \\ &\simeq -0.013 \times 0.04197 + 0.00398 \times 0.18509 \\ &\simeq 1.910482 \times 10^{-4} = W_{\mathrm{PA}}(y_1, y_2)(2\pi), \end{aligned} \tag{44}$$

where $W_{\mathrm{PA}}(y_1, y_2)(2\pi)$ denotes the Puig-Adam's numerical approximation to that quantity. As such, the absolute error obtained when comparing the theoretical value of that Wronskian with respect to $W_{\mathrm{PA}}(y_1, y_2)(2\pi)$, (c.f. Equation (44)) was found to be approximately equal to 1.10349×10^{-4}, quite close to zero. Going beyond, our midpoint-based approach, which approximated $W(y_1, y_2)(2\pi)$ by the quantity 8.0699523×10^{-5}, threw an absolute error approximately equal to 5.33809×10^{-12}.

Furthermore, it is possible to provide a qualitative viewpoint in regard to the stability of the oscillations of the blade of the autogiro in its upcoming turns. In fact, let $\omega = 2\pi$ and consider Equation (35). Applying such initial conditions to both Equations (3) and (36), it holds that the former turns into the following expression:

$$\begin{aligned} y_1(x + 2\pi) &= y_1(2\pi)\, y_1(x) + y_1'(2\pi)\, y_2(x) \\ y_2(x + 2\pi) &= y_2(2\pi)\, y_1(x) + y_2'(2\pi)\, y_2(x). \end{aligned} \tag{45}$$

By recursively applying Equation (45), we have

$$\begin{aligned} y_1(x + 4\pi) &= y_1(2\pi)\, y_1(x + 2\pi) + y_1'(2\pi)\, y_2(x + 2\pi) \\ &= \left(y_1^2(2\pi) + y_1'(2\pi)\, y_2(2\pi)\right)\, y_1(x) \\ &\quad + y_1'(2\pi)\, \left(y_1(2\pi) + y_2'(2\pi)\right)\, y_2(x) \\ y_2(x + 4\pi) &= y_2(2\pi)\, y_1(x + 2\pi) + y_2'(2\pi)\, y_2(x + 2\pi) \\ &= y_2(2\pi)\, \left(y_1(2\pi) + y_2'(2\pi)\right)\, y_1(x) \\ &\quad + \left(y_2'^2(2\pi) + y_2(2\pi)\, y_1'(2\pi)\right)\, y_2(x) \end{aligned} \tag{46}$$

which corresponds to the second turn of the blade. Figure 3 depicts (numerical approximations of) both solutions after two turns of the autogiro's blade. Also, regarding the third turn of the blade, the following expression holds:

$$\begin{aligned} y_1(x+6\pi) &= \alpha_1\, y_1(x) + \alpha_2\, y_2(x) \\ y_2(x+6\pi) &= \beta_1\, y_1(x) + \beta_2\, y_2(x), \end{aligned} \tag{47}$$

where

$$\begin{aligned} \alpha_1 &= \left(y_1^2(2\pi) + y_1'(2\pi)\, y_2(2\pi)\right)\, y_1(2\pi) + y_1'(2\pi)\, \left(y_1(2\pi) + y_2'(2\pi)\right)\, y_2(2\pi) \\ \alpha_2 &= \left(y_1^2(2\pi) + y_1'(2\pi)\, y_2(2\pi)\right)\, y_1'(2\pi) + y_1'(2\pi)\, \left(y_1(2\pi) + y_2'(2\pi)\right)\, y_2'(2\pi) \\ \beta_1 &= y_2(2\pi)\, \left(y_1(2\pi) + y_2'(2\pi)\right)\, y_1(2\pi) + \left(y_2'^2(2\pi) + y_2(2\pi)\, y_1'(2\pi)\right)\, y_2(2\pi) \\ \beta_2 &= y_2(2\pi)\, \left(y_1(2\pi) + y_2'(2\pi)\right)\, y_1'(2\pi) + \left(y_2'^2(2\pi) + y_2(2\pi)\, y_1'(2\pi)\right)\, y_2'(2\pi). \end{aligned} \tag{48}$$

As with Figure 3, (numerical approximations) of the particular solutions of la Cierva's equation (for parameters $m = 0.5$ and $\lambda = 1$) after three turns of the autogiro's blade are illustrated at Figure 4. It can be seen that for angles beyond $\frac{5\pi}{2}$, the graph of the first particular solution of la Cierva's equation at the second turn of the blade becomes indistinguishable from the $x-$axis, as it is the case of the plot of $y_2(x)$ as of the third turn of the blade.

Notice that, as Puig-Adam pointed out, the initial conditions $y_1(0) = 1, y_2'(0) = 1$ (c.f. Equation (35)) are quite extreme. Nevertheless, for k small enough, particular solutions of the form ky_1 and ky_2, which exhibit smaller oscillations than those from y_1 and y_2, and whose graphs can be depicted by a $y-$axis rescaling of those appeared in Figure 2, are possible.

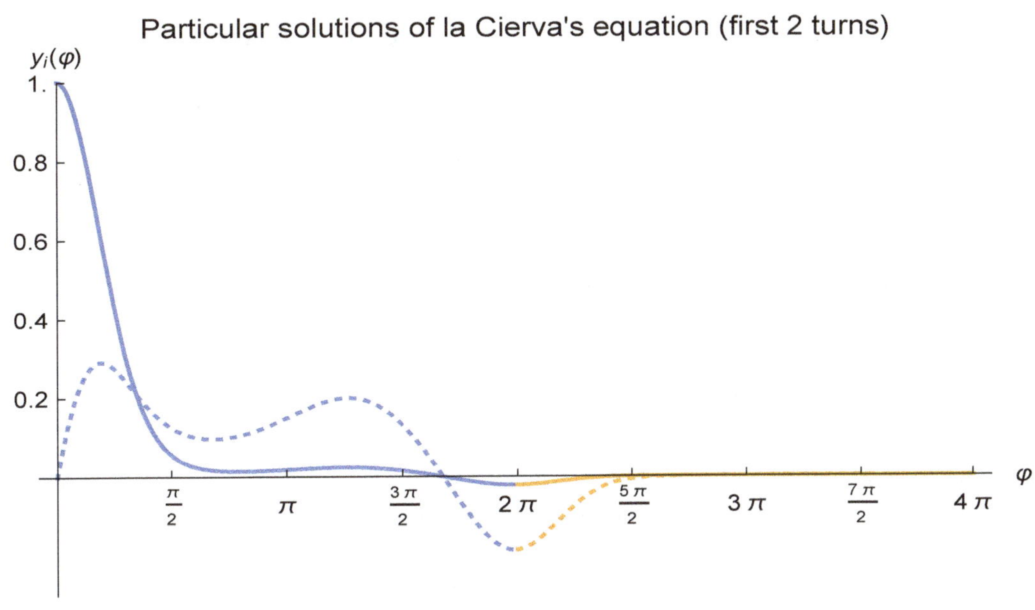

Figure 3. (Numerical approximations to the) particular solutions of la Cierva's equation after two turns of the autogiro's blade (c.f. Equation (46)). In this occassion, the blue lines have been used to distinguish the curves of both particular solutions in regard to the first turn to their prolongations to the second turn of the blade. In addition, notice that the dotted line corresponds to $y_2(\varphi)$.

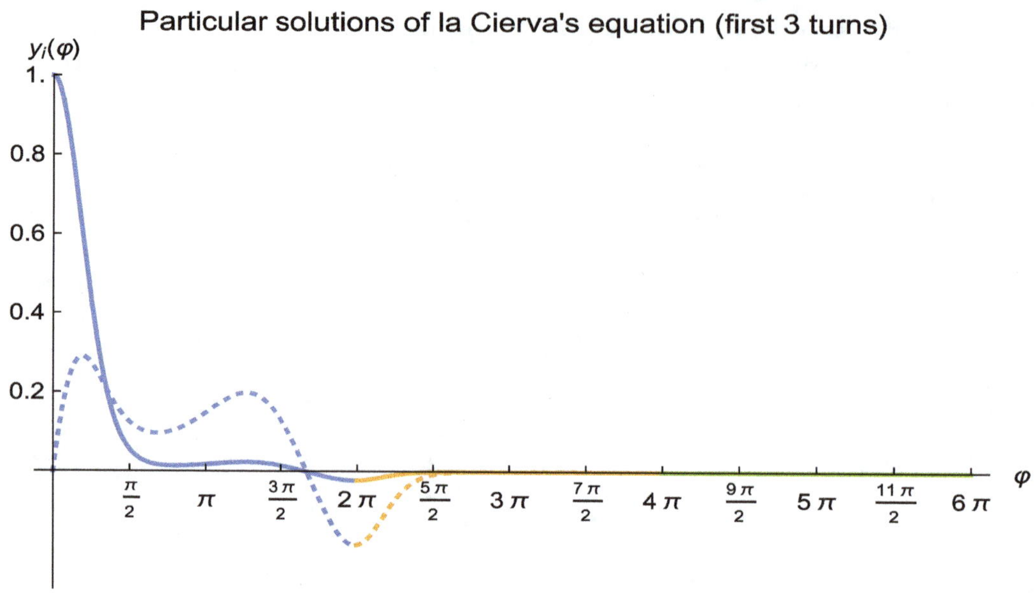

Figure 4. (Numerical approximations to the) particular solutions of la Cierva's equation after the first three turns of the autogiro's blade (c.f. Equations (47) and (48)). The blue lines represent the curves of both particular solutions at the first turn, the orange lines correspond to their prolongations to the second turn of the blade, and the green lines depict the extensions of such solutions to the third turn. As with Figure 3, the dotted line corresponds to $y_2(\varphi)$.

6. La Cierva's Reduced Equation

The aim of this section is to calculate a pair of particular solutions to la Cierva's equation by means of the so-called reduced la Cierva's equation. Furthermore, a comparison of such solutions with those solutions obtained in Section 5 is carried out.

First, recall that in Section 5, it was provided a numerical criterion to determine whether the solutions of la Cierva's equation (c.f. Equation (1)) behave stably for a choice of parameters (λ, m). Specifically, let y_1 and y_2 be the particular solutions of that equation (as provided by a Runge-Kutta method, in this case) in the interval $[0, 2\pi]$, and calculate $y_1(2\pi) + y_2'(2\pi)$. If that quantity stands <1 in absolute value, then the behavior of the oscillations of the autogiro's blade is stable for such parameters.

Firstly, we recall the original expression of la Cierva's equation (c.f. Equation (1)):

$$m\,\Theta'' + \left(\frac{3}{4} + \lambda\,\sin\varphi\right)\,\Theta' + \left(m + \lambda\,\cos\varphi + \frac{3}{4}\,\lambda^2\,\sin(2\varphi)\right)\,\Theta = 0, \tag{49}$$

where φ is the azimuthal angle of the autogiro's blade and Θ is a function of φ that measures the angle of deviation of the blade with respect to its position of dynamic equilibrium when rotating.

Let $\Theta = uv$. Then it is clear that $\Theta' = u'v + uv'$ and $\Theta'' = u''v + 2\,u'v' + uv''$. If we apply that change of variable to Equation (49), then that expression turns into the next one:

$$\begin{aligned} m\,(u''v + 2\,u'v' + uv'') + \left(\frac{3}{4} + \lambda\,\sin\varphi\right)\,(u'v + uv') \\ + \left(m + \lambda\,\cos\varphi + \frac{3}{4}\,\lambda^2\,\sin(2\varphi)\right)\,uv = 0, \end{aligned} \tag{50}$$

which is equivalent to

$$\begin{aligned} &mv\,u'' + \left(2mv' + \left(\frac{3}{4} + \lambda \sin\varphi\right) v\right) u' \\ &+ \left(mv'' + \left(\frac{3}{4} + \lambda \sin\varphi\right) v' + \left(m + \lambda\cos\varphi + \frac{3}{4}\,\lambda^2 \sin(2\varphi)\right) v\right) u \\ &= 0. \end{aligned} \tag{51}$$

The next goal is to cancel the coefficient of u' in Equation (51). In fact,

$$2mv' + \left(\frac{3}{4} + \lambda \sin\varphi\right) v = 0 \Leftrightarrow \frac{v'}{v} = -\frac{1}{2m}\left(\frac{3}{4} + \lambda \sin\varphi\right). \tag{52}$$

The integration of the expression in Equation (52) leads to

$$v = \exp\left[\frac{\lambda}{2m}\cos\varphi - \frac{3}{8m}\,\varphi\right]. \tag{53}$$

As such, Equation (50) has been reduced to the next one:

$$m\,u'' + \left(m\,\frac{v''}{v} + \left(\frac{3}{4} + \lambda\sin\varphi\right)\frac{v'}{v} + m + \lambda\cos\varphi + \frac{3}{4}\,\lambda^2\sin(2\varphi)\right) u = 0. \tag{54}$$

Since

$$\frac{\mathrm{d}}{\mathrm{d}\,\varphi}\left(\frac{v'}{v}\right) = \frac{v''}{v} - \left(\frac{v'}{v}\right)^2, \tag{55}$$

then it is clear that

$$\begin{aligned} \frac{v''}{v} &= \left(\frac{v'}{v}\right)^2 + \frac{\mathrm{d}}{\mathrm{d}\,\varphi}\left(\frac{v'}{v}\right) \\ &= \left[-\frac{1}{2m}\left(\frac{3}{4} + \lambda\,\sin\varphi\right)\right]^2 - \frac{\lambda}{2m}\,\cos\varphi. \end{aligned}$$

Hence, Equation (54) can be rewritten as $u'' = -q(\varphi)\,u$, where

$$\begin{aligned} q(\varphi) &= 1 + \frac{\lambda}{2m}\,\cos\varphi + \frac{3}{4m}\,\lambda^2\,\sin(2\varphi) - \frac{1}{4m^2}\left(\frac{3}{4} + \lambda\,\sin\varphi\right)^2 \\ &= 1 + \frac{\lambda}{2m}\,\cos\varphi + \frac{3}{4m}\,\lambda^2\,\sin(2\varphi) - \frac{9}{64}\,m^2 - \frac{\lambda^2}{4m^2}\,\sin^2\varphi - \frac{3\lambda}{8m^2}\,\sin\varphi. \end{aligned}$$

Firstly, notice that we can write

$$\frac{3\lambda}{8m^2}\,\sin\varphi - \frac{\lambda}{2m}\,\cos\varphi = A\,\sin(\varphi + \varphi_1),$$

where $A = \frac{\lambda}{2m}\sqrt{1 + \left(\frac{3}{4m}\right)^2}$ and $\varphi_1 = \arctan(-\frac{4}{3}\,m)$. In fact, just apply Theorem 2 for $\alpha = \frac{3\lambda}{8m^2} > 0$ and $\beta = -\frac{\lambda}{2m}$.

On the other hand, we also affirm that

$$\frac{\lambda^2}{4m^2}\,\sin^2\varphi - \frac{3}{4m}\,\lambda^2\,\sin(2\varphi) = B\,\sin(2\varphi + \varphi_2),$$

where $B = -\frac{\lambda^2}{4m}\sqrt{9 + \frac{1}{4m^2}}$ and $\varphi_2 = \arctan\left(\frac{1}{6m}\right)$. In this case, it has been used that $\sin^2\varphi = \frac{1}{2}\,(1 - \cos(2\varphi))$, and applied Theorem 2 again for $\alpha = -\frac{3}{4m}\,\lambda^2$ and $\beta = -\frac{\lambda^2}{8m^2}$. Following the above, it holds that Equation (54) is equivalent to the next one, that we shall name as la Cierva's reduced equation, hereafter:

$$u'' = (a + b\,\sin(\varphi + \varphi_1) + c\,\sin(2\varphi + \varphi_2))\,u, \tag{56}$$

where

$$\begin{aligned} a &= \frac{9}{64\,m^2} + \frac{\lambda^2}{8\,m^2} - 1, \quad b = \frac{\lambda}{2m}\sqrt{1 + \left(\frac{3}{4\,m}\right)^2}, \quad c = -\frac{\lambda^2}{4m}\sqrt{9 + \frac{1}{4\,m^2}} \\ \varphi_1 &= \arctan\left(-\frac{4}{3}\,m\right), \quad \varphi_2 = \arctan\left(\frac{1}{6\,m}\right). \end{aligned} \tag{57}$$

Going beyond, it is possible to turn la Cierva's reduced equation into a Riccati type one. In fact, similarly to Equation (55), we have that

$$\frac{u''}{u} = \frac{\mathrm{d}}{\mathrm{d}\,\varphi}\left(\frac{u'}{u}\right) + \left(\frac{u'}{u}\right)^2 = \frac{\mathrm{d}\,\eta}{\mathrm{d}\,\varphi} + \eta^2,$$

where the second equality has been denoted $\eta := \frac{u'}{u}$. Hence, Equation (56) can be even rewritten in terms of a first order Ricatti type equation:

$$\eta' = a + b\,\sin(\varphi + \varphi_1) + c\,\sin(2\varphi + \varphi_2) - \eta^2, \tag{58}$$

where the coefficients a, b, c appear in Equation (57). In this regard, in [7], Puig-Adam realized that a particular solution to la Cierva's equation had been obtained previously by Prof. Aracil in [8]. Despite the form of that particular solution was similar to the one provided in Equation (58) (c.f. Equation (24)), it is worth pointing out that it was obtained for the choice of parameters $\lambda = 0.7249$ and $m = 0.0914 \notin [0.15, 1]$, the range proposed by Mr. la Cierva.

As with the numerical analysis carried out in Section 5 regarding la Cierva's equation, next we shall apply the midpoint method approach to a pair of (independent) particular solutions of the reduced la Cierva's equation (c.f. Equation (56)), namely $u_1(\varphi)$ and $u_2(\varphi)$, with initial conditions $u_1(0) = 1, u_1'(0) = 0$, and $u_2(0) = 0, u_2'(0) = 1$. Also, the same parameters as in Section 5 will be used, i.e., $\lambda = 1$ and $m = 0.5$, and both solutions will be numerically approximated in the subinterval $[0, 2\pi]$ (a turn of the autogiro's blade). In this case, the values of the coefficients and angles in Equation (57) are as follows: $a \simeq 0.0625, b \simeq 80278, c \simeq -1.58114, \varphi_1 \simeq -0.588003$, and $\varphi_2 \simeq 0.321751$. Figure 5 depicts both particular solutions.

Our next goal is to compare the particular solutions (obtained by the midpoint method) of la Cierva's equation (c.f. Figure 2) to the ones of the reduced la Cierva's one. Since $\{u_1(\varphi), u_2(\varphi)\}$ is a fundamental system of solutions of the reduced la Cierva's equation, then $\{u_1(\varphi)\,v(\varphi), u_2(\varphi)\,v(\varphi)\}$ is a fundamental system of solutions of la Cierva's equation, where $v(\varphi) = \exp\left(\frac{\lambda}{2m}\cos\varphi - \frac{3}{8m}\,\varphi\right)$ (c.f. Equation (53)). Hence, each solution of la Cierva's equation, $y(\varphi)$, can be expressed in the following terms:

$$\begin{aligned} y(\varphi) &= v(\varphi)\,\left(C\,u_1(\varphi) + D\,u_2(\varphi)\right) \\ &= e^{\cos\varphi - \frac{3}{4}\varphi}\,\left(C\,u_1(\varphi) + D\,u_2(\varphi)\right) : C, D \in \mathbb{R}, \end{aligned} \tag{59}$$

where the last identity has been used that $\lambda = 1$ and $m = 0.5$. Also, it is clear that

$$y'(\varphi) = e^{\cos\varphi - \frac{3}{4}\varphi}\left[Cu_1' + Du_2' - \left(\frac{3}{4} + \sin\varphi\right)\,(Cu_1 + Du_2)\right]. \tag{60}$$

Since $y_1(0) = 1$, then Equation (59) leads to $C = \frac{1}{e}$. Moreover, the condition $y_1'(0) = 0$ applied to Equation (60) implies that $D = \frac{3}{4e}$. As such, we have

$$y_1(\varphi) = e^{\cos\varphi - \frac{3}{4}\varphi - 1}\left(u_1(\varphi) + \frac{3}{4}u_2(\varphi)\right). \tag{61}$$

On the other hand, the initial condition $y_2(0) = 0$ applied to Equation (59) gives $C = 0$. Furthermore, $y_2'(0) = 1$ implies that $D = \frac{1}{e}$. Thus,

$$y_2(\varphi) = e^{\cos\varphi - \frac{3}{4}\varphi - 1}\,u_2(\varphi). \tag{62}$$

Upcoming Figure 6 displays a graphical comparison involving the particular solutions of la Cierva's equation from both Sections 5 and 6. Specifically, the first particular solutions of that equation are depicted in blue (the dotted line corresponds to the expression in Equation (61)), and the second particular solutions appear in orange (the dashed line corresponds to the expression in Equation (62)). Observe that all the curves behave similarly, especially for angles $\varphi \geq \frac{\pi}{2}$.

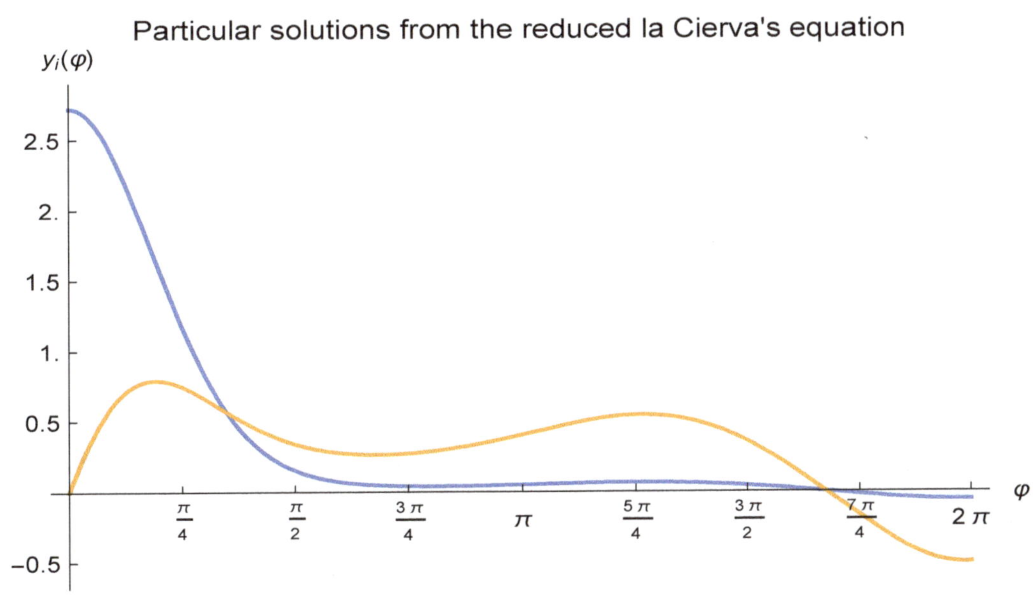

Figure 5. Second order Runge-Kutta approximations (obtained by the explicit midpoint method) to each particular solution from the reduced la Cierva's equation, where the first particular solution, $y_1(\varphi)$ (c.f. Equation (61)), is depicted by a blue line, φ varies in the range $[0, 2\pi]$, and the choice of parameters was as suggested by Mr. la Cierva, i.e., $m = 0.5$ and $\lambda = 1$.

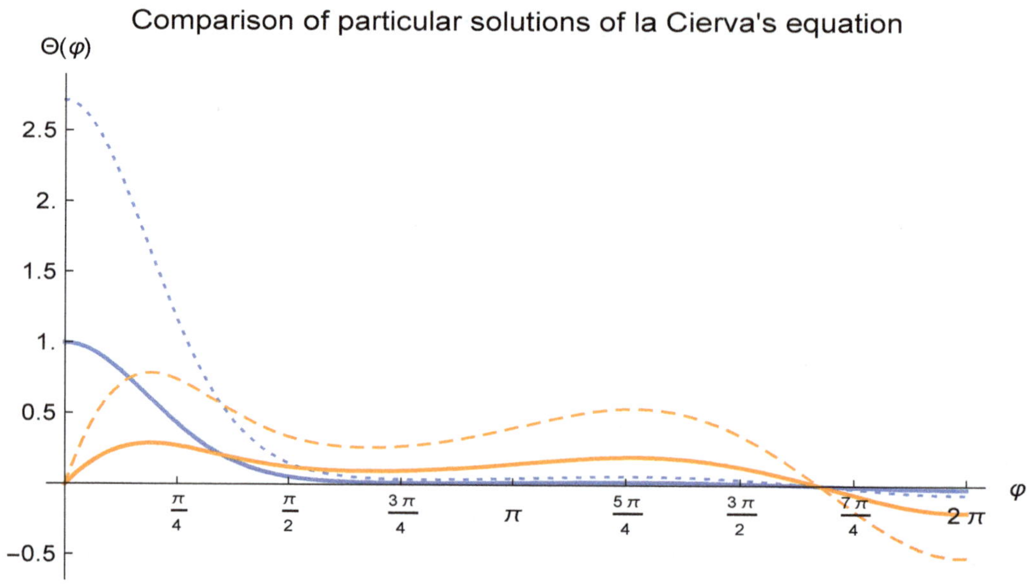

Figure 6. Second order Runge-Kutta approximations (obtained by the explicit midpoint method) to those pairs of particular solutions of la Cierva's equation that were obtained in Sections 5 and 6, respectively. The blue dotted line depicts the first particular solution of that equation as it appears in Equation (61), whereas the orange dashed line illustrates the second particular solution of la Cierva's equation (c.f. Equation (62)). On the other hand, the continuous curves correspond to the particular solutions of la Cierva's equation as they were obtained in Section 5. As with Figure 5, φ varies in the range $[0, 2\pi]$, which means a turn of the blade of the autogiro, and the choice of parameters has been as suggested by Mr. la Cierva, i.e., $m = 0.5$ and $\lambda = 1$. Notice that the y-axis has been labeled as $\Theta(\varphi)$ to denote an approximation to each particular solution of la Cierva's equation for $\varphi \in [0, 2\pi]$.

7. Final Remarks

Next, we provide some additional remarks allowing us to complete our study on the stability of la Cierva's autogiro.

1. We recall that the conditions provided in Section 4 to guarantee the existence of convergent solutions for la Cierva's equation are sufficient but not necessary. In fact, let us consider the reduced la Cierva's equation (c.f. Equation (56)), and define

$$q(\varphi) = a + b\,\sin(\varphi + \varphi_1) + c\,\sin(2\varphi + \varphi_2), \tag{63}$$

where the coefficients a, b, and c are given as in Equation (57). Then for $\lambda = 1$ and $m = 0.5$, i.e., the choice of parameters used in both Sections 5 and 6, it holds that the function $q(\varphi)$ is not positive in the whole interval $[0, 2\pi]$ (c.f., e.g., Figure 7). As such, the Liapounov's condition (c.f. Theorem 3) cannot guarantee the existence of convergent solutions in regard to the reduced la Cierva's equation for that choice of parameters. However, as proved in Section 5, la Cierva's equation behaves stably for such parameters.

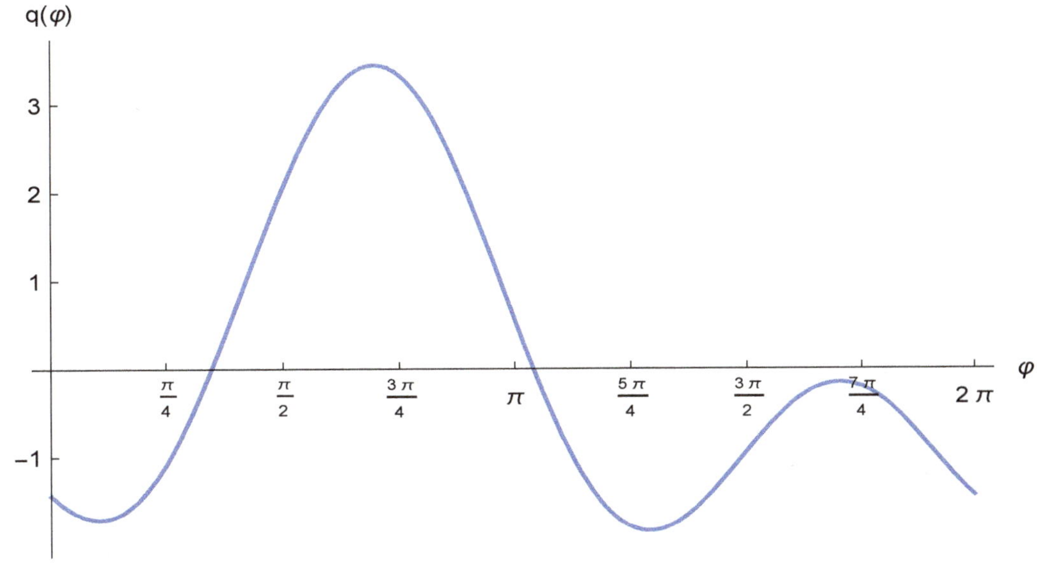

Figure 7. Graph of the function $q(\varphi)$ (as defined in Equation (63)) in the interval $[0, 2\pi]$.

2. Let

$$y''(x) + p_2(x)\,y(x) = 0 \tag{64}$$

be a second order differential equation with $p_1(x) = 0$, as it is the case of the la Cierva's reduced equation (c.f. Equation (56)). Then its associated characteristic equation can be expressed in the following terms (c.f. Equation (27)):

$$s^2 - As + 1 = 0, \tag{65}$$

where the roots of Equation (65) are of the form

$$s_1 = e^{2\pi\,\alpha} \qquad s_2 = e^{-2\pi\,\alpha}.$$

Hence, it is clear that $A = s_1 + s_2 = 2\,\cosh(2\pi\alpha)$, which leads to

$$\alpha = \frac{1}{2\pi}\,\text{arcosh}\left(\frac{A}{2}\right).$$

Let $\Theta = uv$, where $u = e^{\pm\alpha x}$ and v being as in Equation (53). Then the aperiodic part of Θ is given by the next expression:

$$\exp\left[\left(-\frac{3}{8m} \pm \alpha\right)\,x\right]$$

Since $\alpha > 0$, then it is clear that

$$-\frac{3}{8m} - \alpha < \alpha - \frac{3}{8m}.$$

As such, $\alpha - \frac{3}{8m} < 0$ implies $-\frac{3}{8m} - \alpha < 0$. Observe that the stability condition consists of $\alpha - \frac{3}{8m} < 0$, which is satisfied whether $A < 2\,\cosh(\frac{3\pi}{4m})$. On the other hand, the condition $A < 2$ is fulfilled provided that the characteristic exponent $\alpha \in i\,\mathbb{R}$. In that case, the aperiodic part of Θ is of the form $\exp\left(-\frac{3x}{8m}\right)$, which goes to 0 as $x \to +\infty$.
Notice that A could be approximated by the quantity $u_1(2\pi) + u_2'(2\pi)$ through the midpoint approach, for instance, as carried out in both Sections 5 and 6.

3. In Section 4, it was provided a method, first proposed by Liapounov in [10], which allows calculating the coefficient A that appears in characteristic equations of the form Equation (65) that are associated to the next kind of differential equations (c.f. Equation (64)):

$$\frac{\mathrm{d}^2\, y(x)}{\mathrm{d}\, x^2} = \varepsilon\, p(x)\, y(x).$$

In fact, it holds that

$$A = 1 + \frac{1}{2} \sum_{n=1}^{+\infty} \left[F_n(\omega) + f_n'\omega)\right] \varepsilon^n, \tag{66}$$

where $\varepsilon \in (0,1)$, and $F_n(\omega)$ and $f_n(\omega)$ being as in Equation (28). On the other hand, in [11], Goursat applied that method for $\varepsilon = 1$, thus leading to the expressions contained in Equation (28). However, even under the assumption that the series in Equation (66) is convergent, it holds that such a convergence would be quite slow, especially as the period ω increases. As a consequence, that particular expression becomes quite limited to deal with practical applications regarding the calculation of the coefficient A.

4. The reader may think, at least at a first glance, that the form of the reduced la Cierva's equation is similar to the one of the generalized Hill's type equation, whose origins go back to the study of the movement of the Moon under the influence of the gravitational field of the system Earth-Sun. That equation admits the following expression:

$$\frac{\mathrm{d}^2\, y(x)}{\mathrm{d}\, x^2} + [\lambda + \gamma\, \Phi(x)]\; y(x) = 0 : \lambda, \gamma \in \mathbb{R}.$$

However, notice that the parameters at the reduced la Cierva's equation, λ and m, do not appear linearly in Equation (56) (c.f. Equation (57)). As such, the reduced la Cierva's equation cannot be understood as a particular case of the generalized Hill's equation.

5. The stability of la Cierva's autogiro has been proved for the choice of parameters $\lambda = 1$ and $m = 0.5$ (c.f. Sections 5 and 6). Going beyond, observe that the roots (i.e., the characteristic numbers) of the characteristic equation (c.f. Equation (42)) are continuous functions of their coefficients, which, in turn, are continuous functions of both parameters, λ and m. Hence, the stability of la Cierva's equation will be preserved in a neighborhood of such parameters due to ([10] (Theorem, pp. 400)). Moreover, that neighborhood is expected to be wide enough since it is evident that la Cierva's equation behaves stably for those parameters. It is also worth pointing out that if the movement of la Cierva's autogiro is stable for a given speed, then it will be also stable for lower speeds. In other words, the stability will be preserved by decreasing the value of λ. This is a reason for which $\lambda = 1$ was selected to explore the stability of la Cierva's equation in the previous sections. In fact, observe that for $\lambda = 0$, the oscillations are dampened quickly.

6. In [7], Puig-Adam posed to analyze the area of the plane $\lambda - m$ where la Cierva's equation becomes stable. To deal with, we considered the rectangle of the Euclidean plane, $R = [0,1] \times [0.15,1]$, by taking into account the intervals proposed by Mr. la Cierva for each parameter. A partition consisting of 50 points was considered for each subinterval, thus leading to a 2500−point mesh contained in R. As such, for each $(\lambda, m) \in R$, a la Cierva's type equation (c.f. Equation (49)) holds, which was numerically solved as in Section 5 by means of the midpoint approach. Next step was to apply the Puig-Adam criterion to determine whether that equation is stable. Recall that such a condition consists of calculating $|y_1(2\pi) + y_2'(2\pi)|$, where y_1 and y_2 denote the particular solutions of the corresponding la Cierva's equation for a choice of parameters. If $k_{\lambda,m} := |y_1(2\pi) + y_2'(2\pi)| < 1$, then the la Cierva's equation is stable for those

parameters. All the above allowed us to construct a 3D-surface, $S = \{(\lambda, m, k_{\lambda,m}) : (\lambda, m) \in R\}$, we shall refer to as la Cierva's surface. Figure 8 depicts la Cierva's surface, whereas Figure 9 displays the contours of la Cierva's surface. Such figures reveal an overall stable behavior of almost all la Cierva's surface. On the other hand, Figures 10 and 11 depict a neighborhood of Puig-Adam's choice of parameters where la Cierva's surface behaves stably, as stated in remark (5).

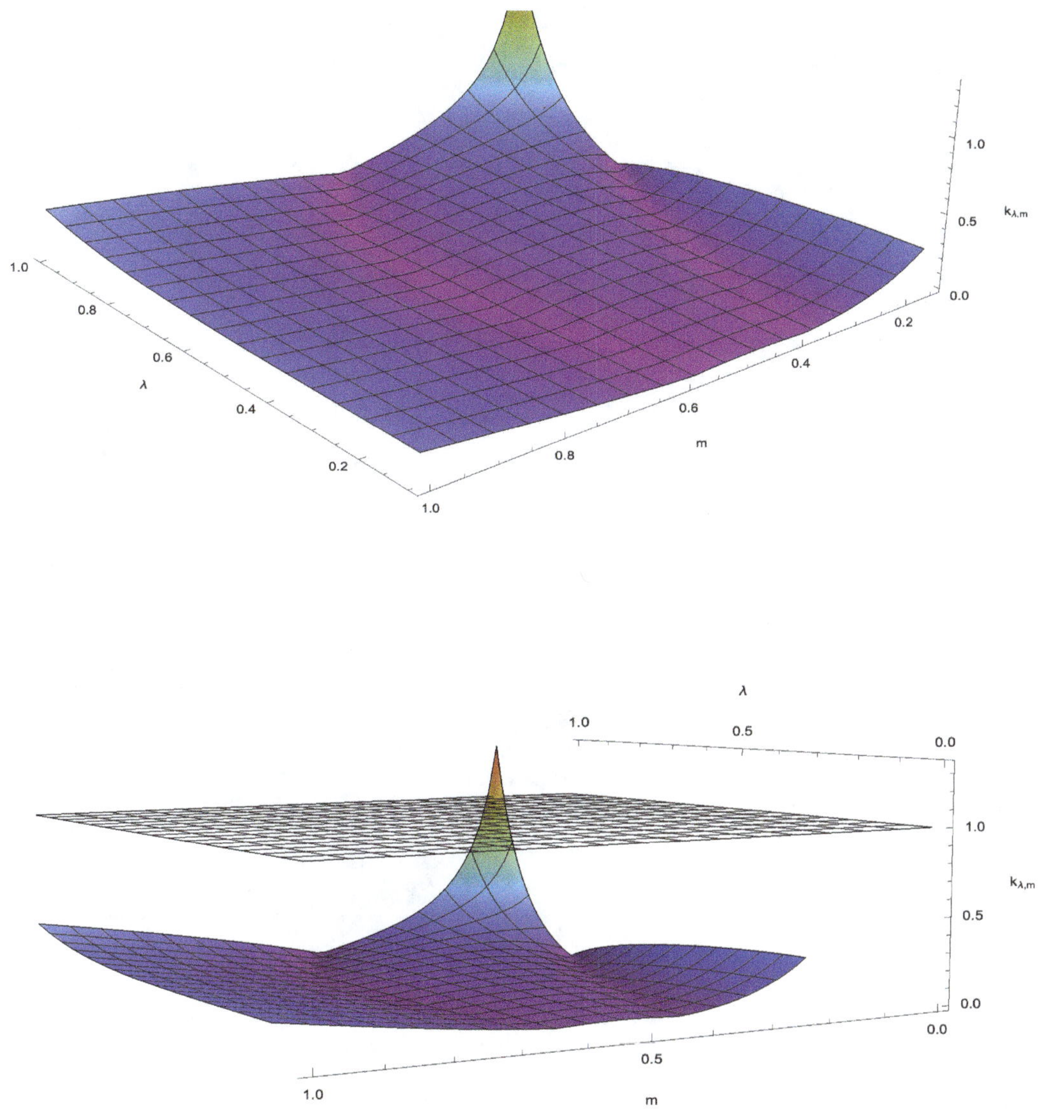

Figure 8. La Cierva's surface, $S = \{(\lambda, m, k_{\lambda,m}) : (\lambda, m) \in R\}$, where $R = [0,1] \times [0.15, 1]$ (above). The plane $\{(\lambda, m, 1) : (\lambda, m) \in R\}$ has been graphically displayed as a benchmark regarding the limit of the stability zone for la Cierva's surface (below).

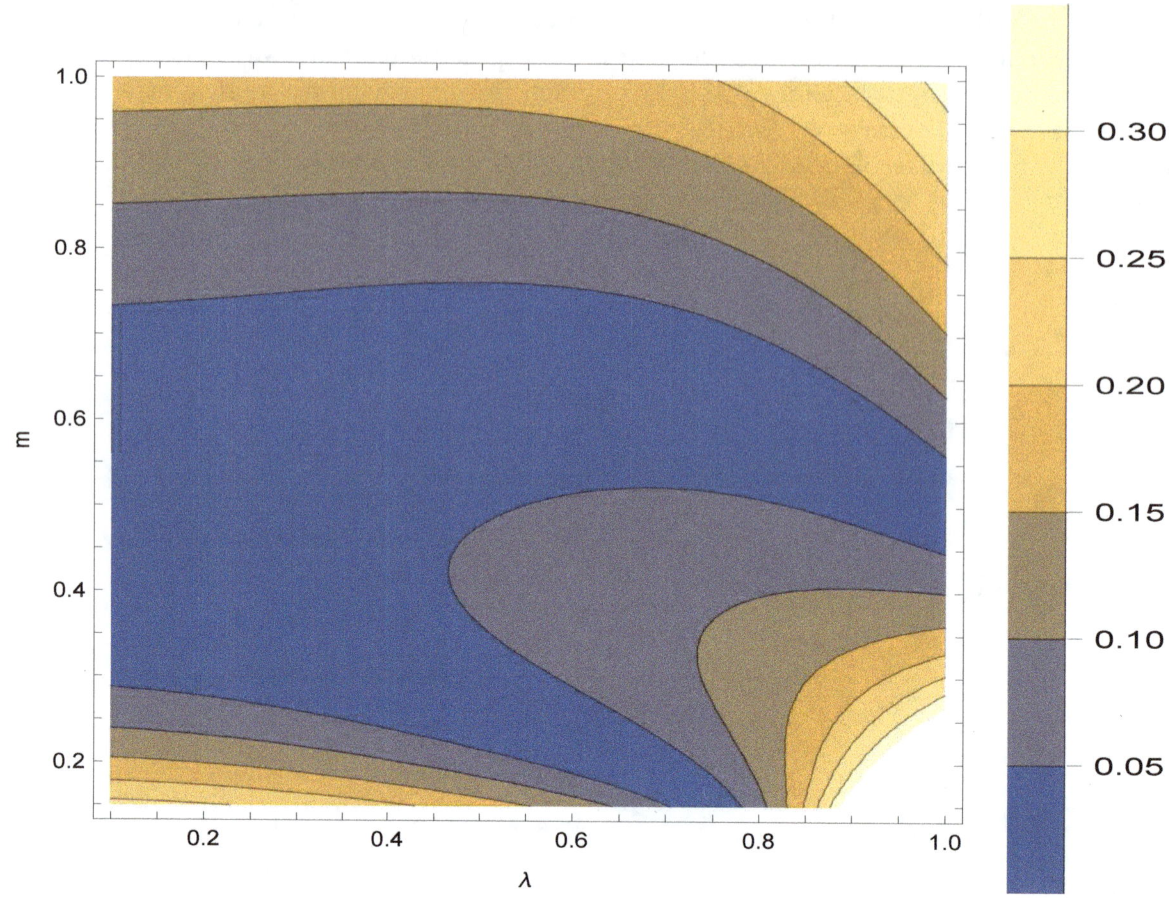

Figure 9. Contours of la Cierva's surface. Observe that the Puig-Adam's choice of parameters, $\lambda = 1, m = 0.5$ is indeed surrounded by a region of points with low $k_{\lambda,m}$ numbers. Notice that almost all the whole surface behaves stably.

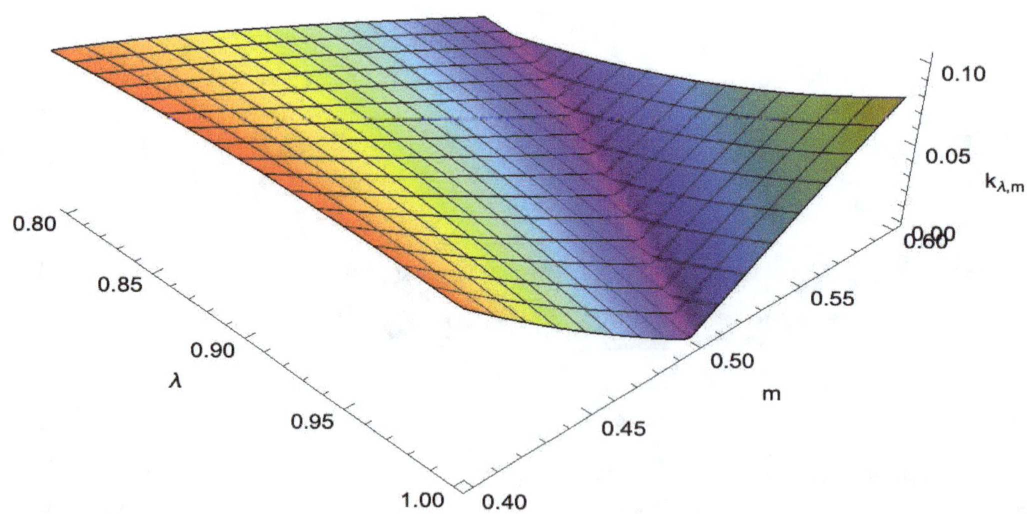

Figure 10. A neighborhood of the Puig-Adam's choice of parameters where la Cierva's surface is stable.

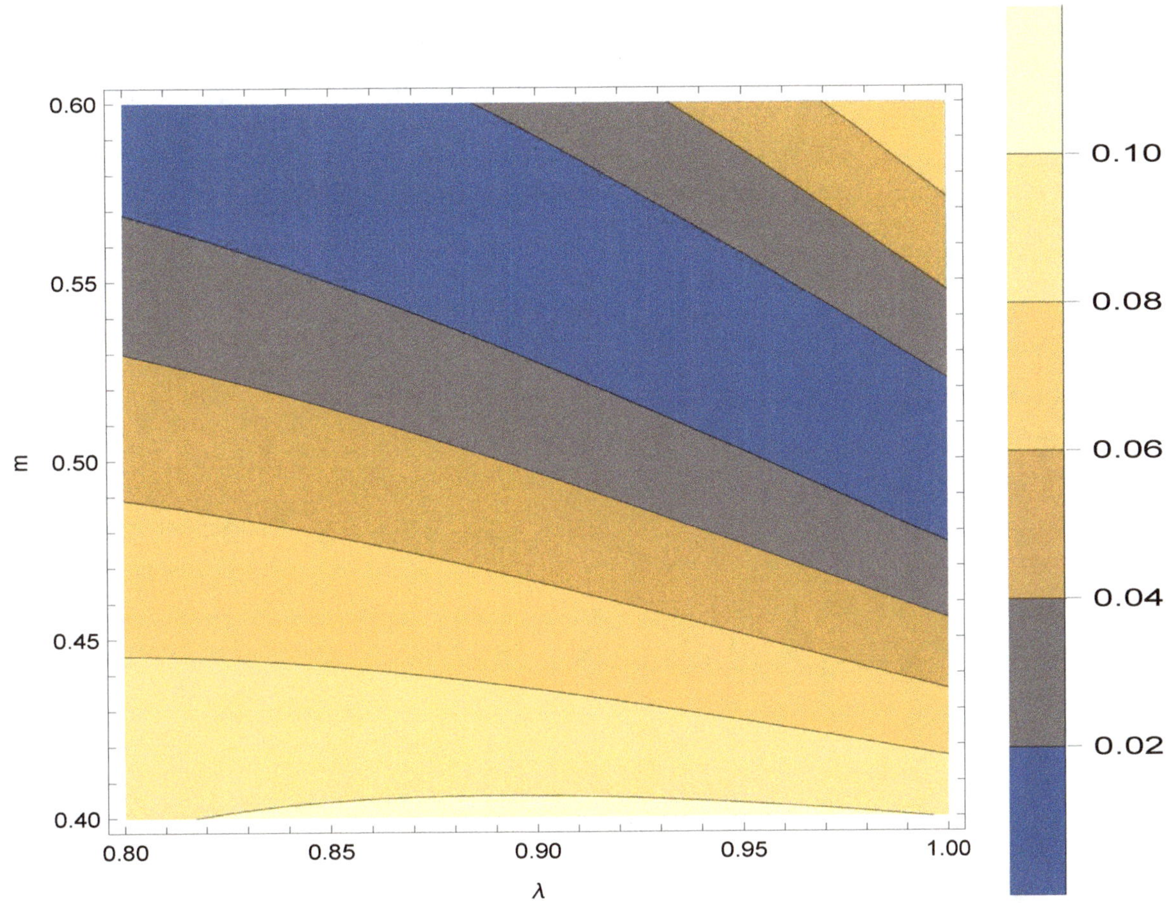

Figure 11. Contours of a neighborhood of the Puig-Adam's choice of parameters where la Cierva's surface behaves stably.

Author Contributions: Conceptualization, M.F.-M. and J.L.G.G.; methodology, M.F.-M. and J.L.G.G.; software, M.F.-M. and J.L.G.G.; validation, M.F.-M. and J.L.G.G.; formal analysis, M.F.-M. and J.L.G.G.; investigation, M.F.-M. and J.L.G.G.; resources, M.F.-M. and J.L.G.G.; data curation, M.F.-M. and J.L.G.G.; writing–original draft preparation, M.F.-M. and J.L.G.G.; writing–review and editing, M.F.-M. and J.L.G.G.; visualization, M.F.-M. and J.L.G.G.; supervision, M.F.-M. and J.L.G.G.; project administration, M.F.-M. and J.L.G.G.; funding acquisition, M.F.-M. and J.L.G.G. All authors have read and agreed to the published version of the manuscript.

Acknowledgments: The authors would also like to express their gratitude to the anonymous reviewers whose suggestions, comments, and remarks have allowed them to enhance the quality of this paper. M.F.M. would like to dedicate this work to the memory of Susi, who passed away while writing this article. The authors sincerely appreciate some insightful comments that were made to this manuscript by Prof. Tareq Saeed from King Abdulaziz University at Saudi Arabia.

References

1. Johnson, W. *Helicopter Theory*; Dover Publications, Inc.: Mineola, NY, USA, 1994.
2. List of NASA Aircraft. 2020. Available online: https://en.wikipedia.org/wiki/List_of_NASA_aircraft (accessed on 16 October 2020).
3. Machado, J.T.; Galhano, A.M. A fractional calculus perspective of distributed propeller design. *Commun. Nonlinear Sci. Numer. Simul.* **2018**, *55*, 174–182.
4. Bing, L.; Ning, Y.; YeHong, D. LAV Path Planning by Enhanced Fireworks Algorithm on Prior Knowledge. *Appl. Math. Nonlinear Sci.* **2016**, *1*, 65–78.
5. Zhang, Q. Fully discrete convergence analysis of non-linear hyperbolic equations based on finite element analysis. *Appl. Math. Nonlinear Sci.* **2019**, *4*, 433–444.
6. Herrera, E. *Sin Alas ni Timones*; Madrid Científico: Madrid, Spain, 1934; pp. 51–53.
7. Puig-Adam, P. Sobre la estabilidad del movimiento de las palas del autogiro. *Rev. Aeronaut.* **1934**, *30*, 478–485.

8. Orts y Aracil, J.M. Nota sobre la ecuación diferencial que plantea la estabilidad del autogiro ultrarrápido. *Ibérica* **1934**, *1024*, 295–296.
9. Orts y Aracil, J.M. Nueva contribución al estudio del problema matemático del autogiro ultrarrápido. *Ibérica* **1934**, *1029*, 376–377.
10. Liapounov, A.M. Problème général de la stabilité du mouvement. *Ann. Fac. Sci. Toulouse* **1903**, *2* ,203–474.
11. Goursat, E. Differential Equations. In *A Course in Mathematical Analysis*; Hedrick, E.R., Dunkel, O., Eds; Ginn and Company: Boston, MA, USA, 1905; Volume II.

17

The Analytical Analysis of Time-Fractional Fornberg–Whitham Equations

A. A. Alderremy [1], Hassan Khan [2,*], Rasool Shah [2], Shaban Aly [1,3] and Dumitru Baleanu [4,5]

1 Department of Mathematics, Faculty of Science, King Khalid University, Abha 61413, Saudi Arabia; aaldramy@kku.edu.sa (A.A.A.); shhaly70@yahoo.com (S.A.)

2 Department of Mathematics, Abdul Wali khan University, Mardan 23200, Pakistan; rasoolshah@awkum.edu.pk

3 Department of Mathematics, Faculty of Science, AL-Azhar University, Assiut 71516, Egypt

4 Department of Mathematics, Faculty of Arts and Sciences, Cankaya University, 06530 Ankara, Turkey; dumitru@cankaya.edu.tr

5 Institute of Space Sciences, 077125 Magurele, Romania

* Correspondence: hassanmath@awkum.edu.pk

Abstract: This article is dealing with the analytical solution of Fornberg–Whitham equations in fractional view of Caputo operator. The effective method among the analytical techniques, natural transform decomposition method, is implemented to handle the solutions of the proposed problems. The approximate analytical solutions of nonlinear numerical problems are determined to confirm the validity of the suggested technique. The solution of the fractional-order problems are investigated for the suggested mathematical models. The solutions-graphs are then plotted to understand the effectiveness of fractional-order mathematical modeling over integer-order modeling. It is observed that the derived solutions have a closed resemblance with the actual solutions. Moreover, using fractional-order modeling various dynamics can be analyzed which can provide sophisticated information about physical phenomena. The simple and straight-forward procedure of the suggested technique is the preferable point and thus can be used to solve other nonlinear fractional problems.

Keywords: Adomian decomposition method; Caputo operator; Natural transform; Fornberg–Whitham equations

---

1. Introduction

It is well known that in many fields of physics, the studies of non-linear wave problems and their effects are of wide significance. Traveling wave solutions are a significant kind of result for the non-linear partial differential condition and numerous non-linear fractional differential equations (FDEs) have been shown to an assortment of traveling wave results. Although water wave are among the extremely important of all-natural phenomena, they have an extraordinarily rich mathematical structure. Water waves are one of the most complicated fields in wave dynamics, including the study in non-linear, electromagnetic waves in 1 space and 3-time dimensions [1–5]. For illustration, the well-known Korteweg–de Vries equation

$$D_t\mu - 6\mu D_x\mu + \mu D_{xxx}\mu = 0,$$

has a simple solitary-wave solution [6]. Camassa-Holm equation

$$D_t\mu - D_{xxt}\mu + 3\mu D_x\mu = 2D_x\mu D_{xx}\mu + \mu D_{xxx}\mu,$$

a model approximation for symmetric non-linear dispersive waves in shallow water, was suggested by Camassa and Holm [7]. Due to its useful mathematical proprieties, this scenario has attracted much attention during the past decade. It has been found that the Camassa Holm equation includes poles, composite wave, stumpons, and cuspons solutions [8]. The specific Camassa–Holm equation solutions were studied by Vakhnenko and Parkes [9]. In mathematical physics the Fornberg-Whitham (FW) model is a significant mathematical equation. The FWE [10,11] is expressed as

$$D_t\mu - D_{xxt}\mu + D_x\mu = \mu D_{xxx}\mu - \mu D_x\mu + 3D_x\mu D_{xx}\mu,$$

where $\mu(x,t)$ is the fluid velocity, x is the spatial co-ordinate and t is the time. In 1978 Fornberg and Whitham derived a $\mu(x,t) = Ce^{\frac{x}{2}-\frac{4t}{3}}$ peaked solution with an arbitrary constant of C [12]. This algorithm was developed to analyze the breakup of dispersive nonlinear water waves. The FWE has been found to require peakon results as a simulation for limiting wave heights as well as the frequency of wave breaks. In fractional calculus (FC) has gained considerable significance and popularity, primarily because of its well-shown applications in a wide range of apparently disparate areas of engineering and science [13]. Many scholars, such as Singh et al. [14], Merdan et al. [15], Saker et al. [16], Gupta and Singh [17] etc., have therefore researched the fractional extensions of the FW model for the Caputo fractional-order derivative [18].

The existence, uniqueness and stability are the important ingredient to show for any mathematical problems in science and engineering. In this connection Li et al. have determine the existance and unique of the solutions for some nonlinear fractional differential equations [19]. Becani et al. have discussed the theory of existence and uniqueness for some singular PDEs [20]. The generalized theorem of existence and uniqueness for nth order fractional DEs was analyzed by Dannan et al. in [21]. Similarly the stability of solutions for the Fornberg-Whitham equation was investigated by Xiujuan Gao et al. in [22]. Shan et al. have discussed the optimal control of the Fornberg-Whitham equation [23].

Recently, the researchers have taken greater interest in FC, i.e., the study of integrals and derivatives of fractional-order non-integer. Major importance have been demonstrated in the analysis of the FC and its various implementations in the field engineering [24–27]. FDEs are widely utilized to model in a variety of fields of study, including an analysis of fractional random walking, kinetic control schemes theory, signal processing, electrical networks, reaction and diffusion procedure [28,29]. FD provides a splendid method for characterizing the memories and genetic properties of different procedures [30,31].

Over the last few years, FDEs have become the subject of several studies owing to their frequent use in numerous implementations in viscoelasticity, biology, fluid mechanics, physics, dynamical schemes, electrical network, physics, signal and optics process, as they can be modelled by linear and nonlinear FDEs [32–36]. FD offer an outstanding method for explaining the memories and inherited properties of specific materials and processes. Fractional-order integrals and derivatives have proven more effective in formulating such electrical and chemical problems than the standard models. Non-linear FPDEs have many applications in various areas of engineering such as heat and mass transfer, thermodynamics and micro-electro mechanics scheme [37–39].

The technique of natural decomposition (NDM) was initially developed by Rawashdeh and Maitama in 2014 [40–42], to solve ODEs and PDEs that appear in different fields of mathematics. The suggested technique is mixing of the Adomian technique (ADM) and natural transformation. The key benefit of this suggested technique is the potential to integrate two important methods of achieving fast convergent series for PDEs. Many scholars have recently solved different types of fractional-order PDEs, for example heat and wave equations [43], coupled Burger equations [44], hyperbolic telegraph equation [45], Harry Dym equation [46] and diffusion equations [47].

2. Preliminaries

Definition 1. *Let $g \in C_\beta$ and $\beta \geq -1$, then the Riemann–Liouville integral of order γ, $\gamma > 0 >$ is given by [48–50]:*

$$J_t^\gamma g(x,t) = \frac{1}{\Gamma(\gamma)} \int_0^t (t-\theta)^{\gamma-1} g(x,\theta) d\theta, \qquad t > 0. \tag{1}$$

Definition 2. *Let $g \in \mathbb{C}_t$ and $t \geq -1$, then Caputo definition of fractional derivative of order γ if $m-1 < \gamma \leq m$ with $m \in \mathbb{N}$ is describe as [48–50]*

$$D_t^\gamma g(t) = \begin{cases} \frac{d^m g(t)}{dt^m}, & \gamma = m \in N, \\ \frac{1}{\Gamma(m-\gamma)} \int_0^t (t-\theta)^{m-\gamma-1} g^{(m)}(\theta) d\theta, & m-1 < \gamma < m, \quad m \in N, \end{cases} \tag{2}$$

Remark 1. *Some basic properties are below [48–50]*

$$\begin{aligned} D_x^\gamma I_x^\gamma g(x) &= g(x), \\ I^\gamma x^\lambda &= \frac{\Gamma(\lambda+1)}{\Gamma(\gamma+\lambda+1)} x^{\gamma+\lambda}, \qquad \gamma > 0, \lambda > -1, \quad x > 0, \\ D_x^\gamma I_x^\gamma g(x) &= g(x) - \sum_{k=0}^{m} g^{(k)}(0^+) \frac{x^k}{k!}, \qquad \text{for} \quad x > 0. \end{aligned}$$

Definition 3. *The natural transform of the function $g(t)$ is expressed by $N[g(t)]$ for $t \in R$ and is given by [40–42,51]*

$$N[g(t)] = \mathcal{G}(s,\omega) = \int_{-\infty}^{\infty} e^{-st} g(\omega t) dt; \quad s,\omega \in (-\infty,\infty), aa$$

where s and ω are the NT variables. If $g(t)H(t)$ is defined for positive real numbers, then NT can be presented as [40–42,51]

$$N[g(t)Q(t)] = N^+[g(t)] = \mathcal{G}^+(s,\omega) = \int_0^\infty e^{-st} g(\omega t) dt; \quad s,\omega \in (0,\infty), \quad \text{and} \quad t \in R, \tag{3}$$

where $Q(t)$ denotes the Heaviside function.

Theorem 1. *The NT of the Caputo derivative of fractional order of any function $g(t)$ is defined as [40–42,51]*

$$N^+[{}^cD^\gamma g(t)] = \frac{s^\gamma}{\omega^\gamma} \mathcal{G}(s,\omega) - \sum_{k=0}^{m-1} \frac{s^{\gamma-(k+1)}}{\omega^{\gamma-k}} [D^k g(t)]_{t=0}, \quad m-1 \leq \gamma < m. \tag{4}$$

where m is the natural number and γ represent the order of the derivative with fractional order.

Remark 2. *Some basic NT properties are listed below [40–42,51]*

$$\begin{aligned} N^+[1] &= \frac{1}{s}, \\ N^+[t^\gamma] &= \frac{\Gamma(\gamma+1)\omega^\gamma}{s^{\gamma+1}}, \\ N^+[g^{(m)}(t)] &= \frac{s^m}{\omega^m} \mathcal{R}(s,\omega) - \sum_{k=0}^{m-1} \frac{s^{m-(k+1)}}{\omega^{m-k}} \frac{\Gamma(\gamma+1)\omega^\gamma}{s^{\gamma+1}}. \end{aligned}$$

3. NDM Procedure

In this section, NDM procedure is introduced to solve general FPDEs of the form [41,42]

$$D^{\gamma}\mu(x,t) + L\mu(x,t) + N\mu(x,t) = P(x,t), \quad x,t \geq 0, \quad \ell - 1 < \gamma < \ell, \tag{5}$$

The fractional derivative in Equation (5) is represented by Caputo operator. The linear and nonlinear terms are denoted by L and N respectively and the source term is $P(x,t)$.

The solution at $t = 0$ is

$$\mu(x,0) = h(x). \tag{6}$$

Using NT to Equation (5), we get [41,42]

$$N^{+}\left[D^{\gamma}\mu(x,t)\right] + N^{+}\left[L\mu(x,t) + N\mu(x,t)\right] = N^{+}\left[P(x,t)\right], \tag{7}$$

Applying the differential property of NT [41,42]

$$\frac{s^{\gamma}}{\omega^{\gamma}}N^{+}\left[\mu(x,t)\right] - \frac{s^{\gamma-1}}{\omega^{\gamma}}\mu(x,0) = N^{+}\left[P(x,t)\right] - N^{+}\left[L\mu(x,t) + N\mu(x,t)\right],$$

$$N^{+}\left[\mu(x,t)\right] = \frac{1}{s}\mu(x,0) + \frac{\omega^{\gamma}}{s^{\gamma}}N^{+}\left[P(x,t)\right] - \frac{\omega^{\gamma}}{s^{\gamma}}N^{+}\left[L\mu(x,t) + N\mu(x,t)\right].$$

Now $\mu(x,0) = k(x)$,

$$N^{+}\left[\mu(x,t)\right] = \frac{h(x)}{s} + \frac{\omega^{\gamma}}{s^{\gamma}}N^{+}\left[P(x,t)\right] - \frac{\omega^{\gamma}}{s^{\gamma}}N^{+}\left[L\mu(x,t) + N\mu(x,t)\right]. \tag{8}$$

The infinite series of NDM $\mu(x,t)$ is shown by

$$\mu(x,t) = \sum_{\ell=0}^{\infty}\mu_{\ell}(x,t). \tag{9}$$

Adomian polynomial for nonlinear terms is

$$N\mu(x,t) = \sum_{\ell=0}^{\infty}A_{\ell}, \tag{10}$$

$$A_{\ell} = \frac{1}{\ell!}\left[\frac{d^{\ell}}{d\lambda^{\ell}}\left[N\sum_{\ell=0}^{\infty}(\lambda^{\ell}\mu_{\ell})\right]\right]_{\lambda=0}, \qquad \ell = 0,1,2\cdots \tag{11}$$

putting Equations (9) and (11) into Equation (8), we have

$$N^{+}\left[\sum_{\ell=0}^{\infty}\mu_{\ell}(x,t)\right] = \frac{h(x)}{s} + \frac{\omega^{\gamma}}{s^{\gamma}}N^{+}\left[P(x,t)\right] - \frac{\omega^{\gamma}}{s^{\gamma}}N^{+}\left[L\sum_{\ell=0}^{\infty}\mu_{\ell}(x,t) + \sum_{\ell=0}^{\infty}A_{\ell}\right]. \tag{12}$$

$$N^{+}\left[\mu_{0}(x,t)\right] = \frac{h(x)}{s} + \frac{\omega^{\gamma}}{s^{\gamma}}N^{+}\left[P(x,t)\right], \tag{13}$$

$$N^{+}\left[\mu_{1}(x,t)\right] = -\frac{\omega^{\gamma}}{s^{\gamma}}N^{+}\left[L\mu_{0}(x,t) + A_{0}\right].$$

We will usually compose

$$N^{+}\left[\mu_{\ell+1}(x,t)\right] = -\frac{\omega^{\gamma}}{s^{\gamma}}N^{+}\left[L\mu_{\ell}(x,t) + A_{\ell}\right], \quad \ell \geq 1. \tag{14}$$

Using the inverse NT to Equations (13) and (14) [41,42].

$$\mu_0(x,t) = h(x) + N^-\left[\frac{\omega^\gamma}{s^\gamma} N^+\left[P(x,t)\right]\right],$$

$$v_{\ell+1}(x,t) = -N^-\left[\frac{\omega^\gamma}{s^\gamma} N^+\left[L\mu_\ell(x,t) + A_\ell\right]\right]. \tag{15}$$

4. NDM Implementation

Example 1. *The following nonlinear Fornberg-Whitham with fractional derivative is considered [14]*

$$D_t^\gamma \mu - D_{xxt}\mu + D_x\mu = \mu D_{xxx}\mu - \mu D_x\mu + 3D_x\mu D_{xx}\mu, \quad t > 0, \quad 0 < \gamma \leq 1, \tag{16}$$

having initial solution as

$$\mu(x,0) = \exp\left(\frac{x}{2}\right). \tag{17}$$

Applying NT to Equation (16), we have

$$\frac{s^\gamma}{\omega^\gamma} N^+\left[\mu(x,t)\right] - \frac{s^{\gamma-1}}{\omega^\gamma}\mu(x,0) = N^+\left[D_{xxt}\mu - D_x\mu + \mu D_{xxx}\mu - \mu D_x\mu + 3D_x\mu D_{xx}\mu\right].$$

$$N^+\left[\mu(x,t)\right] - \frac{1}{s}\mu(x,0) = \frac{\omega^\gamma}{s^\gamma} N^+\left[D_{xxt}\mu - D_x\mu + \mu D_{xxx}\mu - \mu D_x\mu + 3D_x\mu D_{xx}\mu\right].$$

Using inverse natural transformation

$$\mu(x,t) = N^-\left[\frac{\mu(x,0)}{s} - \frac{\omega^\gamma}{s^\gamma} N^+\left[D_{xxt}\mu - D_x\mu + \mu D_{xxx}\mu - \mu D_x\mu + 3D_x\mu D_{xx}\mu\right]\right].$$

Applying the ADM process, we have

$$\mu_0(x,t) = N^-\left[\frac{\mu(x,0)}{s}\right] = N^-\left[\frac{\exp\left(\frac{x}{2}\right)}{s}\right],$$

$$\mu_0(x,t) = \exp\left(\frac{x}{2}\right), \tag{18}$$

$$\sum_{\ell=0}^{\infty}\mu_{\ell+1}(x,t) = N^-\left[\frac{\omega^\gamma}{s^\gamma} N^+\left[\sum_{\ell=0}^{\infty}(D_{xxt}\mu)_\ell - \sum_{\ell=0}^{\infty}(D_x\mu)_\ell + \sum_{\ell=0}^{\infty}A_\ell - \sum_{\ell=0}^{\infty}B_\ell + 3\sum_{\ell=0}^{\infty}C_\ell\right]\right], \quad \ell = 0,1,2,\cdots$$

$$\begin{aligned}
A_0(\mu D_{xxx}\mu) &= \mu_0 D_{xxx}\mu_0,\\
A_1(\mu D_{xxx}\mu) &= \mu_0 D_{xxx}\mu_1 + \mu_1 D_{xxx}\mu_0,\\
A_2(\mu D_{xxx}\mu) &= \mu_1 D_{xxx}\mu_2 + \mu_1 D_{xxx}\mu_1 + \mu_2 D_{xxx}\mu_0,
\end{aligned}$$

$$\begin{aligned}
B_0(\mu D_x\mu) &= \mu_0 D_x\mu_0,\\
B_1(\mu D_x\mu) &= \mu_0 D_x\mu_1 + \mu_1 D_x\mu_0,\\
B_2(\mu D_x\mu) &= \mu_1 D_x\mu_2 + \mu_1 D_x\mu_1 + \mu_2 D_x\mu_0,
\end{aligned}$$

$$\begin{aligned}
C_0(D_x\mu D_{xx}\mu) &= D_x\mu_0 D_{xx}\mu_0,\\
C_1(D_x\mu D_{xx}\mu) &= D_x\mu_0 D_{xx}\mu_1 + D_x\mu_1 D_{xx}\mu_0,\\
C_2(D_x\mu D_{xx}\mu) &= D_x\mu_1 D_{xx}\mu_2 + D_x\mu_1 D_{xx}\mu_1 + D_x\mu_2 D_{xx}\mu_0,
\end{aligned}$$

for $\ell = 1$

$$\mu_1(x,t) = N^- \left[\frac{\omega^\gamma}{s^\gamma} N^+ \left[(D_{xxt}\mu)_0 - (D_x\mu)_0 + A_0 - B_0 + 3C_0\right]\right],$$
$$\mu_1(x,t) = -\frac{1}{2} N^- \left[\frac{\omega^\gamma \exp(\frac{x}{2})}{s^{\gamma+1}}\right] = -\frac{1}{2}\exp\left(\frac{x}{2}\right)\frac{t^\gamma}{\Gamma(\gamma+1)}. \tag{19}$$

for $\ell = 2$

$$\mu_2(x,t) = N^- \left[\frac{\omega^\gamma}{s^\gamma} N^+ \left[(D_{xxt}\mu)_1 - (D_x\mu)_1 + A_1 - B_1 + 3C_1\right]\right],$$
$$\mu_2(x,t) = -\frac{1}{8}\exp\left(\frac{x}{2}\right)\frac{t^{2\gamma-1}}{\Gamma(2\gamma)} + \frac{1}{4}\exp\left(\frac{x}{2}\right)\frac{t^{2\gamma}}{\Gamma(2\gamma+1)}, \tag{20}$$

for $\ell = 3$

$$\mu_3(x,t) = N^- \left[\frac{\omega^\gamma}{s^\gamma} N^+ \left[(D_{xxt}\mu)_2 - (D_x\mu)_2 + A_2 - B_2 + 3C_2\right]\right],$$
$$\mu_3(x,t) = -\frac{1}{32}\exp\left(\frac{x}{2}\right)\frac{t^{3\gamma-2}}{\Gamma(3\gamma-1)} + \frac{1}{8}\exp\left(\frac{x}{2}\right)\frac{\gamma^{3\gamma-1}}{\Gamma(3\gamma)} - \frac{1}{8}\exp\left(\frac{x}{2}\right)\frac{t^{3\gamma}}{\Gamma(3\gamma+1)}, \tag{21}$$

The NDM solution for problem (16) is

$$\mu(x,t) = \mu_0(x,t) + \mu_1(x,t) + \mu_2(x,t) + \mu_3(x,t) + \mu_4(x,t)\cdots.$$
$$\mu(x,t) = \exp\left(\frac{x}{2}\right) - \frac{1}{2}\exp\left(\frac{x}{2}\right)\frac{t^\gamma}{\Gamma(\gamma+1)}$$
$$-\frac{1}{8}\exp\left(\frac{x}{2}\right)\frac{t^{2\gamma-1}}{\Gamma(2\gamma)} + \frac{1}{4}\exp\left(\frac{x}{2}\right)\frac{t^{2\gamma}}{\Gamma(2\gamma+1)} - \frac{1}{32}\exp\left(\frac{x}{2}\right) \tag{22}$$
$$\frac{t^{3\gamma-2}}{\Gamma(3\gamma-1)} + \frac{1}{8}\exp\left(\frac{x}{2}\right)\frac{\gamma^{3\gamma-1}}{\Gamma(3\gamma)} - \frac{1}{8}\exp\left(\frac{x}{2}\right)\frac{t^{3\gamma}}{\Gamma(3\gamma+1)} - \cdots.$$

The simplification of Equation (22);

$$\mu(x,t) = \exp\left(\frac{x}{2}\right)\left[1 - \frac{t^\gamma}{2\Gamma(\gamma+1)} - \frac{1}{8}\frac{t^{2\gamma-1}}{\Gamma(2\gamma)} + \frac{1}{4}\frac{t^{2\gamma}}{\Gamma(2\gamma+1)} - \frac{1}{32}\frac{t^{3\gamma-2}}{\Gamma(3\gamma-1)} + \frac{1}{8}\frac{t^{3\gamma-1}}{\Gamma(3\gamma)} - \frac{1}{8}\frac{t^{3\gamma}}{\Gamma(3\gamma+1)} + \cdots\right]. \tag{23}$$

The exact result of Example 1

$$\mu(x,t) = \exp\left(\frac{x}{2} - \frac{2t}{3}\right), \tag{24}$$

In Table 1, the NDM-solutions at different fractional-order derivatives, $\gamma = 0.5, 0.7$ and 1 are shown. The NDM-solutions at various time level, $t = 0.2, 0.4$ and $t = 1$ are determined. The absolute error of the proposed method at $\gamma = 1$ is also displayed. From Table 1, it is investigated that suggested method has the desire rate of convergence and considered to be the best tool for the analytical solution of FPDEs. In Table 2, the NDM and LADM solutions are compared at various fractional-order of the derivatives. It is observed that the NDM has the higher degree of accuracy as compared to LDM. The comparison has been done at $\gamma = 0.5, 0.7$ and 0.9. It is also investigated that the fractional-order solutions of NDM have the higher accuracy as compared LDM.

Table 1. The NDM solutions and absolute error of Example 1 at $\gamma = 0.5, 0.7$ and 1.

t	x	$\gamma = 0.50$	$\gamma = 0.7$	NDM ($\gamma = 1$)	Exact	NDM (AE) ($\gamma = 1$)
0.2	0.5	1.168497921	1.229840967	1.266952492	1.267018708	6.620×10^{-5}
	1.0	1.500381030	1.579147061	1.626799201	1.626884224	8.500×10^{-5}
	1.5	1.926527377	2.027664962	2.088851523	2.088960694	1.090×10^{-4}
	2.0	2.473710118	2.603573348	2.682138447	2.682278626	1.400×10^{-4}
0.4	0.5	1.123744786	1.197168588	1.250129766	1.25023725	1.070×10^{-4}
	1.0	1.442916867	1.537194895	1.605198393	1.60533640	1.380×10^{-4}
	1.5	1.852741931	1.973797315	2.061115536	2.06129274	1.770×10^{-4}
	2.0	2.378967730	2.534405920	2.646524735	2.64675227	2.270×10^{-4}
1	0.5	1.039959208	1.124099772	1.201030155	1.20114746	1.84×10^{-4}
	1.0	1.335334055	1.443372678	1.542153245	1.542390265	2.370×10^{-4}
	1.5	1.714602867	1.853327204	1.980163963	1.980468303	3.004×10^{-4}
	2.0	2.201593661	2.379719235	2.542580858	2.542971638	3.900×10^{-4}

Table 2. Two terms comparison of NDM and LDM [16] of different fractional-order at $\gamma = 0.5, 0.7$ and 0.9 of Example 1.

x	t	NDM	LDM	NDM	LDM	NDM	LDM
		$\gamma = 0.5$	$\gamma = 0.5$	$\gamma = 0.7$	$\gamma = 0.7$	$\gamma = 0.9$	$\gamma = 0.9$
0.2	0.1	1.259×10^{-1}	4.021×10^{-1}	5.006×10^{-2}	2.158×10^{-1}	2.833×10^{-3}	4.973×10^{-2}
	0.2	1.408×10^{-1}	4.171×10^{-1}	5.916×10^{-2}	2.249×10^{-1}	2.978×10^{-3}	5.227×10^{-2}
	0.3	1.411×10^{-1}	4.174×10^{-1}	6.148×10^{-2}	2.272×10^{-1}	3.131×10^{-3}	5.496×10^{-2}
	0.4	1.356×10^{-1}	4.119×10^{-1}	6.153×10^{-2}	2.273×10^{-1}	3.291×10^{-3}	5.777×10^{-2}
	0.5	1.276×10^{-1}	4.039×10^{-1}	6.107×10^{-2}	2.268×10^{-1}	3.460×10^{-3}	6.074×10^{-2}
	0.6	1.186×10^{-1}	3.949×10^{-1}	6.096×10^{-2}	2.267×10^{-1}	3.638×10^{-3}	6.385×10^{-2}
	0.7	1.095×10^{-1}	3.858×10^{-1}	6.164×10^{-2}	2.274×10^{-1}	3.824×10^{-3}	6.712×10^{-2}
	0.8	1.007×10^{-1}	3.771×10^{-1}	6.337×10^{-2}	2.291×10^{-1}	4.020×10^{-3}	7.057×10^{-2}
	0.9	9.288×10^{-2}	3.691×10^{-1}	6.626×10^{-2}	2.320×10^{-1}	4.226×10^{-3}	7.418×10^{-2}
	1.0	8.576×10^{-2}	3.622×10^{-1}	7.038×10^{-2}	2.361×10^{-1}	4.443×10^{-3}	7.799×10^{-2}

In Figures 1 and 2, the NDM and actual solution of Example 1 are plotted. It is observed that NDM solutions are in closed contact with the exact solutions of Example 1. In Figures 3 and 4, the solutions of Example 1 at various fractional-order of the derivatives are plotted. The graphical representation has shown the convergence phenomena of fractional-order solution towards the solution at integer order of Example 1.

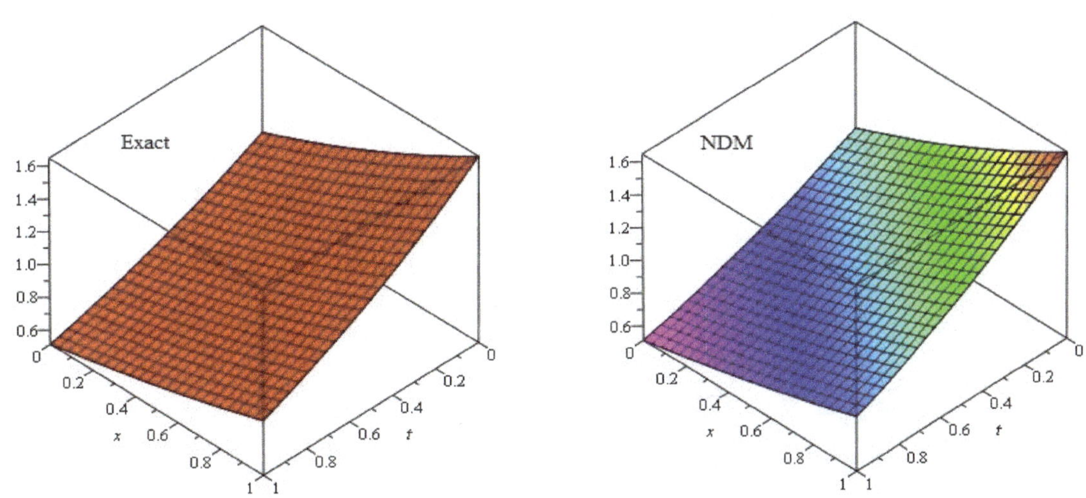

Figure 1. Exact and NDM solutions at $\gamma = 1$ of Example 1.

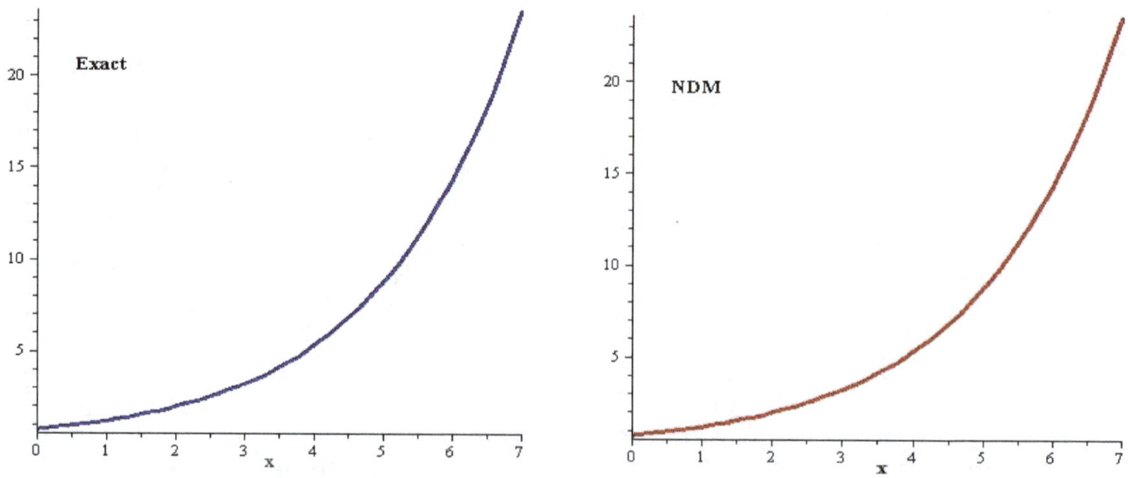

Figure 2. Exact and NDM solutions at $\gamma = 1$ of Example 1.

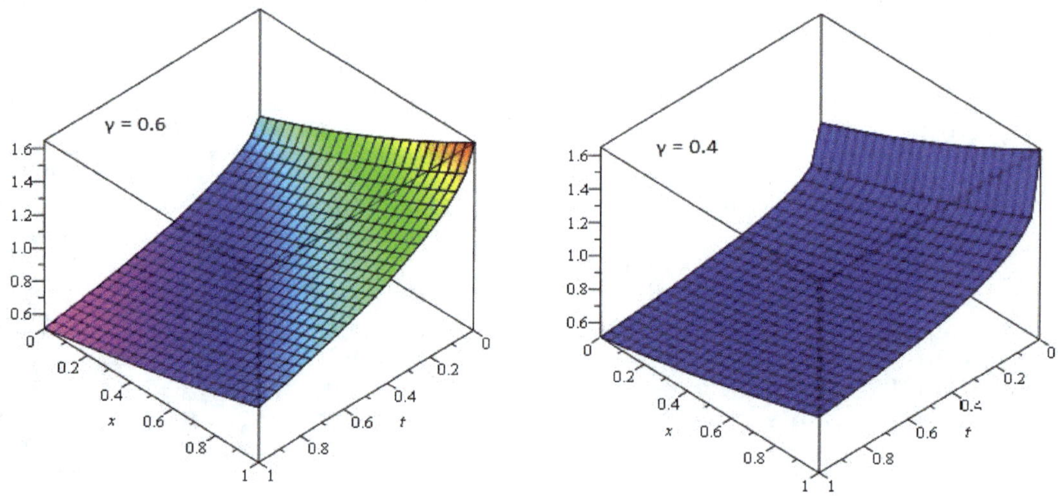

Figure 3. The NDM solutions of different valve of γ of Example 1.

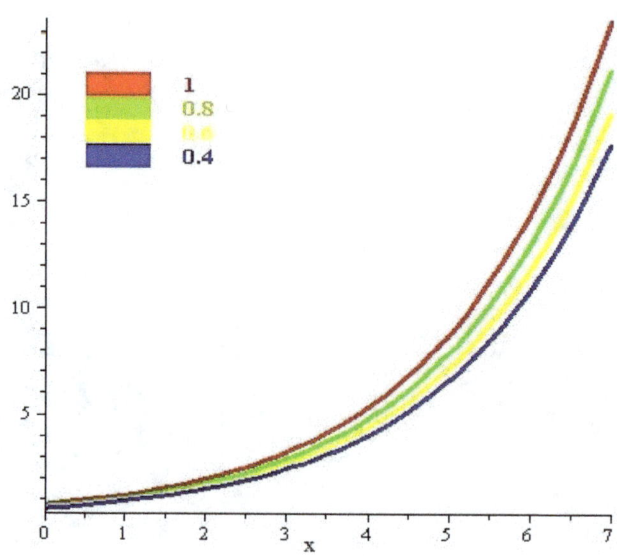

Figure 4. Solution graph of Example 1, at various value of γ.

Example 2. *Consider the following nonlinear time-fractional Fornberg–Whitham equation [18]*

$$D_t^{\gamma}\mu - D_{xxt}\mu + D_x\mu = \mu D_{xxx}\mu - \mu D_x\mu + 3D_x\mu D_{xx}\mu, \quad t>0, \quad 0<\gamma\leq 1, \tag{25}$$

with initial condition

$$\mu(x,0) = \cosh^2\left(\frac{x}{4}\right), \tag{26}$$

Applying natural transformation of Equation (25),

$$\frac{s^{\gamma}}{\omega^{\gamma}}N^+\left[\mu(x,t)\right] - \frac{s^{\gamma-1}}{\omega^{\gamma}}\mu(x,0) = N^+\left[D_{xxt}\mu - D_x\mu + \mu D_{xxx}\mu - \mu D_x\mu + 3D_x\mu D_{xx}\mu\right].$$

$$N^+\left[\mu(x,t)\right] - \frac{1}{s}\mu(x,0) = \frac{\omega^{\gamma}}{s^{\gamma}}N^+\left[D_{xxt}\mu - D_x\mu + \mu D_{xxx}\mu - \mu D_x\mu + 3D_x\mu D_{xx}\mu\right].$$

Using inverse natural transformation

$$\mu(x,t) = N^-\left[\frac{\mu(x,0)}{s} - \frac{\omega^{\gamma}}{s^{\gamma}}N^+\left[D_{xxt}\mu - D_x\mu + \mu D_{xxx}\mu - \mu D_x\mu + 3D_x\mu D_{xx}\mu\right]\right].$$

Applying the ADM process, we have

$$\mu_0(x,t) = N^-\left[\frac{\mu(x,0)}{s}\right] = N^-\left[\frac{\cosh^2\left(\frac{x}{4}\right)}{s}\right],$$

$$\mu_0(x,t) = \cosh^2\left(\frac{x}{4}\right), \tag{27}$$

$$\sum_{\ell=0}^{\infty}\mu_{\ell+1}(x,t) = N^-\left[\frac{\omega^{\gamma}}{s^{\gamma}}N^+\left[\sum_{\ell=0}^{\infty}(D_{xxt}\mu)_{\ell} - \sum_{\ell=0}^{\infty}(D_x\mu)_{\ell} + \sum_{\ell=0}^{\infty}A_{\ell} - \sum_{\ell=0}^{\infty}B_{\ell} + 3\sum_{\ell=0}^{\infty}C_{\ell}\right]\right], \quad \ell = 0,1,2,\cdots$$

$$\begin{aligned}
A_0(\mu D_{xxx}\mu) &= \mu_0 D_{xxx}\mu_0,\\
A_1(\mu D_{xxx}\mu) &= \mu_0 D_{xxx}\mu_1 + \mu_1 D_{xxx}\mu_0,\\
A_2(\mu D_{xxx}\mu) &= \mu_1 D_{xxx}\mu_2 + \mu_1 D_{xxx}\mu_1 + \mu_2 D_{xxx}\mu_0,
\end{aligned}$$

$$\begin{aligned}
B_0(\mu D_x\mu) &= \mu_0 D_x\mu_0,\\
B_1(\mu D_x\mu) &= \mu_0 D_x\mu_1 + \mu_1 D_x\mu_0,\\
B_2(\mu D_x\mu) &= \mu_1 D_x\mu_2 + \mu_1 D_x\mu_1 + \mu_2 D_x\mu_0.
\end{aligned}$$

$$\begin{aligned}
C_0(D_x\mu D_{xx}\mu) &= D_x\mu_0 D_{xx}\mu_0,\\
C_1(D_x\mu D_{xx}\mu) &= D_x\mu_0 D_{xx}\mu_1 + D_x\mu_1 D_{xx}\mu_0,\\
C_2(D_x\mu D_{xx}\mu) &= D_x\mu_1 D_{xx}\mu_2 + D_x\mu_1 D_{xx}\mu_1 + D_x\mu_2 D_{xx}\mu_0,
\end{aligned}$$

for $\ell = 1$

$$\begin{aligned}
\mu_1(x,t) &= N^-\left[\frac{\omega^{\gamma}}{s^{\gamma}}N^+\left[(D_{xxt}\mu)_0 - (D_x\mu)_0 + A_0 - B_0 + 3C_0\right]\right],\\
\mu_1(x,t) &= -\frac{11}{32}N^-\left[\frac{\omega^{\gamma}\sinh\left(\frac{x}{2}\right)}{s^{\gamma+1}}\right] = -0.3437\sinh\left(\frac{x}{4}\right)\frac{t^{\gamma}}{\Gamma(\gamma+1)},
\end{aligned} \tag{28}$$

for $\ell = 2$

$$\begin{aligned} \mu_2(x,t) &= N^-\left[\frac{\omega^\gamma}{s^\gamma}N^+\left[(D_{xxt}\mu)_1-(D_x\mu)_1+A_1-B_1+3C_1\right]\right], \\ \mu_2(x,t) &= -0.08593\sinh\left(\frac{x}{4}\right)\frac{t^\gamma}{\Gamma(\gamma+1)}+0.11816\cosh\left(\frac{x}{4}\right)\frac{t^{2\gamma}}{\Gamma(2\gamma+1)}, \end{aligned} \tag{29}$$

for $\ell = 3$

$$\begin{aligned} \mu_3(x,t) &= N^-\left[\frac{\omega^\gamma}{s^\gamma}N^+\left[(D_{xxt}\mu)_2-(D_x\mu)_2+A_2-B_2+3C_2\right]\right], \\ \mu_3(x,t) &= -0.08593\sinh\left(\frac{x}{4}\right)\frac{t^\gamma}{\Gamma(\gamma+1)}+0.11816\cosh\left(\frac{x}{4}\right)\frac{t^{2\gamma}}{\Gamma(2\gamma+1)}-0.02707\sinh\left(\frac{x}{4}\right)\frac{t^{3\gamma}}{\Gamma(3\gamma+1)}. \end{aligned} \tag{30}$$

The NDM result for problem 2 is

$$\mu(x,t) = \mu_0(x,t)+\mu_1(x,t)+\mu_2(x,t)+\mu_3(x,t)+\mu_4(x,t)\cdots.$$

$$\begin{aligned} \mu(x,t) = &\cosh^2\left(\frac{x}{4}\right)-0.3437\sinh\left(\frac{x}{4}\right)\frac{t^\gamma}{\Gamma(\gamma+1)}-0.08593\sinh\left(\frac{x}{4}\right)\frac{t^\gamma}{\Gamma(\gamma+1)} \\ &+0.11816\cosh\left(\frac{x}{4}\right)\frac{t^{2\gamma}}{\Gamma(2\gamma+1)}-0.08593\sinh\left(\frac{x}{4}\right)\frac{t^\gamma}{\Gamma(\gamma+1)} \\ &+0.11816\cosh\left(\frac{x}{4}\right)\frac{t^{2\gamma}}{\Gamma(2\gamma+1)}-0.02707\sinh\left(\frac{x}{4}\right)\frac{t^{3\gamma}}{\Gamma(3\gamma+1)}-\cdots. \end{aligned}$$

The exact result is;

$$\mu(x,t) = \cosh^2\left(\frac{x}{4}-\frac{11t}{24}\right).$$

In Figures 5 and 6, the solution graph of exact and NDM of Example 2 at integer-order are plotted. The closed relation is observed between NDM and exact solution of Example 2. In Figures 7 and 8, the fractional-order solutions of Example 2 are presented. The graphical representation have confirmed the different dynamics of Example 2, which are correlated with each other.

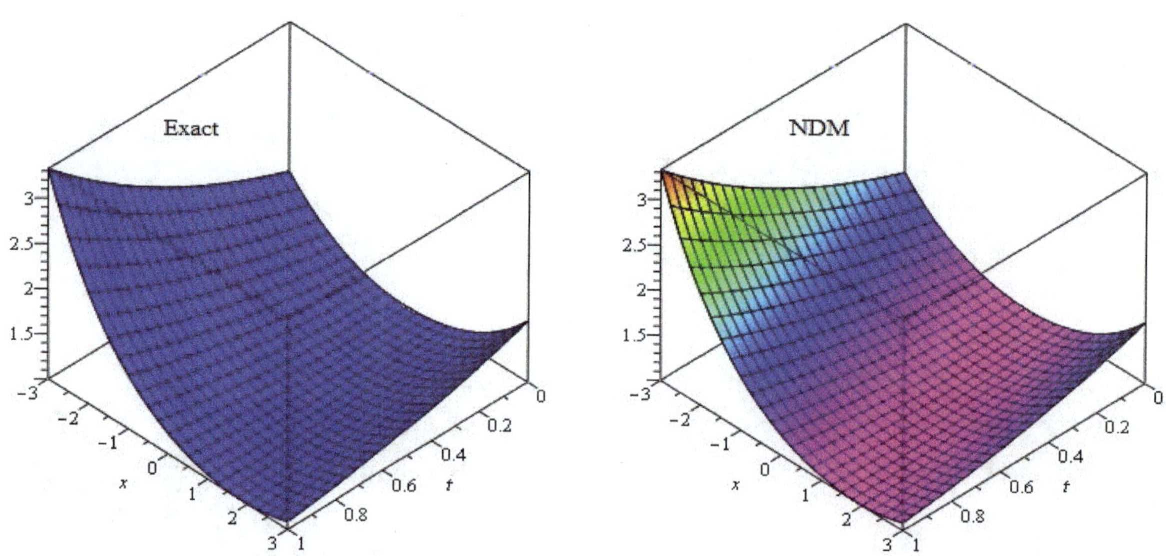

Figure 5. The graph of exact and approximate solution of Example 2.

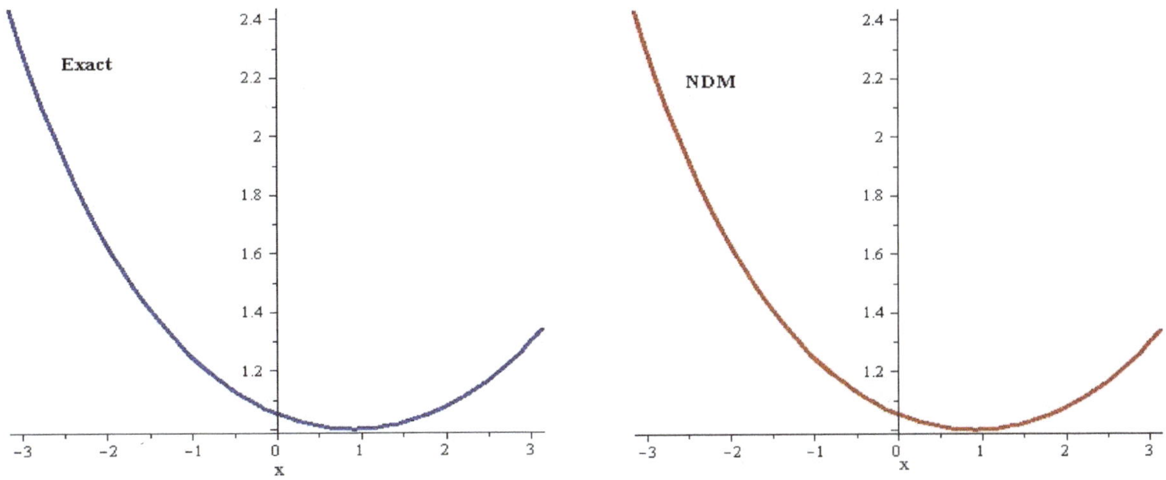

Figure 6. The graph of exact and approximate solution of Example 2.

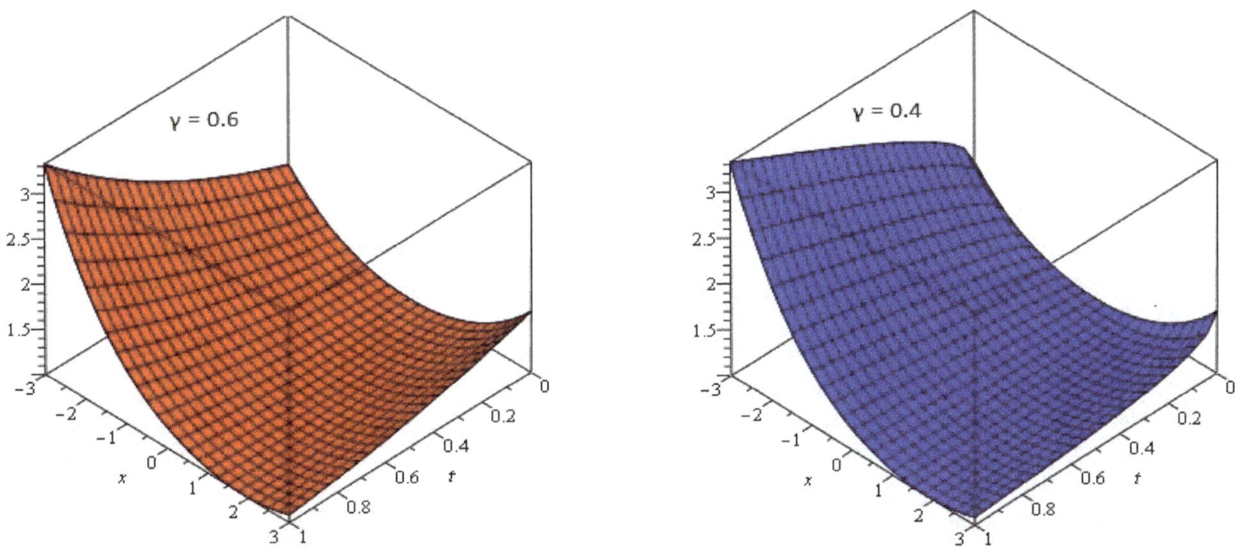

Figure 7. The NDM solutions of different valve of γ of Example 2.

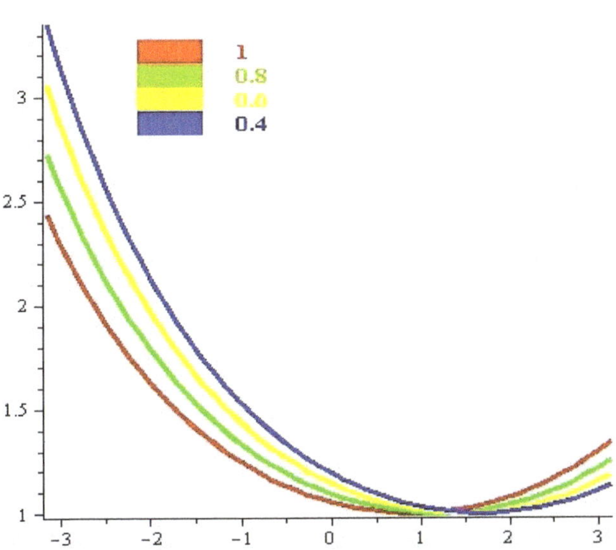

Figure 8. The graph of Example 2, for different value of γ.

5. Conclusions

In the current work. an innovative technique is used to find the solution of fractional Fornberg-Whitham equations. The fractional-derivatives are discussed within Caputo operator. The solutions are determined for fractional-order problems and an aesthetically a strong relation is found. The fractional models have shown convergence to the ordinary model as the order of the derivative tends towards to an integer. The graphical representation has provided similar behavior of actual and derived results. It is also noted the current method needs small calculation and higher convergence to achieve the solution of the targeted problems.

Author Contributions: Conceptualization, R.S. and H.K.; Methodology, R.S.; Software, A.A.A.; Validation, D.B. and S.A; Formal Analysis, H.K.; Investigation, R.S. and A.A.A.; Resources, H.K. and R.S.; Writing—Original Draft Preparation, R.S.; Writing—Review and Editing, H.K., D.B. and S.A.; Visualization, H.K.; Supervision, H.K., H.K.; Project Administration, D.B.; Funding Acquisition, A.A.A. and S.A. All authors have read and agreed to the published version of the manuscript.

Acknowledgments: The authors extend their appreciation to the Deanship of Scientific Research at King Khalid University, Saudi Arabia for founding this work through Research Groups program under grant number (R.G.P2./99/41).

References

1. Acan, O.; Firat, O.; Keskin, Y. Conformable variational iteration method, conformable fractional reduced differential transform method and conformable homotopy analysis method for non-linear fractional partial differential equations. *Waves Random Complex Media* **2020**, *30*, 250–268. [CrossRef]
2. Najafi, R. Group-invariant solutions for time-fractional Fornberg-Whitham equation by Lie symmetry analysis. *Comput. Methods Differ. Equ.* **2020**, *8*, 251–258.
3. Hörmann, G.; Okamoto, H. Weak periodic solutions and numerical case studies of the Fornberg-Whitham equation. *arXiv* **2018**, arXiv:1807.02320.
4. Zhou, J.; Tian, L. A type of bounded traveling wave solutions for the Fornberg–Whitham equation. *J. Math. Anal. Appl.* **2008**, *346*, 255–261. [CrossRef]
5. Moldabayev, D.; Kalisch, H.; Dutykh, D. The Whitham equation as a model for surface water waves. *Phys. D Nonlinear Phenom.* **2015**, *309*, 99–107. [CrossRef]
6. Lenells, J. Traveling wave solutions of the Camassa–Holm and Korteweg–de Vries Equations. *J. Nonlinear Math. Phys.* **2004**, *11*, 508–520. [CrossRef]
7. Camassa, R.; Holm, D. An integrable shallow wave equation with peaked solitons. *Phys. Rev. Lett.* **1993**, *71*, 1661–1664. [CrossRef]
8. Lenells, J. Traveling wave solutions of the Camassa–Holm equation. *J. Differ. Equ.* **2005**, *217*, 393–430. [CrossRef]
9. Liu, Z.; Chen, C. Compactons in a general compressible hyperelastic rod. *Chaos Soliton Fractals* **2004**, *22*, 627–640. [CrossRef]
10. Parkes, E.J.; Vakhnenko, V.O. Explicit solutions of the Camassa–Holm equation. *Chaos Solitons Fractals* **2005**, *26*, 1309–1316. [CrossRef]
11. Whitham, G.B. Variational methods and applications to water waves. *Philos. Trans. R. Soc. Lond. Ser. A Math. Phys. Sci.* **1967**, *299*, 6–25.
12. Fornberg, B.; Whitham, G.B. A numerical and theoretical study of certain nonlinear wave phenomena. *Philos. Trans. R. Soc. Lond. Ser. A Math. Phys. Sci.* **1978**, *289*, 373–404.
13. Purohit, S.D. Solutions of fractional partial differential equations of quantum mechanics. *Adv. Appl. Math. Mech.* **2013**, *5*, 639–651. [CrossRef]
14. Singh, J.; Kumar, D.; Kumar, S. New treatment of fractional Fornberg–Whitham equation via Laplace transform. *Ain Shams Eng. J.* **2013**, *4*, 557–562. [CrossRef]
15. Iyiola, O.S.; Ojo, G.O. On the analytical solution of Fornberg–Whitham equation with the new fractional derivative. *Pramana* **2015**, *85*, 567–575. [CrossRef]
16. Kumar, D.; Singh, J.; Baleanu, D. A new analysis of the Fornberg-Whitham equation pertaining to a fractional derivative with Mittag-Leffler-type kernel. *Eur. Phys. J. Plus* **2018** *133*, 70. [CrossRef]

17. Gupta, P.K.; Singh, M. Homotopy perturbation method for fractional Fornberg–Whitham equation. *Comput. Math. Appl.* **2011**, *61*, 250–254. [CrossRef]
18. Abidi, F.; Omrani, K. Numerical solutions for the nonlinear Fornberg-Whitham equation by He's method. *Int. J. Mod. Phys. B* **2011**, *25*, 4721–4732. [CrossRef]
19. Li, Q.; Sun, S.; Han, Z.; Zhao, Y. On the existence and uniqueness of solutions for initial value problem of nonlinear fractional differential equations. In Proceedings of the 2010 IEEE/ASME International Conference on Mechatronic and Embedded Systems and Applications, Qingdao, China, 15–17 July 2010; pp. 452–457.
20. Bacani, D.B.; Tahara, H. Existence and uniqueness theorem for a class of singular nonlinear partial differential equations. *Publ. Res. Inst. Math. Sci.* **2012**, *48*, 899–917. [CrossRef]
21. Dannan, F.M.; Saleeby, E.G. An existence and uniqueness theorem for n-th order functional differential equations. *Int. J. Pure Appl. Math.* **2013**, *84*, 193–200. [CrossRef]
22. Gao, X.; Lai, S.; Chen, H. The stability of solutions for the Fornberg–Whitham equation in $L^1(\mathbb{R})$ space. *Bound. Value Probl.* **2018**, *2018*, 142. [CrossRef]
23. Shen, C.; Gao, A. Optimal distributed control of the Fornberg–Whitham equation. *Nonlinear Anal. Real World Appl.* **2015**, *21*, 127–141. [CrossRef]
24. Jajarmi, A.; Yusuf, A.; Baleanu, D.; Inc, M. A new fractional HRSV model and its optimal control: A non-singular operator approach. *Phys. A Stat. Mech. Its Appl.* **2019**, *547*, 123860. [CrossRef]
25. Tuan, N.H.; Baleanu, D.; Thach, T.N.; O'Regan, D.; Can, N.H. Approximate solution for a 2-D fractional differential equation with discrete random noise. *Chaos Solitons Fractals* **2020**, *133*, 109650. [CrossRef]
26. Anh Triet, N.; Van Au, V.; Dinh Long, L.; Baleanu, D.; Huy Tuan, N. Regularization of a terminal value problem for time fractional diffusion equation. *Math. Methods Appl. Sci.* **2020**, *43*, 3850–3878. [CrossRef]
27. Tuan, N.H.; Baleanu, D.; Thach, T.N.; O'Regan, D.; Can, N.H. Final value problem for nonlinear time fractional reaction–diffusion equation with discrete data. *J. Comput. Appl. Math.* **2020**, *376*, 112883. [CrossRef]
28. Şenol, M.; Iyiola, O.S.; Kasmaei, H.D.; Akinyemi, L. Efficient analytical techniques for solving time-fractional nonlinear coupled Jaulent–Miodek system with energy-dependent Schrödinger potential. *Adv. Differ. Equ.* **2019**, *2019*, 462. [CrossRef]
29. Akinyemi, L.; Iyiola, O. Exact and approximate solutions of time-fractional models arising from physics via Shehu transform. *Math. Methods Appl. Sci.* **2020**. [CrossRef]
30. Akinyemi, L.; Iyiola, O.S. A reliable technique to study nonlinear time-fractional coupled Korteweg–de Vries equations. *Adv. Differ. Equ.* **2020**, *2020*, 1–27. [CrossRef]
31. Shah, R.; Khan, H.; Baleanu, D.; Kumam, P.; Arif, M. The analytical investigation of time-fractional multi-dimensional Navier–Stokes equation. *Alex. Eng. J.* **2020**, in press. [CrossRef]
32. Jan, R.; Xiao, Y. Effect of partial immunity on transmission dynamics of dengue disease with optimal control. *Math. Methods Appl. Sci.* **2019**, *42*, 1967–1983. [CrossRef]
33. Jan, R.; Xiao, Y. Effect of pulse vaccination on dynamics of dengue with periodic transmission functions. *Adv. Differ. Equ.* **2019**, *1*, 368. [CrossRef]
34. Mahmood, S.; Shah, R.; Arif, M. Laplace Adomian Decomposition Method for Multi Dimensional Time Fractional Model of Navier-Stokes Equation. *Symmetry* **2019**, *11*, 149. [CrossRef]
35. Shah, R.; Khan, H.; Farooq, U.; Baleanu, D.; Kumam, P.; Arif, M. A New Analytical Technique to Solve System of Fractional-Order Partial Differential Equations. *IEEE Access* **2019**, *7*, 150037–150050. [CrossRef]
36. Shah, R.; Khan, H.; Kumam, P.; Arif, M. An analytical technique to solve the system of nonlinear fractional partial differential equations. *Mathematics* **2019**, *7*, 505. [CrossRef]
37. Shah, R.; Khan, H.; Baleanu, D.; Kumam, P.; Arif, M. A novel method for the analytical solution of fractional Zakharov–Kuznetsov equations. *Adv. Differ. Equ.* **2019**, *2019*, 1–14. [CrossRef]
38. Khan, H.; Farooq, U.; Shah, R.; Baleanu, D.; Kumam, P.; Arif, M. Analytical Solutions of (2+ Time Fractional Order) Dimensional Physical Models, Using Modified Decomposition Method. *Appl. Sci.* **2020**, *10*, 122. [CrossRef]
39. Shah, R.; Farooq, U.; Khan, H.; Baleanu, D.; Kumam, P.; Arif, M. Fractional view analysis of third order Kortewege-De Vries equations, using a new analytical technique. *Front. Phys.* **2020**, *7*, 244. [CrossRef]
40. Khan, Z.H.; Khan, W.A. N-transform-properties and applications. *NUST J. Eng. Sci.* **2008**, *1*, 127–133.
41. Rawashdeh, M.S.; Maitama, S. Solving nonlinear ordinary differential equations using the NDM. *J. Appl. Anal. Comput.* **2015**, *5*, 77–88.

42. Rawashdeh, M.S.; Maitama, S. Solving PDEs using the natural decomposition method. *Nonlinear Stud.* **2016**, *23*, 63–72.
43. Khan, H.; Shah, R.; Kumam, P.; Arif, M. Analytical Solutions of Fractional-Order Heat and Wave Equations by the Natural Transform Decomposition Method. *Entropy* **2019**, *21*, 597. [CrossRef]
44. Rawashdeh, M.S.; Maitama, S. Solving coupled system of nonlinear PDE's using the natural decomposition method. *Int. J. Pure Appl. Math.* **2014**, *92*, 757–776. [CrossRef]
45. Khan, H.; Shah, R.; Baleanu, D.; Kumam, P.; Arif, M. Analytical Solution of Fractional-Order Hyperbolic Telegraph Equation, Using Natural Transform Decomposition Method. *Electronics* **2019**, *8*, 1015. [CrossRef]
46. Rawashdeh, M.S. The fractional natural decomposition method: Theories and applications. *Math. Methods Appl. Sci.* **2017**, *40*, 2362–2376. [CrossRef]
47. Shah, R.; Khan, H.; Mustafa, S.; Kumam, P.; Arif, M. Analytical Solutions of Fractional-Order Diffusion Equations by Natural Transform Decomposition Method. *Entropy* **2019**, *21*, 557. [CrossRef]
48. Miller, K.S.; Ross, B. *An Introduction to the Fractional Calculus and Fractional Differential Equations*; Wiley: New York, NY, USA, 1993.
49. Hilfer, R. *Applications of Fractional Calculus in Physics*; World Sci. Publishing: River Edge, NJ, USA, 2000.
50. Agarwal, R.; Lakshmikantham, V.; Nieto, J. On the concept of solution for fractional differential equations with uncertainty. *Nonlinear Anal. Theory Methods Appl.* **2010**, *72*, 2859–2862. [CrossRef]
51. Belgacem, F.B.M.; Silambarasan, R. Maxwell's equations solutions by means of the natural transform. *Int. J. Math. Eng. Sci. Aerosp.* **2012**, *3*, 313–323.

Permissions

All chapters in this book were first published by MDPI; hereby published with permission under the Creative Commons Attribution License or equivalent. Every chapter published in this book has been scrutinized by our experts. Their significance has been extensively debated. The topics covered herein carry significant findings which will fuel the growth of the discipline. They may even be implemented as practical applications or may be referred to as a beginning point for another development.

The contributors of this book come from diverse backgrounds, making this book a truly international effort. This book will bring forth new frontiers with its revolutionizing research information and detailed analysis of the nascent developments around the world.

We would like to thank all the contributing authors for lending their expertise to make the book truly unique. They have played a crucial role in the development of this book. Without their invaluable contributions this book wouldn't have been possible. They have made vital efforts to compile up to date information on the varied aspects of this subject to make this book a valuable addition to the collection of many professionals and students.

This book was conceptualized with the vision of imparting up-to-date information and advanced data in this field. To ensure the same, a matchless editorial board was set up. Every individual on the board went through rigorous rounds of assessment to prove their worth. After which they invested a large part of their time researching and compiling the most relevant data for our readers.

The editorial board has been involved in producing this book since its inception. They have spent rigorous hours researching and exploring the diverse topics which have resulted in the successful publishing of this book. They have passed on their knowledge of decades through this book. To expedite this challenging task, the publisher supported the team at every step. A small team of assistant editors was also appointed to further simplify the editing procedure and attain best results for the readers.

Apart from the editorial board, the designing team has also invested a significant amount of their time in understanding the subject and creating the most relevant covers. They scrutinized every image to scout for the most suitable representation of the subject and create an appropriate cover for the book.

The publishing team has been an ardent support to the editorial, designing and production team. Their endless efforts to recruit the best for this project, has resulted in the accomplishment of this book. They are a veteran in the field of academics and their pool of knowledge is as vast as their experience in printing. Their expertise and guidance has proved useful at every step. Their uncompromising quality standards have made this book an exceptional effort. Their encouragement from time to time has been an inspiration for everyone.

The publisher and the editorial board hope that this book will prove to be a valuable piece of knowledge for researchers, students, practitioners and scholars across the globe.

List of Contributors

Liana Eneeva, Arsen Pskhu and Sergo Rekhviashvili
Institute of Applied Mathematics and Automation, Kabardino-Balkarian Scientific Center of Russian Academy of Sciences, 89-A Shortanov Street, 360000 Nalchik, Russia

Omar Bazighifan
Department of Mathematics, Faculty of Science, Hadhramout University, Hadhramout 50512, Yemen
Department of Mathematics, Faculty of Education, Seiyun University, Hadhramout 50512, Yemen

Mihai Postolache
Center for General Education, China Medical University, Taichung 40402, Taiwan
Department of Mathematics and Informatics, University Politehnica of Bucharest, 060042 Bucharest, Romania
Romanian Academy, Gh. Mihoc-C. Iacob Institute of Mathematical Statistics and Applied Mathematics, 050711 Bucharest, Romania

Antonio Barrera
Departamento de Análisis Matemático, Estadística e Investigación Operativa y Matemática Aplicada, Facultad de Ciencias, Universidad de Málaga, Bulevar Louis Pasteur, 31, 29010 Málaga, Spain

Patricia Román-Román and Francisco Torres-Ruiz
Departamento de Estadística e Investigación Operativa, Facultad de Ciencias, Universidad de Granada, Avenida Fuente Nueva s/n, 18071 Granada, Spain
Instituto de Matemáticas de la Universidad de Granada (IEMath-GR), Calle Ventanilla 11, 18001 Granada, Spain

Ioannis Dassios
AMPSAS, University College Dublin, D4 Dublin, Ireland

Taher S. Hassan
Department of Mathematics, Faculty of Science, University of Ha'il, Ha'il 2440, Saudi Arabia
Department of Mathematics, Faculty of Science, Mansoura University, Mansoura 35516, Egypt

Yuangong Sun
School of Mathematical Sciences, University of Jinan, Jinan 250022, China

Amir Abdel Menaem
Department of Automated Electrical Systems, Ural Power Engineering Institute, Ural Federal University, 620002 Yekaterinburg, Russia

Jan Awrejcewicz
Department of Automation, Biomechanics and Mechatronics, Lodz University of Technology, 1/15 Stefanowski St., 90-924 Lodz, Poland

Ali Muhib
Department of Mathematics, Faculty of Science, Mansoura University, Mansoura 35516, Egypt
Department of Mathematics, Faculty of Education (Al-Nadirah), University, Ibb Yemen

Dumitru Baleanu
Department of Mathematics and Computer Science, Faculty of Arts and Sciences, Çankaya University Ankara, 06790 Etimesgut, Turkey
Department of Medical Research, China Medical University Hospital, China Medical University, Taichung 40402, Taiwan
Department of Mathematics, Faculty of Arts and Sciences, Cankaya University, 06530 Ankara, Turkey
Institute of Space Sciences, 077125 Magurele, Romania

Hoang Viet Long
Division of Computational Mathematics and Engineering, Institute for Computational Science, Ton Duc Thang University, Ho Chi Minh City 70000, Vietnam
Faculty of Mathematics and Statistics, Ton Duc Thang University, Ho Chi Minh City 70000, Vietnam

Haifa Bin Jebreen
Department of Mathematics, College of Science, King Saud University, Riyadh 11451, Saudi Arabia

Stefania Tomasiello
Institute of Computer Science, University of Tartu, 50090 Tartu, Estonia

Asıf Yokus
Department of Actuary, Faculty of Science, Firat University, Elazig 23200, Turkey

Hülya Durur
Department of Computer Engineering, Faculty of Engineering, Ardahan University, Ardahan 75000, Turkey

Hijaz Ahmad
Department of Basic Sciences, University of Engineering and Technology, Peshawar 25000, Pakistan

Shao-Wen Yao
School of Mathematics and Information Science, Henan Polytechnic University, Jiaozuo 454000, China

Zhen Yang and Junjie Ma
School of Mathematics and Statistics, Guizhou University, Guiyang 550025, Guizhou, China

Osama Moaaz and Waad Muhsin
Department of Mathematics, Faculty of Science, Mansoura University, Mansoura 35516, Egypt

Rami Ahmad El-Nabulsi
Athens Institute for Education and Research, Mathematics and Physics Divisions, 10671 Athens, Greece

Thabet Abdeljawad
Department of Mathematics and General Sciences, Prince Sultan University, Riyadh 11586, Saudi Arabia
Department of Medical Research, China Medical University, Taichung 40402, Taiwan
Department of Computer Science and Information Engineering, Asia University, Taichung 40402, Taiwan

Jehad Alzabut
Department of Mathematics and General Sciences, Prince Sultan University, Riyadh 11586, Saudi Arabia

James Viji and Velu Muthulakshmi
Department of Mathematics, Periyar University, Salem 636 011, Tamilnadu, India

Weerawat Sudsutad
Department of General Education, Navamindradhiraj University, Bangkok 10300, Thailand

Rabha W. Ibrahim
Informetrics Research Group, Ton Duc Thang University, Ho Chi Minh City 758307, Vietnam
Faculty of Mathematics & Statistics, Ton Duc Thang University, Ho Chi Minh City 758307, Vietnam

Rafida M. Elobaid
Department of General Sciences, Prince Sultan University, Riyadh 12345, Saudi Arabia

Suzan J. Obaiys
School of Mathematical and Computer Sciences, Heriot-Watt University Malaysia, Putrajaya 62200, Malaysia

M. Fernández-Martínez
MDE-UPCT, University Centre of Defence at the Spanish Air Force Academy, 30720 Santiago de la Ribera, Región de Murcia, Spain

Juan L.G. Guirao
Departamento de Matematica Aplicada y Estadística, Universidad Politécnica de Cartagena, 30203 Cartagena, Spain
Department of Mathematics, Faculty of Science, King Abdulaziz University, 80203, Jeddah 21589, Saudi Arabia

A. A. Alderremy
Department of Mathematics, Faculty of Science, King Khalid University, Abha 61413, Saudi Arabia

Hassan Khan and Rasool Shah
Department of Mathematics, Abdul Wali khan University, Mardan 23200, Pakistan

Shaban Aly
Department of Mathematics, Faculty of Science, AL-Azhar University, Assiut 71516, Egypt

Index

www.ingramcontent.com/pod-product-compliance
Lightning Source LLC
LaVergne TN
LVHW080930210926
848485LV00004B/764